D0558542

PRELUDE TO
CALCULUS

Student Solutions Manual

Warren L. Ruud
Santa Rosa Junior College

Terry L. Shell
Santa Rosa Junior College

Wadsworth Publishing Company
Belmont, California
A Division of Wadsworth, Inc

© 1990 by Wadsworth, Inc. All rights reserved. No part of this book may be reproduced, stored in a retrieval system, or transcribed, in any form or by any means, electronic, mechanical, photocopying, recording, or otherwise, without the prior written permission of the publisher, Wadsworth Publishing Company, Belmont, California 94002, a division of Wadsworth, Inc.

Printed in the United States of America   49

2 3 4 5 6 7 8 9 10—94 93 92 91

CONTENTS

Chapter 1

Section 1.1

1. Because $\frac{5}{3}$ is the ratio of two integers, it is a rational number. In reduced form, $-\frac{12}{4}$ is -3, an integer (and therefore a rational number). The next two numbers, $\sqrt{7}$ and $\sqrt{19}$, are both irrational, since neither 7 nor 19 are perfect squares of rational numbers. Because $2^2 < 7 < 3^2$ and $4^2 < 19 < 5^2$, it follows that $2 < \sqrt{7} < 3$ and $4 < \sqrt{19} < 5$.
In summary:

integers: $-\frac{12}{4}$ rational numbers: $-\frac{12}{4}, \frac{5}{3}$ irrational numbers: $\sqrt{7}, \sqrt{19}$

increasing order: $-\frac{12}{4}, \frac{5}{3}, \sqrt{7}, \sqrt{19}$

5. In Example 1 on page 3, $\frac{\pi}{2}$ is shown to be irrational; its value is approximately 1.57.
Because $-\frac{3}{2}$ is the ratio of two integers, it is a rational number. When simplified, $\sqrt{4}$ is 2 and $\sqrt[3]{-8}$ is -2; these are integers (and therefore rational numbers).
In summary:

integers: $\sqrt{4}, \sqrt[3]{-8}$ rational numbers: $-\frac{3}{2}, \sqrt{4}, \sqrt[3]{-8}$ irrational numbers: $\frac{\pi}{2}$

increasing order: $\sqrt[3]{-8}, -\frac{3}{2}, \frac{\pi}{2}, \sqrt{4},$

9. From the table on page 4, this is equivalent to the interval notation $(-2, 4)$.

13. From the table on page 4, this is equivalent to the interval notation $(-\infty, -2]$ or $(2, \infty)$.

17. From the table on page 4, this is equivalent to the interval notation $(-3, 1]$ or $(4, \infty)$.

21. From the table on page 4, this is equivalent to the inequality notation $4 \le x \le 9$.

25. From the table on page 4, this is equivalent to the inequality notation $x < -2$ or $x > 2$.

29. From the table on page 4, this is equivalent to the inequality notation $-5 < x < -2$ or $x > 0$.

33. It follows from $3^2 < 13 < 4^2$ that $3 < \sqrt{13} < 4$. It follows that $\sqrt{13} < 6$, or $\sqrt{13} - 6$ is negative. From page 5, we know that if r is a negative real number, then $|r| = -r$.
So, $\left|\sqrt{13} - 6\right| = -\left(\sqrt{13} - 6\right) = 6 - \sqrt{13}$

37. The square of a real number is nonnegative. From page 5, we know that if r is a nonnegative real number, then $|r| = r$. So, $|x^2| = x^2$.

41. From page 5, we know that if r is a nonnegative real number, then $|r| = r$. It follows from this that $\dfrac{x}{|x|} = \dfrac{x}{x} = 1$.

45. By the table on page 7, the graph of the numbers described by $|x| = 6$ is the set of points that are 6 units from the origin. These are the points that correspond to 6 and -6.

$$-6\ -5\ -4\ -3\ -2\ -1\ \ 0\ \ 1\ \ 2\ \ 3\ \ 4\ \ 5\ \ 6$$

49. By the table on page 7, the graph of the numbers described by $|x-3| < 0.5$ is the set of points that are 0.5 units from 3. This is the interval $(2.5, 3.5)$.

$$-1\qquad 0\qquad 1\qquad 2\qquad 3\qquad 4$$

53. By the table on page 7, the graph of the numbers described by $|x+1| < 0.2$ is the set of points that are 0.2 units from -1. This is the interval $(-1.2, -0.8)$.

$$-1.5\qquad\qquad -1\qquad\qquad -0.5$$

57. The keystrokes for $\frac{1}{2}(\sqrt{2} + \sqrt{6})$: 2 $\boxed{\sqrt{\ }}$ $\boxed{+}$ 6 $\boxed{\sqrt{\ }}$ $\boxed{=}$ $\boxed{\div}$ 2 $\boxed{=}$ Display: 1.9319

The keystrokes for $\sqrt{2 + \sqrt{3}}$: 2 $\boxed{+}$ 3 $\boxed{\sqrt{\ }}$ $\boxed{=}$ $\boxed{\sqrt{\ }}$ Display: 1.9319

The numbers appear to be equal. If we square each of these positive numbers, and arrive at the same number exactly, then they are equal. Proceeding, we find that

$$\left(\sqrt{2 + \sqrt{3}}\right)^2 = 2 + \sqrt{3}$$

and

$$\left(\tfrac{1}{2}(\sqrt{2} + \sqrt{6})\right)^2 = \tfrac{1}{4}(\sqrt{2} + \sqrt{6})^2 = \tfrac{1}{4}(2 + 2\sqrt{12} + 6) = 2 + \sqrt{3}$$

They are the same number.

61. From the table on page 7, a is the midpoint of $a-\delta$ and $a+\delta$. For this problem, $a-\delta = 3$ and $a+\delta = 7$. It follows that $a = 5$. The value of δ is the distance from 3 to 5 or 5 to 7, so $\delta = 2$. Thus the inequality we seek is $|x-2| < 5$.

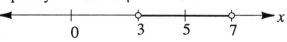

$$0\qquad\quad 3\quad\ 5\quad\ 7$$

Section 1.2

1. Each part of this problem uses the property that $x^{-n} = \dfrac{1}{x^n}$.

   a) $\quad x^{-2/3} = \dfrac{1}{x^{2/3}}$
                  b) $\quad 2y^{-1} = 2 \cdot \dfrac{1}{y^1} = \dfrac{2}{y}$

   c) $\quad (2y)^{-1} = \dfrac{1}{(2y)^1} = \dfrac{1}{2y}$
            d) $\quad x^5 y^{-2} = x^5 \dfrac{1}{y^2} = \dfrac{x^5}{y^2}$

5. Each part of this problem uses the property that $x^{p/q} = \sqrt[q]{x^p}$.

   a) $\quad x^{2/5} = \sqrt[5]{x^2}$
                  b) $\quad x^{-3/7} = \sqrt[7]{x^{-3}}$

   c) $\quad \dfrac{1}{x^{3/2}} = x^{-3/2} = \sqrt{x^{-3}}$
            d) $\quad \dfrac{1}{x^{-2/9}} = x^{2/9} = \sqrt[9]{x^2}$

9. $\left(-5x^3\right)^2 (4x)^{-1} = (-5)^2 \left(x^3\right)^2 \cdot 4^{-1} x^{-1} = 25x^6 \cdot \dfrac{1}{4} x^{-1} = \dfrac{25x^5}{4}$

13. $\left(25x^4\right)^{-1/2} = 25^{-1/2} \left(x^4\right)^{-1/2} = 25^{-1/2} \left(x^4\right)^{-1/2} = 25^{-1/2} x^{-2} = \dfrac{1}{25^{1/2}} \dfrac{1}{x^2} = \dfrac{1}{5x^2}$

17. $\left(x^4 y^2 z^{-6}\right)^{5/2} = \left(x^4\right)^{5/2} \left(y^2\right)^{5/2} \left(z^{-6}\right)^{5/2} = x^{10} y^5 z^{-15} = x^{10} y^5 \dfrac{1}{z^{15}} = \dfrac{x^{10} y^5}{z^{15}}$

21. $\sqrt[3]{8x^{12}} = \sqrt[3]{8} \, \sqrt[3]{x^{12}} = 2\sqrt[3]{\left(x^4\right)^3} = 2x^4$

25. $\left(2x^2 + 4x - 7\right) - 2x(x - 8) = 2x^2 + 4x - 7 - 2x^2 + 16x = 20x - 7$

29. Using the second product-factoring formula in the table on page 14, we get
$$(2x - 1)(3x + 1) = 6x^2 - x - 1$$

33. Multiplying directly yields
$$\left(x^2 - x\right)(x + 1) = \left(x^2 - x\right)x + \left(x^2 - x\right)1 = x^3 - x^2 + x^2 - x = x^3 - x.$$
There is an alternate solution. We use the factorization $x^2 - x = x(x - 1)$.
$$\left(x^2 - x\right)(x + 1) = [x(x - 1)](x + 1) = x[(x - 1)(x + 1)] = x\left[x^2 - 1\right] = x^3 - x$$

37. Multiplying directly yields
$$(x - 2)(x + 4) + (x + 4)^2 = \left(x^2 + 2x - 8\right) + \left(x^2 + 8x + 16\right) = 2x^2 + 10x + 8.$$
There is an alternate solution. We factor $(x + 4)$ from the sum first:
$$(x - 2)(x + 4) + (x + 4)^2 = (x + 4)[(x - 2) + (x + 4)] = (x + 4)[2x + 2] = 2x^2 + 10x + 8$$

41. The factors can be rearranged to take advantage of the third product-factoring formula in the table on page 14:

$$(x-2)(x^2+4)(x+2) = [(x-2)(x+2)](x^2+4)$$

$$= (x^2-4)(x^2+4) = (x^2)^2 - 4^2 = x^4 - 16$$

45. Using the second product-factoring formula in the table on page 14, we get

$$3x^2 - 2x - 1 = (3x+1)(x-1)$$

49. Using the third product-factoring formula in the table on page 14, we get

$$25x^2 - 36 = (5x)^2 - 6^2 = (5x+6)(5x-6)$$

53. First we factor an x from the difference, then we use the ninth product-factoring formula in the table on page 14: $x^4 - 27x = x(x^3 - 27) = x(x-3)(x^2 + 3x + 9)$

57. Using the third product-factoring formula in the table on page 14, we get

$$(\sqrt{x}+2)^2 = (\sqrt{x})^2 + 2(\sqrt{x})(2) + 2^2 = x + 4\sqrt{x} + 4$$

61. Multiplying directly and simplifying gives

$$x^{1/2}(x^{3/2} + 2x^{1/2} - x^{-1/2}) = x^{1/2}x^{3/2} + 2x^{1/2}x^{1/2} - x^{1/2}x^{-1/2} = x^2 + 2x - 1$$

65. This fits the pattern of the eighth product-factoring formula in the table on page 14:

$$(x^2+y^2)(x^4 - x^2y^2 + y^4) = (x^2+y^2)((x^2)^2 - x^2y^2 + (y^2)^2) = (x^2)^3 + (y^2)^3 = x^6 + y^6$$

69. Multiplying directly and simplifying gives

$$2x^{4/3} - x^{2/3}(x^{2/3} - 3) = 2x^{4/3} - x^{2/3}x^{2/3} + 3x^{2/3} = 2x^{4/3} - x^{4/3} + 3x^{2/3} = x^{4/3} + 3x^{2/3}$$

73. Multiplying directly and simplifying gives

$$(x-1)(x^6 + x^5 + x^4 + x^3 + x^2 + x + 1)$$

$$= x(x^6 + x^5 + x^4 + x^3 + x^2 + x + 1) - (x^6 + x^5 + x^4 + x^3 + x^2 + x + 1)$$

$$= x^7 + x^6 + x^5 + x^4 + x^3 + x^2 + x - x^6 - x^5 - x^4 - x^3 - x^2 - x - 1$$

$$= x^7 - 1$$

77. The polynomial is factored by grouping the terms by pairs.

$$2x^3 + x^2 + 8x + 4 = (2x^3 + x^2) + (8x+4)$$

$$= x^2(2x+1) + 4(2x+1) = (x^2+4)(2x+1)$$

81. The polynomial is factored using the third product-factoring formula in the table on page 14.
$$x^4 - 16y^8 = \left(x^2\right)^2 - \left(4y^4\right)^2 = \left(x^2 + 4y^4\right)\left(x^2 - 4y^4\right)$$
$$= \left(x^2 + 4y^4\right)\left(x^2 - \left(2y^2\right)^2\right) = \left(x^2 + 4y^4\right)\left(x + 2y^2\right)\left(x - 2y^2\right)$$

85. Simplifying and then factoring, we get
$$(x+2)^2 + 3(x+2) - 4 = x^2 + 4x + 4 + 3x + 6 - 4 = x^2 + 7x + 6 = (x+6)(x+1)$$
There is an alternate solution. Consider the polynomial $u^2 + 3u - 4$. This is factored as $(u+4)(u-1)$. The polynomial in this problem is equivalent to this polynomial in u with $u = x + 2$. So, we get
$$(x+2)^2 + 3(x+2) - 4 = ((x+2)+4)((x+2)-1) = (x+6)(x+1)$$

89. $\dfrac{x^{5/2} - 2x^{8/3} + 4x - 2}{x^2} = \dfrac{1}{x^2}\left(x^{5/2} - 2x^{8/3} + 4x - 2\right) = x^{-2}\left(x^{5/2} - 2x^{8/3} + 4x - 2\right)$

$$= x^{-2}x^{5/2} - 2x^{-2}x^{8/3} + 4x^{-2}x - 2x^{-2} = x^{1/2} - 2x^{2/3} + 4x^{-1} - 2x^{-2}$$

Section 1.3

---

1. Factoring the numerator and denominator and eliminating the common factor yields
$$\frac{2x^4 + x^2}{1 + 2x^2} = \frac{x^2\left(2x^2 + 1\right)}{\left(2x^2 + 1\right)} = \frac{x^2}{1}\frac{\left(2x^2 + 1\right)}{\left(2x^2 + 1\right)} = x^2$$

5. Factoring the numerator and denominator and eliminating the common factor yields
$$\frac{\left(x^3 + 3x^2 + 4x\right)(2x+4)}{\left(x^2 + 2\right)\left(x^2 + 3x + 4\right)} = \frac{x\left(x^2 + 3x + 4\right)\cdot 2(x+2)}{\left(x^2 + 2\right)\left(x^2 + 3x + 4\right)} = \frac{2x(x+2)}{x^2 + 2}$$

9. Factoring the numerator and denominator and eliminating the common factor yields
$$\frac{x^2 - y^2}{(x-y)^2} = \frac{(x+y)(x-y)}{(x-y)(x-y)} = \frac{(x+y)}{(x-y)}\frac{(x-y)}{(x-y)} = \frac{x+y}{x-y}$$

13. The numerators and denominators are written in factored form before multiplying.
$$\frac{4x^2 + 4x + 1}{4x^2 - 1}\cdot\frac{2x^2 + x - 1}{2x^2 - x - 1} = \frac{(2x+1)^2}{(2x-1)(2x+1)}\cdot\frac{(2x-1)(x+1)}{(2x+1)(x-1)} = \frac{x+1}{x-1}$$

17. The least common multiple of $x-2$ and $x+1$ is $(x-2)(x+1)$.
$$\frac{3}{x+1} + \frac{4}{x-2} = \frac{3}{(x+1)}\frac{(x-2)}{(x-2)} + \frac{4}{(x-2)}\frac{(x+1)}{(x+1)} = \frac{3(x-2) + 4(x+1)}{(x-2)(x+1)} = \frac{7x-2}{(x-2)(x+1)}$$

21. Factoring the denominators, we get
$$x^2 - 7x + 6 = (x-6)(x-1) \text{ and } x^2 - 2x - 24 = (x-6)(x+4) \,.$$
The least common denominator is $(x-6)(x-1)(x+4)$. So,
$$\frac{x}{x^2-7x+6} - \frac{x}{x^2-2x-24} = \frac{x}{(x-6)(x-1)}\frac{(x+4)}{(x+4)} - \frac{x}{(x-6)(x+4)}\frac{(x-1)}{(x-1)}$$
$$= \frac{x(x+4)-x(x-1)}{(x-6)(x+4)(x-1)} = \frac{5x}{(x-6)(x+4)(x-1)}$$

25. The least common denominator of the rational expressions in the numerator and the denominator is $x(x+2)$. We multiply the numerator and the denominator by this least common denominator and simplify.
$$\frac{\frac{1}{x+2}-\frac{1}{x}}{2} \cdot \frac{x(x+2)}{x(x+2)} = \frac{\frac{1}{x+2}\cdot x(x+2) - \frac{1}{x}\cdot x(x+2)}{2x(x+2)}$$
$$= \frac{x-(x+2)}{2x(x+2)} = \frac{-2}{2x(x+2)} = \frac{-1}{x(x+2)}$$

29. The least common denominator of the rational expressions in the numerator and the denominator is $(x+1)(x+h+1)$. We multiply the numerator and the denominator by this least common denominator and simplify.
$$\frac{\frac{x+h-1}{x+h+1}-\frac{x-1}{x+1}}{h} \cdot \frac{(x+1)(x+h+1)}{(x+1)(x+h+1)}$$
$$= \frac{\frac{x+h-1}{x+h+1}\cdot(x+1)(x+h+1) - \frac{x-1}{x+1}\cdot(x+1)(x+h+1)}{h(x+1)(x+h+1)}$$
$$= \frac{(x+h-1)\cdot(x+1)-(x-1)(x+h+1)}{h(x+1)(x+h+1)}$$
$$= \frac{(x+h-1)x+(x+h-1)-x(x+h+1)+(x+h+1)}{h(x+1)(x+h+1)}$$
$$= \frac{x^2+xh-x+x+h-1-x^2-xh-x+x+h+1}{h(x+1)(x+h+1)}$$
$$= \frac{2h}{h(x+1)(x+h+1)} = \frac{2}{(x+1)(x+h+1)}$$

33. There is only one fraction, $\dfrac{1}{\sqrt{x}}$, in the numerator or the denominator, so we multiply the

numerator and the denominator by $\sqrt{x}$ and simplify.

$$\frac{\dfrac{1}{\sqrt{x}}-\sqrt{x}}{\sqrt{x}}=\frac{\dfrac{1}{\sqrt{x}}-\sqrt{x}}{\sqrt{x}}\,\frac{\sqrt{x}}{\sqrt{x}}=\frac{\dfrac{1}{\sqrt{x}}\cdot\sqrt{x}-\sqrt{x}\sqrt{x}}{\sqrt{x}\sqrt{x}}=\frac{1-x}{x}$$

37. The numerators and denominators are written in factored form before multiplying

$$\left(x^2-4\right)\left(\frac{x^2-2x-8}{x-2}\right)=\frac{(x-2)(x+2)}{1}\frac{(x-4)(x+2)}{(x-2)}$$

$$=\frac{(x+2)^2(x-4)(x-2)}{(x-2)}=(x+2)^2(x-4)\ \text{ or }\ x^3-12x-16$$

41. Because $\sqrt{x}+2$ and $\sqrt{x}-2$ are conjugates, their product is $x-4$. We factor the denominator, but their are no common factors in the numerator and denominator.

$$\left(\frac{\sqrt{x}+2}{x^2-4x-12}\right)\left(\sqrt{x}-2\right)=\frac{\left(\sqrt{x}+2\right)\left(\sqrt{x}-2\right)}{(x-6)(x+2)}=\frac{x-4}{(x-6)(x+2)}$$

45. We use $\sqrt{1-x^2}$ as the least common denominator.

$$\frac{x^2}{\sqrt{1-x^2}}+\sqrt{1-x^2}=\frac{x^2}{\sqrt{1-x^2}}+\frac{\sqrt{1-x^2}}{1}\frac{\sqrt{1-x^2}}{\sqrt{1-x^2}}=\frac{x^2}{\sqrt{1-x^2}}+\frac{1-x^2}{\sqrt{1-x^2}}=\frac{1}{\sqrt{1-x^2}}$$

Next, we rationalize the denominator:

$$\frac{1}{\sqrt{1-x^2}}=\frac{1}{\sqrt{1-x^2}}\frac{\sqrt{1-x^2}}{\sqrt{1-x^2}}=\frac{\sqrt{1-x^2}}{1-x^2}$$

49. Each term is simplified and then added. The least common denominator is $2\sqrt{x-4}\sqrt{3x-5}$.

$$\sqrt{3x+5}\,\frac{1}{2\sqrt{x-4}}+\sqrt{x-4}\,\frac{3}{2\sqrt{3x+5}}=\frac{\sqrt{3x+5}}{2\sqrt{x-4}}+\frac{3\sqrt{x-4}}{2\sqrt{3x+5}}$$

$$=\frac{\sqrt{3x+5}}{2\sqrt{x-4}}\frac{\sqrt{3x+5}}{\sqrt{3x+5}}+\frac{3\sqrt{x-4}}{2\sqrt{3x+5}}\frac{\sqrt{x-4}}{\sqrt{x-4}}$$

$$=\frac{3x+5}{2\sqrt{x-4}\sqrt{3x+5}}+\frac{3(x-4)}{2\sqrt{x-4}\sqrt{3x+5}}=\frac{6x-7}{2\sqrt{x-4}\sqrt{3x+5}}$$

Next, we rationalize the denominator. The rationalizing factor is $\sqrt{x-4}\sqrt{3x-5}$.

$$\frac{6x-7}{2\sqrt{x-4}\sqrt{3x+5}}\cdot\frac{\sqrt{x-4}\sqrt{3x+5}}{\sqrt{x-4}\sqrt{3x+5}}=\frac{(6x-7)\sqrt{x-4}\sqrt{3x+5}}{2(x-4)(3x+5)}$$

53. First we write the expression without negative exponents and then simplify the resulting compound fraction.

$$\frac{3x^2\left(1-x^2\right)^{1/3}-x^3\cdot\frac{1}{3}\left(1-x^2\right)^{-2/3}(-2x)}{\left(1-x^2\right)^{2/3}}$$

$$=\frac{3x^2\left(1-x^2\right)^{1/3}-x^3\cdot\dfrac{1}{3\left(1-x^2\right)^{2/3}}(-2x)}{\left(1-x^2\right)^{2/3}}\cdot\frac{3\left(1-x^2\right)^{2/3}}{3\left(1-x^2\right)^{2/3}}$$

$$=\frac{3x^2\left(1-x^2\right)^{1/3}\cdot3\left(1-x^2\right)^{2/3}+\dfrac{2x^4}{3\left(1-x^2\right)^{2/3}}\cdot3\left(1-x^2\right)^{2/3}}{\left(1-x^2\right)^{2/3}\cdot3\left(1-x^2\right)^{2/3}}$$

$$=\frac{9x^2\left(1-x^2\right)+2x^4}{3\left(1-x^2\right)^{4/3}}=\frac{9x^2-9x^4+2x^4}{3\left(1-x^2\right)^{4/3}}=\frac{9x^2-7x^4}{3\left(1-x^2\right)^{4/3}}$$

57. First we simplify the radical and then factor:

$$\sqrt{\left(\frac{3}{x^4}-2\right)^2+\frac{24}{x^4}}=\sqrt{\frac{9}{x^8}-\frac{12}{x^4}+4+\frac{24}{x^4}}=\sqrt{\frac{9}{x^8}+\frac{12}{x^4}+4}=\sqrt{\left(\frac{3}{x^4}+2\right)^2}=\frac{3}{x^4}+2$$

61. The rationalizing factor of $\sqrt{x}+1$ is its conjugate, $\sqrt{x}-1$.

$$\frac{1}{\sqrt{x}+1}=\frac{1}{\sqrt{x}+1}\frac{\sqrt{x}-1}{\sqrt{x}-1}=\frac{\sqrt{x}-1}{\left(\sqrt{x}\right)^2-1^2}=\frac{\sqrt{x}-1}{x-1}$$

65. The rationalizing factor of $x-\sqrt{1-x^2}$ is its conjugate, $x+\sqrt{1-x^2}$.

$$\frac{x+\sqrt{1-x^2}}{x-\sqrt{1-x^2}}=\frac{x+\sqrt{1-x^2}}{x-\sqrt{1-x^2}}\frac{x+\sqrt{1-x^2}}{x+\sqrt{1-x^2}}=\frac{x^2+2x\sqrt{1-x^2}+\left(\sqrt{1-x^2}\right)^2}{x^2-\left(\sqrt{1-x^2}\right)^2}$$

$$=\frac{x^2+2x\sqrt{1-x^2}+1-x^2}{x^2-\left(1-x^2\right)}=\frac{1+2x\sqrt{1-x^2}}{2x^2-1}$$

69. Performing the long division, we get

$$
\begin{array}{r}
x-1 \\
x+4\overline{)x^2+3x+7} \\
\underline{x^2+4x} \\
-x+7 \\
\underline{-x-4} \\
11
\end{array}
\qquad \Rightarrow \quad x^2+3x+7 = x-1+\dfrac{11}{x+4}
$$

73. Performing the long division, we get

$$
\begin{array}{r}
1 \\
x^2-2\overline{)x^2-6x-3} \\
\underline{x^2-2} \\
-6x-1
\end{array}
\qquad \Rightarrow \quad x^2-6x-3 = 1+\dfrac{-6x-1}{x^2-2} \ \text{ or } \ 1-\dfrac{6x+1}{x^2-2}
$$

77. Performing the long division, we get

$$
\begin{array}{r}
2x^2+x+3 \\
x^2-4\overline{)2x^4+x^3-5x^2-x-18} \\
\underline{2x^4-8x^2} \\
x^3+3x^2-x \\
\underline{x^3-4x} \\
3x^2+3x-18 \\
\underline{3x^2-12} \\
3x-6
\end{array}
\quad \Rightarrow
$$

$$
\begin{aligned}
2x^4+x^3-5x^2-x-18 &= 2x^2+x+3+\dfrac{3x-6}{x^2-4} \\
&= 2x^2+x+3+\dfrac{3}{x+2}
\end{aligned}
$$

Section 1.4

1. To eliminate the fractions, we multiply each side of the equation by 6, the least common multiple of 2 and 3.

$$
6\left(\frac{x}{2}-\frac{x}{3}\right)=(4)6 \ \Rightarrow\ 3x-2x=24 \ \Rightarrow\ x=24
$$

5. To eliminate the fractions, we multiply each side of the equation by x, the denominator of the fractions in the equation.

$$
x\left(\frac{2}{x}-3\right)=\left(\frac{5}{x}-2\right)x \ \Rightarrow\ 2-3x=5-2x \ \Rightarrow\ -x=3 \ \Rightarrow\ x=-3
$$

9. To solve for m, we divide each side by v^2 to get $m=\dfrac{FR}{v^2}$.

13. We subtract 12 from each side and factor.
$$12x^2 + 7x = 12 \quad \Rightarrow \quad 12x^2 + 7x - 12 = 0 \quad \Rightarrow \quad (4x-3)(3x+4) = 0$$
$$\Rightarrow \quad 4x - 3 = 0 \text{ or } 3x + 4 = 0 \quad \Rightarrow \quad x = \tfrac{3}{4} \text{ or } x = -\tfrac{4}{3}$$
The solutions are $\tfrac{3}{4}$ and $-\tfrac{4}{3}$.

17. We subtract x^2 from each side and factor.
$$2x^3 - x = x^2 \quad \Rightarrow \quad 2x^3 - x^2 - x = 0 \quad \Rightarrow \quad x\left(2x^2 - x - 1\right) = 0$$
$$\Rightarrow \quad x(2x+1)(x-1) = 0 \quad \Rightarrow \quad x = 0 \text{ or } 2x + 1 = 0 \text{ or } x - 1 = 0$$
$$\Rightarrow \quad x = 0 \text{ or } x = -\tfrac{1}{2} \text{ or } x = 1$$
The solutions are $0, -\tfrac{1}{2}$ and 1 .

21. We extract roots and then add 4 to each side.
$$(x-4)^2 = 12 \quad \Rightarrow \quad x - 4 = \pm\sqrt{12} \quad \Rightarrow \quad x - 4 = \pm 2\sqrt{3} \quad \Rightarrow \quad x = 4 \pm 2\sqrt{3}$$
The solutions are $4 + 2\sqrt{3}$ and $4 - 2\sqrt{3}$.

25. We use $a = 2$, $b = -5$, $c = -42$ in the quadratic formula
$$x = \frac{-(-5) \pm \sqrt{(-5)^2 - 4(2)(-42)}}{2(2)} = \frac{5 \pm \sqrt{361}}{4} = \frac{5 \pm 19}{4} = 6 \text{ or } -\tfrac{7}{2}$$
The solutions are 6 and $-\tfrac{7}{2}$.

29. The equation can be rewritten as $0 = x^2 - 2x + 5$. We use $a = 1$, $b = -2$, $c = 5$ in the quadratic formula
$$x = \frac{-(-2) \pm \sqrt{(-2)^2 - 4(1)(5)}}{2(1)} = \frac{2 \pm \sqrt{-16}}{2}$$
The value of the discriminant is negative, so there are no real solutions.

33. The equation can be rewritten as
$$x^2 - 2x = -4.$$
We take half of the coefficient of x , square it, and add it to each side of the equation:
$$x^2 - 2x + 1 = -4 + 1$$
$$(x - 1)^2 = -3$$
$$x - 1 = \pm\sqrt{-3}$$
Because $\sqrt{-3}$ is undefined in the real number system, there are no real solutions to this equation.

37. Using the property that $|u| = a \Rightarrow u = a$ or $u = -a$, we get
$$|2x - 1| = 7 \quad \Rightarrow \quad 2x - 1 = 7 \text{ or } 2x - 1 = -7 \quad \Rightarrow \quad x = 4 \text{ or } x = -3$$
The solutions are -3 and 4 .

41. Using the property that $|u| = a \Rightarrow u = a$ or $u = -a$, we get
$$\left|x^3 + 14\right| = 13 \quad \Rightarrow \quad x^3 + 14 = 13 \text{ or } x^3 + 14 = -13.$$

In each of these equations, we can solve for x^3 and take cube roots to find x.
$$x^3 + 14 = 13 \quad \Rightarrow \quad x^3 = -1 \quad \Rightarrow \quad x = -1$$
and
$$x^3 + 14 = -13 \quad \Rightarrow \quad x^3 = -27 \quad \Rightarrow \quad x = -3$$
The real solutions to the equation are -1 and -3.

45. This equation can be solved by taking cube roots and solving for x:
$$(x+2)^3 = 24 \quad \Rightarrow \quad x+2 = \sqrt[3]{24} \quad \Rightarrow \quad x+2 = 2\sqrt[3]{3} \quad \Rightarrow \quad x = -2 + 2\sqrt[3]{3}$$
The real solution to the equation is $-2 + \sqrt[3]{3}$.

49. The least common denominator of the fractions $\dfrac{1}{x+1}$ and $\dfrac{1}{x}$ is $x(x+1)$. The fractions can be eliminated by multiplying each side of the equation by this factor.
$$x(x+1)\left(\frac{1}{x+1} + \frac{1}{x}\right) = (1)x(x+1) \quad \Rightarrow \quad x(x+1)\left(\frac{1}{x+1}\right) + x(x+1)\left(\frac{1}{x}\right) = x(x+1)$$
$$\Rightarrow \quad x + (x+1) = x(x+1) \quad \Rightarrow \quad 2x+1 = x^2 + x \quad \Rightarrow \quad 0 = x^2 - x - 1$$
This quadratic equation does not yield to factoring, so we use the quadratic formula, with $a = 1$, $b = -1$, and $c = -1$:
$$x = \frac{-(-1) \pm \sqrt{(-1)^2 - 4(1)(-1)}}{2(1)} = \frac{1 \pm \sqrt{5}}{2}.$$ The real solutions are $\dfrac{1+\sqrt{5}}{2}$ and $\dfrac{1-\sqrt{5}}{2}$.

53. First we solve the equation for $|x|$:
$$|x|^2 - 8|x| = 0 \quad \Rightarrow \quad |x|(|x| - 8) = 0 \quad \Rightarrow \quad |x| = 0 \text{ or } |x| - 8 = 0$$
$$\Rightarrow \quad |x| = 0 \text{ or } |x| = 8$$
Using the property that $|u| = a \Rightarrow u = a$ or $u = -a$, we get
$$x = 0 \quad \text{or} \quad x = 8 \quad \text{or} \quad x = -8 .$$ These are the real solutions to the equation.

57. First we rewrite the equation as a quadratic equation of x^2:
$$x^4 - 2x^2 = 3 \quad \Rightarrow \quad x^4 - 2x^2 - 3 = 0 \quad \Rightarrow \quad \left(x^2\right)^2 - 2\left(x^2\right) - 3 = 0$$
This is similar to the equation $u^2 - 2u - 3 = 0$, which can be solved by factoring:
$$u^2 - 2u - 3 = 0 \quad \Rightarrow \quad (u+1)(u-3) = 0 \quad \Rightarrow \quad u = -1 \text{ or } u = 3.$$
So,
$$\left(x^2\right)^2 - 2\left(x^2\right) - 3 = 0 \quad \Rightarrow \quad \left(x^2 + 1\right)\left(x^2 - 3\right) = 0 \quad \Rightarrow \quad x^2 = -1 \text{ or } x^2 = 3$$
The first equation of this pair, $x^2 = -1$, has no real solutions. Solving the second equation, we get $x^2 = 3 \Rightarrow x = \pm\sqrt{3}$. The real solutions to the equation are $\sqrt{3}$ and $-\sqrt{3}$.

61. Notice that this equation is quadratic in $\dfrac{1}{x-3}$. Instead of simplifying and eliminating the fractions as we normally do, we can first solve for $\dfrac{1}{x-3}$ and then for x. This is similar to the equation $u^2-u=12$, which can be solve by factoring:

$$u^2-u=12 \implies u^2-u-12=0 \implies (u+3)(u-4)=0 \implies u=-3 \text{ or } u=4.$$

So,

$$\left(\frac{1}{x-3}\right)^2-\left(\frac{1}{x-3}\right)=12 \implies \left(\frac{1}{x-3}\right)^2-\left(\frac{1}{x-3}\right)-12=0$$

$$\implies \left(\frac{1}{x-3}+3\right)\left(\frac{1}{x-3}-4\right)=0 \implies \frac{1}{x-3}=-3 \text{ or } \frac{1}{x-3}=4$$

Solving each of these equations yields

$$\frac{1}{x-3}=-3 \implies 1=-3(x-3) \implies 1=-3x+9 \implies x=\frac{8}{3}$$

and

$$\frac{1}{x-3}=4 \implies 1=4(x-3) \implies 1=4x-12 \implies 13=4x \implies x=\frac{13}{4}$$

The real solutions are $\dfrac{8}{3}$ and $\dfrac{13}{4}$.

65. First we square each side, simplify, and square again to eliminate the radicals.

$$\sqrt{x+12}=2+\sqrt{x} \implies \left(\sqrt{x+12}\right)^2=\left(2+\sqrt{x}\right)^2$$

$$\implies x+12=4+4\sqrt{x}+x \implies 8=4\sqrt{x} \implies 2=\sqrt{x} \implies 4=x$$

Apparently, 4 is a root of the equation. This needs to be checked, however, since squaring each side of the equation may introduce extraneous roots.

$$\sqrt{4+12} \overset{?}{=} 2+\sqrt{4} \implies \sqrt{16} \overset{?}{=} 2+2 \text{ True}$$

The real solution is 4.

69. Multiplying each side of the equation by the least common denominator, $2\sqrt{3x+1}\sqrt{x-2}$, and simplify:

$$\sqrt{3x+1}\cdot\frac{1}{2\sqrt{x-2}}+\sqrt{x-2}\cdot\frac{3}{2\sqrt{3x+1}}=0$$

$$\implies 2\sqrt{x-2}\sqrt{3x+1}\left(\sqrt{3x+1}\cdot\frac{1}{2\sqrt{x-2}}+\sqrt{x-2}\cdot\frac{3}{2\sqrt{3x+1}}\right)=(0)2\sqrt{x-2}\sqrt{3x+1}$$

$$\implies \left(\sqrt{3x+1}\right)^2+3\left(\sqrt{x-2}\right)^2=0 \implies 3x+1+3(x-2)=0$$

$$\implies 3x+1+3x-6=0 \implies x=\frac{5}{6}$$

Apparently, $\dfrac{5}{6}$ is a root of the equation. Checking, however, shows this not to be the case, since for $x=\dfrac{5}{6}$, the radical $\sqrt{x-2}$ is not a real number. The equation has no real solutions.

73. a) We subtract c from each side and divide each side by a.

$$ax^2 + bx + c = 0 \quad \Rightarrow \quad ax^2 + bx = -c$$

$$\Rightarrow \quad \frac{ax^2 + bx}{a} = \frac{-c}{a} \quad \Rightarrow \quad x^2 + \frac{b}{a}x = -\frac{c}{a}$$

b) To complete the square, we take half of $\dfrac{b}{a}$, the coefficient of x, square it, and add this quantity to each side.

$$x^2 + \frac{b}{a}x = -\frac{c}{a} \quad \Rightarrow \quad x^2 + \frac{b}{a}x + \left(\frac{b^2}{4a^2}\right) = -\frac{c}{a} + \left(\frac{b^2}{4a^2}\right)$$

$$\Rightarrow \quad \left(x + \frac{b}{2a}\right)^2 = \frac{b^2 - 4ac}{4a^2}$$

c) We solve this equation by extracting roots

$$\left(x + \frac{b}{2a}\right)^2 = \frac{b^2 - 4ac}{4a^2} \Rightarrow x + \frac{b}{2a} = \pm\sqrt{\frac{b^2 - 4ac}{4a^2}}$$

$$\Rightarrow x + \frac{b}{2a} = \pm\frac{\sqrt{b^2 - 4ac}}{2a} \Rightarrow x = -\frac{b}{2a} \pm \frac{\sqrt{b^2 - 4ac}}{2a} = \frac{-b \pm \sqrt{b^2 - 4ac}}{2a}$$

The roots of the equation are $\dfrac{-b + \sqrt{b^2 - 4ac}}{2a}$ or $\dfrac{-b - \sqrt{b^2 - 4ac}}{2a}$.

Section 1.5

1. We simplify the left side and solve for x.

$$2(x - 3) - 3(x + 2) < 5 \quad \Rightarrow \quad 2x - 6 - 3x - 6 < 5$$

$$\Rightarrow \quad -x - 12 < 5 \quad \Rightarrow \quad -x < 17 \Rightarrow x > -17$$

The solution is $(-17, \infty)$.

5. We multiply each side of the inequality by 12 to eliminate the fractions, simplify the left side and solve for x.

$$\frac{x + 2}{3} + \frac{x - 3}{4} \le 2 \quad \Rightarrow \quad 12\left(\frac{x + 2}{3} + \frac{x - 3}{4}\right) \le (2)12$$

$$\Rightarrow \quad 4(x + 2) + 3(x - 3) \le 24 \quad \Rightarrow \quad 7x - 1 \le 24 \Rightarrow 7x \le 25 \Rightarrow x \le \frac{25}{7}$$

The solution is $\left(-\infty, \frac{25}{7}\right]$.

9. We multiply each member of the inequality by 5 to eliminate the fraction, then subtract 6 from each member of the inequality.

$$4 \le \frac{6+x}{5} \le 14 \quad \Rightarrow \quad 5(4) \le 5\left(\frac{6+x}{5}\right) \le 5(14)$$

$$\Rightarrow \quad 20 \le 6+x \le 70 \quad \Rightarrow \quad 14 \le x \le 64$$

The solution is $[14, 64]$.

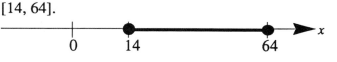

13. The sign graph is shown below.

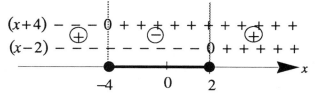

The solution is $[-4, 2]$.

17. First we rewrite the inequality so that there is a zero on the right side.

$$x^2 - 6 \le x \quad \Rightarrow \quad x^2 - x - 6 \le 0 \quad \Rightarrow \quad (x-3)(x+2) \le 0$$

The sign graph is shown below.

$$
\begin{array}{c}
(x+2) \;-\;\underset{\oplus}{-}\;-\;0\;+\;+\underset{\ominus}{+}\;+\;+\;+\;+\;+\underset{\oplus}{+}\;+\;+ \\
(x-3) \;-\;-\;-\;-\;-\;-\;-\;-\;0\;+\;+\;+\;+\;+ \\
\hline
\quad\quad -2 \quad\quad 0 \quad\quad 3
\end{array}
$$

The solution is $[-2, 3]$.

21. This inequality can be expressed as $3x+2 \le -3$ or $3x+2 \ge 3$
Solving each of these we obtain

$$3x+2 \le -3 \quad \Rightarrow \quad x \le -\frac{5}{3} \qquad \text{and} \qquad 3x+2 \ge 3 \quad \Rightarrow \quad x \ge \frac{1}{3}$$

The solution is $(-\infty, -\frac{5}{3}]$ or $[\frac{1}{3}, \infty)$

25. The sign graph is shown below.

$$
\begin{array}{c}
(x-2) \;-\;-\;-\;-\;-\;-\;-\;-\;-\;-\;0\;+\;+\;+ \\
(x+4) \;-\;-\;-\;-\;-\;0\;+\;+\;+\;+\;+\;+\;+\;+\;+ \\
(x+7) \;-\;-\;0\;+\;+\;+\;+\;+\;+\;+\;+\;+\;+\;+\;+\;+\;+ \\
\hline
\quad -7 \quad -4 \quad 0 \quad 2
\end{array}
$$

The solution is $(-\infty, -7)$ or $(-4, 2)$.

29. The sign graph is shown below.

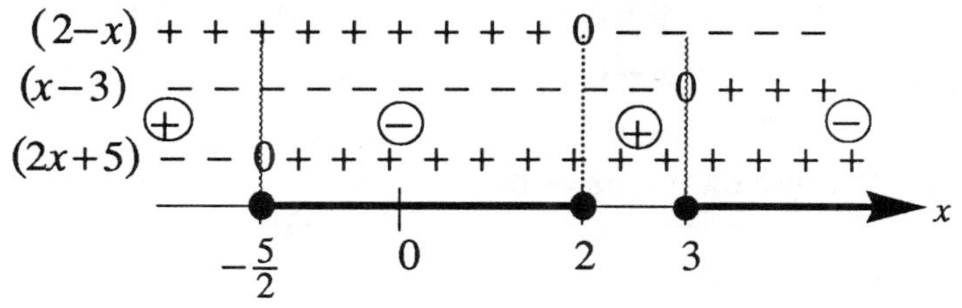

The solution is $[-\frac{5}{2}, 2]$ or $[3, \infty)$.

33. The sign graph is shown below. Notice that 5 is not in the solution set because it makes the denominator zero.

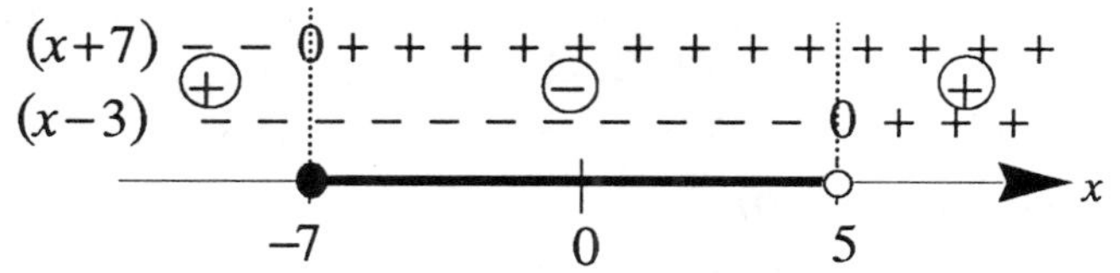

The solution is $[-7, 5)$.

37. First we factor the denominator:

$$\frac{2x}{x^2-25} \geq 0 \Rightarrow \frac{2x}{(x-5)(x+5)} \geq 0$$

The sign graph is shown below. Notice that neither −5 or 5 is in the solution set because it makes the denominator zero.

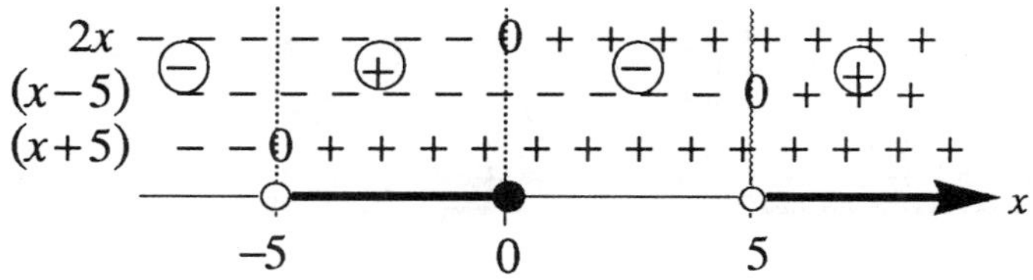

The solution is $(-5, 0]$ or $(5, \infty)$

41. We rewrite the inequality with a zero on the right side and then simplify the left side:

$$\frac{6}{x+1} < \frac{2}{x-2} \Rightarrow \frac{6}{x+1} - \frac{2}{x-2} < 0 \quad \Rightarrow \quad \frac{6(x-2)-2(x+1)}{(x+1)(x-2)} < 0$$

$$\Rightarrow \quad \frac{4x-14}{(x+1)(x-2)} < 0 \quad \Rightarrow \quad \frac{2(2x-7)}{(x+1)(x-2)} < 0$$

Next, we construct a sign graph.

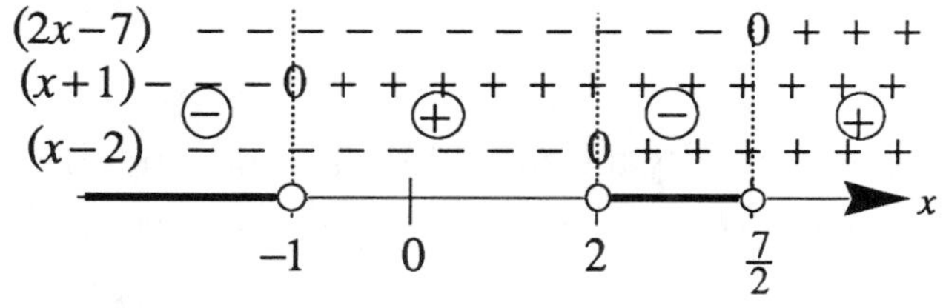

The solution is $(-\infty, -1)$ or $(2, \frac{7}{2})$

45. A real number is a solution to the continued inequality $0 < 3+x < 4x$ if and only if it is a solution to both $0 < 3+x$ and $3+x < 4x$. Solving each of these we obtain

$$0 < 3+x \Rightarrow x > -3 \qquad \text{and} \qquad 3+x < 4x \Rightarrow x > 1$$

The real numbers that are elements of both of these solution sets is $x > 1$

The solution is $(1, \infty)$.

49. **a)** This inequality is equivalent to $-1.2 < 2x-6 < 1.2$. We solve by adding 6 to each member of the inequality, and then dividing by 2.

$$-1.2 < 2x-6 < 1.2 \quad \Rightarrow \quad 4.8 < 2x < 7.2 \quad \Rightarrow \quad 2.4 < x < 3.6$$

The solution is $(2.4, 3.6)$.

b) Consider again the equivalent inequality $-1.2 < 2x-6 < 1.2$. If we divide each member of the inequality by 2, we get

$$-1.2 < 2x-6 < 1.2 \quad \Rightarrow \quad -0.6 < x-3 < 0.6$$

This is in turn equivalent to $|x-3| < 0.6$. This implies that $\delta = 0.6$.

53. **a)** We subtract $9x$ from each side of the equation and then factor:

$$x^3 < 9x \quad \Rightarrow \quad x^3 - 9x < 0 \quad \Rightarrow \quad x(x-3)(x+3) < 0$$

The sign graph is constructed below.

The solution is $(-\infty, -3)$ or $(0, 3)$.

b) We subtract 9 from each side of the equation and then factor:

$$x^2 < 9 \quad \Rightarrow \quad x^2 - 9 < 0 \quad \Rightarrow \quad (x-3)(x+3) < 0$$

The sign graph is constructed below.

The solution is $(-3, 3)$.

57. Suppose that $a > b$ and $c < 0$. By the definition of inequalities on page 43, it follows that $a-b$ is a positive number. Because $a-b$ is a positive number and c is a negative number, the product $(a-b)c = ac - bc$ is a negative number. This, from the definitions on page 433, implies that $ac < bc$. This proves the property.

Section 1.6

1. The second sentence of the problem can be expressed as an equation.
$$\left(\begin{array}{c}\text{cost of}\\\text{calculator}\end{array}\right)+\left(\begin{array}{c}\text{cost of}\\\text{textbook}\end{array}\right)=36$$
Suppose that we let x represent the cost (in dollars) of the textbook. Because the calculator costs \$18.50 less then the textbook, we can represent the cost of the calculator by $x-18.50$. Our equation becomes
$$x+(x-18.50)=36.$$
Solving this linear equation, we obtain
$$x+(x-18.50)=36 \;\Rightarrow\; 2x-18.50=36 \;\Rightarrow\; 2x=54.50 \;\Rightarrow\; x=27.25$$
The cost of the textbook is \$27.25. The cost of the calculator is \$8.75 .

5. For each driver, we are given information about time, distance, and rate traveled. The relation between these quantities is given by
$$\text{Distance} = \text{Rate} \times \text{Time, or } \text{Time} = \frac{\text{Distance}}{\text{Rate}}$$
The first sentence of the problem can be expressed as an equation.
$$\left(\begin{array}{c}\text{number of hours}\\\text{driven by Mike}\end{array}\right)=\left(\begin{array}{c}\text{number of hours}\\\text{driven by Sherry}\end{array}\right)$$
It follows then that
$$\frac{\left(\begin{array}{c}\text{number of miles}\\\text{driven by Mike}\end{array}\right)}{\left(\begin{array}{c}\text{rate (mph)}\\\text{driven by Mike}\end{array}\right)}=\frac{\left(\begin{array}{c}\text{number of miles}\\\text{driven by Sherry}\end{array}\right)}{\left(\begin{array}{c}\text{rate (mph)}\\\text{driven by Sherry}\end{array}\right)}$$
Of these four quantities, we know the distances, and we know a relation between the rates. Suppose that we let r represent the rate at which Sherry drives. Because Mike drives 10 mph faster than Sherry, we can represent Mike's rate by $r+10$. Our equation becomes
$$\frac{135}{r+10}=\frac{125}{r}$$
Solving this equation yields
$$\frac{135}{r+5}=\frac{125}{r} \;\Rightarrow\; 135r=125(r+5)$$
$$\Rightarrow\; 135r=125r+625 \;\Rightarrow\; 10r=625 \;\Rightarrow\; r=62.5$$
Sherry drives at a rate of 62.5 mph and Mike drives at 67.5 mph.

9. In much the same fashion as Example 2 on page 54, we derive this equation from the problem:

$$\left(\begin{array}{c}\text{amount of}\\\text{chlorine remaining}\\\text{after draining}\end{array}\right)+\left(\begin{array}{c}\text{amount of}\\\text{chlorine}\\\text{added}\end{array}\right)=\left(\begin{array}{c}\text{amount of}\\\text{chlorine in}\\\text{final mixture}\end{array}\right)$$

or,

$$0.40\left(\begin{array}{c}\text{amount of original}\\\text{chlorine mixture}\\\text{remaining after draining}\end{array}\right)+\left(\begin{array}{c}\text{amount of}\\\text{chlorine}\\\text{added}\end{array}\right)=0.70\left(\begin{array}{c}\text{amount of final}\\\text{chlorine mixture}\end{array}\right)$$

Suppose that we let x represent the number of gallons that are drained from the original mixture. It follows then that there are $50-x$ gallons remaining and that x gallons of pure liquid chlorine are added. So, our equation becomes

$$0.40(50-x)+x=0.70(50)$$

Solving this equation gives us

$$0.40(50-x)+x=0.70(50) \quad\Rightarrow\quad 20-0.4x+x=35$$
$$\Rightarrow\quad 20+0.6x=35 \quad\Rightarrow\quad 0.6x=15 \quad\Rightarrow\quad x=25$$

This means that 25 gallons must be drained and then added from the tank.

13. The problem suggests this equation:

$$\left(\begin{array}{c}\text{cost of}\\\text{peanuts in mix}\end{array}\right)+\left(\begin{array}{c}\text{cost of}\\\text{cashews in mix}\end{array}\right)=\left(\begin{array}{c}\text{total cost}\\\text{of mix}\end{array}\right)$$

The cost of each ingredient in the mix is the product of the cost per pound times the the number of pounds. So, our equation becomes

$$\left(\begin{array}{c}\text{cost / lb of}\\\text{peanuts}\end{array}\right)\left(\begin{array}{c}\text{number}\\\text{of lbs}\\\text{of peanuts}\end{array}\right)+\left(\begin{array}{c}\text{cost / lb of}\\\text{cashews}\end{array}\right)\left(\begin{array}{c}\text{number}\\\text{of lbs}\\\text{of cashews}\end{array}\right)=\left(\begin{array}{c}\text{cost / lb}\\\text{of mix}\end{array}\right)\left(\begin{array}{c}\text{number}\\\text{of lbs}\\\text{of mix}\end{array}\right)$$

We know the costs/lb, and the number of lbs of mix. Suppose that we let x represent the number of lbs of peanuts used in the mix. Then $10-x$ represents the number of lbs of cashews used in the mix. Our equation becomes

$$0.90x+1.80(10-x)=1.50(10) \quad\Rightarrow\quad 0.9x+18-1.8x=15$$
$$\Rightarrow\quad -0.9x+18=15 \quad\Rightarrow\quad -0.9x=-3 \quad\Rightarrow\quad x=\frac{10}{3}$$

The number of lbs of cashews is $10-\frac{10}{3}=\frac{20}{3}$.

The store owner should use $3\frac{1}{3}$ lbs of peanuts and $6\frac{2}{3}$ lbs of cashews in the mix.

17. The box is constructed from a square (for the bottom) and four identical rectangles (for the sides). The material used must be less than or equal to the material available:

$$\begin{pmatrix} \text{material} \\ \text{for bottom} \end{pmatrix} + 4 \begin{pmatrix} \text{material} \\ \text{for side} \end{pmatrix} \le \begin{pmatrix} \text{total} \\ \text{material} \end{pmatrix}$$

or

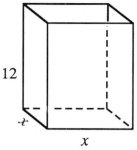

12

x

$$\begin{pmatrix} \text{length} \\ \text{of base} \end{pmatrix}^2 + 4 \begin{pmatrix} \text{length} \\ \text{of base} \end{pmatrix}\begin{pmatrix} \text{length of} \\ \text{height} \end{pmatrix} \le \begin{pmatrix} \text{total} \\ \text{material} \end{pmatrix}$$

We assign x to represent the length of the base. The equation becomes

$$x^2 + 4x(12) \le 448 \implies x^2 + 48x \le 448$$

$$\implies x^2 + 48x - 448 \le 0 \implies (x-8)(x+56) \le 0$$

The sign graph for this quadratic inequality is shown

$(x+56)$ $\oplus$ $- \;\; 0 + + + + + + + + + + + + +$

$(x-8)$ $\ominus$ $- \; - - - - \ominus - - - - \; 0 + \; + +$

                              $\oplus$

                 -56 0 8 x

The solution of the inequality is $[-56, 8]$. A solution to the problem being modeled must be a positive number however, so the side of the square base is in the interval $(0, 8]$. Thus the side of the square cannot exceed 8 cm.

21. The problem suggests this equation:

$$\begin{pmatrix} \text{Fraction of} \\ \text{pool filled by} \\ \text{first pump} \end{pmatrix} + \begin{pmatrix} \text{Fraction of} \\ \text{pool filled by} \\ \text{second pump} \end{pmatrix} = 1$$

For each pump, we are given information about the time it takes to fill the pool. The relation between these quantities is given by

$$\begin{pmatrix} \text{Fraction of} \\ \text{pool filled} \end{pmatrix} = \text{Rate of pump} \times \text{Time pump runs}, \quad \text{or Rate of pump} = \frac{\begin{pmatrix} \text{Fraction of} \\ \text{pool filled} \end{pmatrix}}{\text{Time pump runs}}$$

The equation becomes

$$\begin{pmatrix} \text{Rate of} \\ \text{first pump} \end{pmatrix}\begin{pmatrix} \text{Time of} \\ \text{first pump} \end{pmatrix} + \begin{pmatrix} \text{Rate of} \\ \text{second pump} \end{pmatrix}\begin{pmatrix} \text{Time of} \\ \text{second pump} \end{pmatrix} = 1$$

The rate of the first pump is $\dfrac{1 \, \text{pool filled}}{1\frac{1}{2} \, \text{days}} = \dfrac{2}{3}$ pools/day. The rate of the second pump

is $\dfrac{1 \, \text{pool filled}}{2 \, \text{days}} = \dfrac{1}{2}$ pools/day. We choose x to represent the number of days that the pumps are on.(They are both on the same amount of time.) Our equation now becomes

$$\frac{2}{3}x + \frac{1}{2}x = 1$$

Solving, we obtain $\quad \dfrac{2}{3}x + \dfrac{1}{2}x = 1 \implies \dfrac{7}{6}x = 1 \implies x = \dfrac{6}{7}$

The pumps require $\dfrac{6}{7}$ of a day to fill the pool. This is approximately 20 hours, 34 minutes.

25. Consider the figure at right. From the problem, we determine the equation to be

$$AE + 1 = ED$$

Using the Pythagorean theorem, we determine that

$$AE = \sqrt{49 + x^2} \text{ and } ED = \sqrt{100 + (48 - x)^2}$$

The equation becomes

$$\sqrt{49 + x^2} + 1 = \sqrt{100 + (48 - x)^2}$$

We square both sides and simplify:

$$\left(\sqrt{49 + x^2} + 1\right)^2 = \left(\sqrt{100 + (48 - x)^2}\right)^2$$

$$49 + x^2 + 2\sqrt{49 + x^2} + 1 = 100 + (48 - x)^2$$

$$50 + x^2 + 2\sqrt{49 + x^2} = 100 + 2304 - 96x + x^2$$

$$\sqrt{49 + x^2} = 1177 - 48x$$

We square both sides again and solve the resulting equation by factoring:

$$\sqrt{49 + x^2} = 1177 - 48x \quad \Rightarrow \quad \left(\sqrt{49 + x^2}\right)^2 = (1177 - 48x)^2$$

$$\Rightarrow \quad 49 + x^2 = 2304x^2 - 112992x + 1385329 \quad \Rightarrow \quad 0 = 2303x^2 - 112992x + 138280$$

$$\Rightarrow \quad 0 = (x - 24)(2303x - 57720) \quad \Rightarrow \quad x = 24 \quad \text{(the other root is extraneous)}$$

The distance from B to E is 24 miles.

29. The problem suggests this equation:

$$\left(\begin{array}{c}\text{height} \\ \text{of box}\end{array}\right)\left(\begin{array}{c}\text{length of} \\ \text{base of box}\end{array}\right)^2 = \left(\begin{array}{c}\text{volume} \\ \text{of box}\end{array}\right)$$

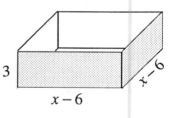

Let x represent the length of the side of the sheet The base of the box is the 6 in less than the side of the sheet, since two squares are cut from the corners. Our equation becomes

$$3(x - 6)^2 = 363$$

Solving, we obtain

$$3(x - 6)^2 = 363 \quad \Rightarrow \quad (x - 6)^2 = 121 \quad \Rightarrow \quad x - 6 = \pm 11 \quad \Rightarrow \quad x = 17 \text{ or } x = -5$$

We discard -5 because the length of the side must be positive. The size of the sheet is 17 inches by 17 inches.

Section 1.7

1. The equation is solved by extracting roots: $x^2 = -25 \Rightarrow x = \pm\sqrt{-25} \Rightarrow x = \pm 5i$
The solutions to the equation are $0 + 5i$ and $0 - 5i$.

5. The equation is solved by extracting roots.
$$(x-3)^2 = -4 \Rightarrow x - 3 = \pm\sqrt{-4} \Rightarrow x = 3 \pm \sqrt{-4} \Rightarrow x = 3 \pm 2i$$
The solutions to the equation are $3 + 2i$ and $3 - 2i$.

9. The equation is solved by extracting roots.
$$4x^2 = -7 \Rightarrow x^2 = -\frac{7}{4} \Rightarrow x = \pm\sqrt{-\frac{7}{4}} \Rightarrow x = \pm\frac{\sqrt{-7}}{2} \Rightarrow x = \pm\frac{\sqrt{7}}{2}i$$
The solutions to the equation are $0 + \frac{\sqrt{7}}{2}i$ and $0 - \frac{\sqrt{7}}{2}i$

13. $(2+i) + (4-7i) = (2+4) + (i-7i) = 6 - 6i$

17. $(2+2i)(1+3i) = 2 + 8i + 6i^2 = 2 + 8i - 6 = -4 + 8i$

21. $(4-i)^2 = 16 - 8i + i^2 = 16 - 8i - 1 = 15 - 8i$

25. $z_1 z_2 = (2-i)(3+7i) = 6 + 11i - 7i^2 = 6 + 11i + 7 = 13 + 11i$

29. $z_1 z_3 - z_2 z_3 = (2-i)(-2i) - (3+7i)(-2i) = \left(-4i + 2i^2\right) - \left(-6i - 14i^2\right)$
$$= (-4i - 2) - (-6i + 14) = -4i - 2 + 6i - 14 = -16 + 2i$$

33. $\left(z_3\right)^{-2} = (-2i)^{-2} = \dfrac{1}{(-2i)^2} = \dfrac{1}{4i^2} = \dfrac{1}{-4} = -\dfrac{1}{4}$

37. $\overline{z_3} z_2 = \overline{(-2i)}(3+7i) = (2i)(3+7i) = 6i + 14i^2 = -14 + 6i$

41. First we write the expression without the negative exponent:
$$\left(z_1\right)^{-3} = (2-i)^{-3} = \frac{1}{(2-i)^3}$$
The next step is to simplify $(2-i)^3$, using a product-factoring formula from page 14:
$$(2-i)^3 = 2^3 - 3 \cdot 2^2 i + 3 \cdot 2i^2 - i^3 = 8 - 12i + 6i^2 - i^3 = 8 - 12i - 6 + i = 2 - 11i$$
Thus $\dfrac{1}{(2-i)^3} = \dfrac{1}{2-11i} = \dfrac{1}{2-11i}\dfrac{2+11i}{2+11i} = \dfrac{2+11i}{2^2+11^2} = \dfrac{2}{125} + \dfrac{11}{125}i$

45. $\overline{z_1} + z_1 = \overline{(2-i)} + (2-i) = (2+i) + (2-i) = 4$

49. We use the quadratic equation with $a = 1$, $b = -2$, and $c = 12$ to find the roots.

$$x = \frac{-(-2) \pm \sqrt{(-2)^2 - 4(1)(3)}}{2(1)} = \frac{2 \pm \sqrt{-8}}{2} = \frac{2 \pm 2\sqrt{-2}}{2} = 1 \pm \sqrt{-2} = 1 \pm \sqrt{2}\, i$$

The complex solutions are $1 + \sqrt{2}\, i$ and $1 - \sqrt{2}\, i$.

53. This equation yields to solution by factoring over the integers:

$$x^2 + 4x - 5 = 0 \quad \Rightarrow \quad (x+5)(x-1) = 0$$
$$\Rightarrow \quad x + 5 = 0 \text{ or } x - 1 = 0 \quad \Rightarrow \quad x = -5 \text{ or } x = 1$$

The complex solutions are -5 and 1 .

57. Consider this as a quadratic equation in x^2:

$$x^4 + 10x^2 + 9 = 0 \quad \Rightarrow \quad \left(x^2 + 1\right)\left(x^2 + 9\right) = 0 \quad \Rightarrow \quad x^2 + 1 = 0 \text{ or } x^2 + 9 = 0$$

Solving each of these equations gives

$$x^2 + 1 = 0 \quad \Rightarrow \quad x^2 = -1 \quad \Rightarrow \quad x = \pm i$$
$$x^2 + 9 = 0 \quad \Rightarrow \quad x^2 = -9 \quad \Rightarrow \quad x = \pm 3i$$

The complex solutions are $0 - i$, $0 + i$, $0 - 3i$, and $0 + 3i$.

61. This equation can be solved using factoring by grouping:

$$x^4 + 5x^3 + x + 5 = 0 \quad \Rightarrow \quad x^3(x+5) + (x+5) = 0$$
$$\Rightarrow \quad \left(x^3 + 1\right)(x+5) = 0 \quad \Rightarrow \quad x^3 + 1 = 0 \text{ or } x + 5 = 0$$

Solving the first equation gives $x + 5 = 0 \quad \Rightarrow \quad x = -5$.
The second of these equations is solved by first factoring:

$$x^3 + 1 = 0 \quad \Rightarrow \quad (x+1)\left(x^2 - x + 1\right) = 0 \quad \Rightarrow \quad x + 1 = 0 \text{ or } x^2 - x + 1 = 0$$

The first of these two equations yields $x + 1 = 0 \quad \Rightarrow \quad x = -1$.
The second of these equations is solved using the quadratic formula:

$$x = \frac{-(-1) \pm \sqrt{(-1)^2 - 4(1)(1)}}{2(1)} = \frac{1 \pm \sqrt{-3}}{2} = \frac{1}{2} \pm \frac{\sqrt{3}}{2}\, i$$

The complex solutions are -5, -1, $\dfrac{1}{2} + \dfrac{\sqrt{3}}{2}\, i$ and $\dfrac{1}{2} - \dfrac{\sqrt{3}}{2}\, i$.

65. This equation yields to factoring:

$$27x^3 - 64 = 0 \implies (3x)^3 - 4^3 = 0 \implies (3x-4)\left((3x)^2 + 4(3x) + 16\right) = 0$$

$$\implies (3x-4)\left(9x^2 + 12x + 16\right) = 0 \implies 3x-4 = 0 \text{ or } 9x^2 + 12x + 16 = 0$$

The first equation yields $3x - 4 = 0 \implies x = \dfrac{4}{3}$.

The second of these equations is solved using the quadratic formula:

$$x = \frac{-(12) \pm \sqrt{(12)^2 - 4(9)(16)}}{2(9)} = \frac{-12 \pm \sqrt{-432}}{18} = \frac{-12 \pm 12\sqrt{-3}}{18} = -\frac{2}{3} \pm \frac{2\sqrt{3}}{3} i$$

The complex solutions are $\dfrac{4}{3}$, $-\dfrac{2}{3} + \dfrac{2\sqrt{3}}{3} i$ and $-\dfrac{2}{3} - \dfrac{2\sqrt{3}}{3} i$.

69. These solutions are checked directly. First, consider $\pm\sqrt{3}$:

$$(\pm\sqrt{3})^4 + 13(\pm\sqrt{3})^2 - 48 \overset{?}{=} 0$$

$$\implies 9 + 13 \cdot 3 - 48 \overset{?}{=} 0 \implies 9 + 39 - 48 \overset{?}{=} 0 \quad \text{True}$$

Next, consider $\pm 4i$:

$$(\pm 4i)^4 + 13(\pm 4i)^2 - 48 \overset{?}{=} 0 \implies 256i^4 + 13 \cdot 16i^2 - 48 \overset{?}{=} 0$$

$$\implies 256 + 13 \cdot 16(-1) - 48 \overset{?}{=} 0 \implies 256 - 208 - 48 \overset{?}{=} 0 \quad \text{True}$$

73. From Problem 72, the roots of the equation we seek are $3 - i$ (the given root) and $3 + i$ (its conjugate). So,

$$x = 3 - i \text{ or } x = 3 + i \implies x - (3-i) = 0 \text{ or } x - (3+i) = 0$$

$$\implies \left(x - (3-i)\right)\left(x - (3+i)\right) = 0 \implies x^2 - 6x + 10 = 0$$

This can be checked by directly solving the equation for x.

Miscellaneous Exercises for Chapter 1

1. Because $\dfrac{3}{5}$ is the ratio of two integers, it is a rational number. Since π is an irrational number, it follows that $-\dfrac{\pi}{4}$ is an irrational number (see Example 1 on page 3); its value is approximately -0.78. The next number, $\sqrt{3}$, is irrational, since 3 is a perfect square of rational number; because $1^2 < 3 < 2^2$ it follows that $1 < \sqrt{3} < 2$. On the other hand, $\sqrt[3]{8}$ is an integer, since $\sqrt[3]{8} = 2$. In summary:

integers: $\sqrt[3]{8}$ rational numbers: $\sqrt[3]{8}, \dfrac{3}{5}$ irrational numbers: $-\dfrac{\pi}{4}, \sqrt{3}$

increasing order: $-\dfrac{\pi}{4}, \dfrac{3}{5}, \sqrt{3}, \sqrt[3]{8}$

5. From the table on page 4, this is equivalent to the interval notation $[-6, 1)$.

9. From the table on page 4, this is equivalent to the interval notation $(-\infty, -2)$ or $(-1, 5]$.

13. From the table on page 4, this is equivalent to the inequality notation $x < -3$ or $-1 \le x < 0$.

17. The square of a number is nonnegative, so $x^2 > 0$ and therefore $x^2 + 1 > 0$. From page 5, we know that if $r > 0$, then $|r| = r$. From this, it follows that $|x^2 + 1| = x^2 + 1$.

21. By the table on page 7, the graph of the numbers described by $|x + 4| < \frac{1}{2}$ is the set of points that are $\frac{1}{2}$ units from -4. This is the interval $\left(-4\frac{1}{2}, -3\frac{1}{2}\right)$.

25. Each part of this problem uses the property that $x^{-n} = \dfrac{1}{x^n}$.

    a) $4x^{-3} = 4 \cdot \dfrac{1}{x^3} = \dfrac{4}{x^3}$

    b) $8x^{-1/3} = 8 \cdot \dfrac{1}{x^{1/3}} = \dfrac{8}{x^{1/3}}$

    c) $(8x)^{-1/3} = \dfrac{1}{(8x)^{1/3}} = \dfrac{1}{8^{1/3}x^{1/3}} = \dfrac{1}{2x^{1/3}}$

    d) $p^{-5}q^2 = \dfrac{1}{p^5}q^2 = \dfrac{q^2}{p^5}$

29. $5x^{-4}\left(-x^{-5}\right)^{-3} = 5x^{-4} \cdot \left(-x^{15}\right) = -5x^{11}$

33. $\left(a^3bc^{-4}\right)^2 \left(\dfrac{a^2c^6}{b^2}\right)^{1/4}$ $\left(a^3bc^{-4}\right)^2 \left(a^2c^6b^{-2}\right)^{1/4} = a^6b^2c^{-8} \cdot a^{1/2}c^{3/2}b^{-1/2}$

$$= a^{13/2}b^{3/2}c^{-13/2} = a^{13/2}b^{3/2}\dfrac{1}{c^{13/2}} = \dfrac{a^{13/2}b^{3/2}}{c^{13/2}}$$

37. $\left(2x^3 + 3x^2 - 10x + 4\right) - 8x\left(x^2 - x + 3\right)$

$$= 2x^3 + 3x^2 - 10x + 4 - 8x^3 + 8x^2 - 24x = -6x^3 + 11x^2 - 34x + 4$$

41. This fits the pattern of the last product-factoring formula in the table on page 14:

$$\left(x^{1/6} - 5\right)\left(x^{1/3} + 5x^{1/6} + 25\right) = \left(x^{1/6}\right)^3 - 5^3 = x^{1/2} - 125 \text{ or } \sqrt{x} - 125$$

45. $24xy - 3x^4y = 3xy\left(8 - x^3\right) = 3xy\left(2^3 - x^3\right) = 3xy(2 - x)\left(4 + 2x + x^2\right)$

49. Factoring $x^{2/5}$ from the expression makes the first term $3x^2$.

$$3x^{12/5} - x^{7/5} + 6x^{2/5} = x^{2/5}\left(3x^2 - x + 6\right)$$

53.
$$\frac{(x+h)^2+5(x+h)-x^2-5x}{h}=\frac{x^2+2xh+h^2+5x+5h-x^2-5x}{h}$$

$$=\frac{2xh+h^2+5h}{h}=\frac{h(2x+h+5)}{h}=2x+h+5$$

57. The least common multiple of $x+2$ and $2x-1$ is $(x+2)(2x-1)$.

$$\frac{3}{x+2}+\frac{1}{2x+1}=\frac{3}{(x+2)}\frac{(2x+1)}{(2x+1)}+\frac{1}{(2x+1)}\frac{(x+2)}{(x+2)}$$

$$=\frac{3(2x+1)+1(x+2)}{(2x+1)(x+2)}=\frac{7x+5}{(2x+1)(x+2)}$$

61. The least common denominator of the rational expressions in the numerator and the denominator is $x^2(x+1)$. We multiply the numerator and the denominator by this least common denominator and simplify.

$$\frac{\dfrac{2}{x+1}+\dfrac{1}{x}}{\dfrac{3}{x^2}}\cdot\frac{x^2(x+1)}{x^2(x+1)}=\frac{\dfrac{2}{x+1}\cdot x^2(x+1)+\dfrac{1}{x}\cdot x^2(x+1)}{\dfrac{3}{x^2}x^2(x+1)}$$

$$=\frac{2x^2+x(x+1)}{3(x+1)}=\frac{3x^2+x}{3(x+1)}\quad\text{or}\quad\frac{x(3x+1)}{3(x+1)}$$

65. Each term is simplified and then added.

$$10x\sqrt{3-4x}-\left(5x^2+1\right)\frac{2}{\sqrt{3-4x}}=\frac{10x\sqrt{3-4x}}{1}\cdot\frac{\sqrt{3-4x}}{\sqrt{3-4x}}-\frac{2\left(5x^2+1\right)}{\sqrt{3-4x}}$$

$$=\frac{10x(3-4x)}{\sqrt{3-4x}}-\frac{2\left(5x^2+1\right)}{\sqrt{3-4x}}=\frac{10x(3-4x)-2\left(5x^2+1\right)}{\sqrt{3-4x}}$$

$$=\frac{30x-40x^2-10x^2-2}{\sqrt{3-4x}}=\frac{-50x^2+30x-2}{\sqrt{3-4x}}$$

The final step is to rationalize the denominator:

$$\frac{-50x^2+30x-2}{\sqrt{3-4x}}=\frac{-50x^2+30x-2}{\sqrt{3-4x}}\frac{\sqrt{3-4x}}{\sqrt{3-4x}}=\frac{\left(-50x^2+30x-2\right)\sqrt{3-4x}}{3-4x}$$

69. Performing the long division, we get

$$
\begin{array}{r}
3\\
x^2+x+1\overline{)3x^2+5x+9}\\
3x^2+3x+3\\
\hline
2x+6
\end{array}
$$

$$\Rightarrow\quad x^2+5x+9=3+\frac{2x+6}{x^2+x+1}$$

73. The equation can be solved by factoring after first simplifying and arranging a zero on the right side:
$$(x+1)(x+2) = 12 \implies x^2 + 3x + 2 = 12 \implies x^2 + 3x - 10 = 0$$
$$\implies (x+5)(x-2) = 0 \implies x+5=0 \text{ or } x-2=0 \implies x=-5 \text{ or } x=2$$
The real solutions to the equation are −5 and 2.

77. We square both sides and then solve the resulting linear equation:
$$(4x-3)^{1/2} = 2 \implies \left[(4x-3)^{1/2}\right]^2 = [2]^2 \implies 4x-3=4 \implies x=\tfrac{7}{4}$$
The only real solution to the equation is $\tfrac{7}{4}$.

81. We rewrite the equation with a zero on the right side.
$$x^2 - 3x = 5 \implies x^2 - 3x - 5 = 0$$
The equation does not yield to solution by factoring, so the quadratic formula is required.
$$x = \frac{-(-3) \pm \sqrt{(-3)^2 - 4(1)(-5)}}{2(1)} = \frac{3 \pm \sqrt{29}}{2}$$
The real solutions to the equation are $\dfrac{3+\sqrt{29}}{2}$ and $\dfrac{3-\sqrt{29}}{2}$.

85. Using the property that $|u| = a \implies u = a$ or $u = -a$, we get
$$\left|\frac{1}{x-2}\right| = 4x \implies \frac{1}{x-2} = 4x \text{ or } \frac{1}{x-2} = -4x$$
Consider the first equation. It can be rewritten as a quadratic equation that is solved using the quadratic formula with $a = 4$, $b = -8$, and $c = -1$.
$$\frac{1}{x-2} = 4x \implies 1 = 4x(x-2) \implies 1 = 4x^2 - 8x \implies 0 = 4x^2 - 8x - 1$$
$$\implies x = \frac{-(-8) \pm \sqrt{(-8)^2 - 4(4)(-1)}}{2(4)} = \frac{8 \pm \sqrt{80}}{8} = \frac{2 \pm \sqrt{5}}{2}$$
The second equation is solved in the same manner as the first.
$$\frac{1}{x-2} = -4x \implies 1 = -4x(x-2) \implies 1 = -4x^2 + 8x \implies 0 = 4x^2 - 8x + 1$$
$$\implies x = \frac{-(-8) \pm \sqrt{(-8)^2 - 4(4)(1)}}{2(4)} = \frac{8 \pm \sqrt{48}}{8} = \frac{2 \pm \sqrt{3}}{2}$$
The real solutions to the equation apparently are $\dfrac{2+\sqrt{5}}{2}$, $\dfrac{2-\sqrt{5}}{2}$, $\dfrac{2+\sqrt{3}}{2}$, and $\dfrac{2-\sqrt{3}}{2}$.

However, when $\dfrac{2-\sqrt{5}}{2}$ is checked in the original equation, we get a positive number on the right side (it is an absolute value) and a negative number on the left. This implies that this number is not a root. The real solutions are $\dfrac{2+\sqrt{5}}{2}$, $\dfrac{2+\sqrt{3}}{2}$, and $\dfrac{2-\sqrt{3}}{2}$.

89. First we rewrite the inequality so that there is a zero on the right side.
$$4x^2 + 7x \geq 2 \quad \Rightarrow \quad 4x^2 + 7x - 2 \geq 0 \quad \Rightarrow \quad (4x - 1)(x + 2) \geq 0$$
The sign graph is shown below.

The solution is $(-\infty, -2]$ or $[\frac{1}{4}, +\infty)$.

93. The problem suggests this equation:
$$\left(\begin{array}{c} \text{Number of labels} \\ \text{by first machine} \end{array}\right) + \left(\begin{array}{c} \text{Number of labels} \\ \text{by second machine} \end{array}\right) = \left(\begin{array}{c} \text{Total number} \\ \text{of labels} \end{array}\right)$$
For each machine, we are given information about the time it takes to label 250 packages. The relation between these quantities is given by

Number of labels = Rate of machine × Time machine works,

or   Rate of machine = $\dfrac{\text{Number of labels}}{\text{Time machine works}}$

The equation becomes
$$\left(\begin{array}{c} \text{Rate of} \\ \text{first} \\ \text{machine} \end{array}\right)\left(\begin{array}{c} \text{Time of first} \\ \text{machine} \end{array}\right) + \left(\begin{array}{c} \text{Rate of} \\ \text{second} \\ \text{machine} \end{array}\right)\left(\begin{array}{c} \text{Time of second} \\ \text{machine} \end{array}\right) = \left(\begin{array}{c} \text{Total number} \\ \text{of labels} \end{array}\right)$$
The rate of the first machine is $\dfrac{250\,\text{labels}}{10\,\text{minutes}} = 25$ labels/minute. The rate of the second

machine is $\dfrac{250\,\text{labels}}{8\,\text{minutes}} = \dfrac{125}{4}$ labels/minute. We choose x to represent the time that the

machines are working. (They both work the same amount of time.)   Our equation now becomes
$$25x + \frac{125}{4}x = 250$$
Solving, we obtain
$$25x + \frac{125}{4}x = 250 \quad \Rightarrow \quad \frac{225}{4}x = 250 \quad \Rightarrow \quad x = \frac{40}{9}$$
The machines require $4\frac{4}{9}$ minutes to label 250 packages. This is approximately 4 minutes, 27 seconds.

97. The equation for this problem comes from the Pythagorean theorem.

$$\left(\begin{array}{c}\text{Distance traveled}\\ \text{by}\\ \text{westbound boat}\end{array}\right)^2 + \left(\begin{array}{c}\text{Distance traveled}\\ \text{by}\\ \text{northbound boat}\end{array}\right)^2 = \left(\begin{array}{c}\text{Distance}\\ \text{between}\\ \text{boats}\end{array}\right)^2$$

For each boat, we are given information about time, distance, and rate traveled. The relation between these quantities is given by

$$\text{Distance} = \text{Rate} \times \text{Time, or Time} = \frac{\text{Distance}}{\text{Rate}}$$

Our equation becomes

$$\left[\left(\begin{array}{c}\text{time of}\\ \text{westbound}\\ \text{boat}\end{array}\right)\left(\begin{array}{c}\text{rate of}\\ \text{westbound}\\ \text{boat}\end{array}\right)\right]^2 + \left[\left(\begin{array}{c}\text{time of}\\ \text{northbound}\\ \text{boat}\end{array}\right)\left(\begin{array}{c}\text{rate of}\\ \text{northbound}\\ \text{boat}\end{array}\right)\right]^2 = \left(\begin{array}{c}\text{Distance}\\ \text{between}\\ \text{boats}\end{array}\right)^2$$

If we assign r to represent the rate of the westbound boat, then the rate of the northbound boat is $r+5$. Because each boat travels for 2 hours, the equation becomes

$$(2r)^2 + (2(r+5))^2 = 100$$

Solving this equation yields

$$(2r)^2 + (2(r+5))^2 = 100 \quad\Rightarrow\quad 4r^2 + 4r^2 + 20r + 50 = 100 \Rightarrow 8r^2 + 20r - 50 = 0$$

We use the quadratic formula to solve this:

$$x = \frac{-(20) \pm \sqrt{(20)^2 - 4(8)(-50)}}{2(8)} = \frac{-20 \pm \sqrt{2000}}{16} = \frac{-5 \pm 5\sqrt{5}}{4}$$

The roots of this equation are $\dfrac{-5 + 5\sqrt{5}}{4}$ and $\dfrac{-5 - 5\sqrt{5}}{4}$. We discard $\dfrac{-5 - 5\sqrt{5}}{4}$ because the rate of the westbound boat is positive, and this number is negative. To the nearest mile per hour, $\dfrac{-5 + 5\sqrt{5}}{4} = 33$. The rate of the westbound boat is 33 mph and the rate of the northbound boat is 38 mph. Keystrokes: $5\;\boxed{+/-}\;\boxed{+}\;5\;\boxed{\sqrt{\ }}\;\boxed{=}\;\boxed{\div}\;4\;\boxed{=}$

101. $(3 + 4i)(-2 + 8i) = -6 + 16i + 32i^2 = -6 + 16i - 32 = -38 + 16i$

105. $\dfrac{3 - 2i}{4 + 5i} = \dfrac{3 - 2i}{4 + 5i}\dfrac{4 - 5i}{4 - 5i} = \dfrac{12 - 23i + 10i^2}{16 - 25i^2} = \dfrac{12 - 23i - 10}{16 + 25} = \dfrac{2 - 23i}{41} = \dfrac{2}{41} - \dfrac{23}{41}i$

109. We use the quadratic equation with $a = 1$, $b = -2$, and $c = 2$ to find the roots.

$$x = \frac{-(-2) \pm \sqrt{(-2)^2 - 4(1)(2)}}{2(1)} = \frac{2 \pm \sqrt{4 - 8}}{2} = \frac{2 \pm \sqrt{-4}}{2} = \frac{2 \pm 2i}{2} = 1 \pm i$$

The complex solutions are $1 + i$ and $1 - i$.

Chapter 2

Section 2.1

1. By the distance formula, we get
$$d = \sqrt{(\Delta x)^2 + (\Delta y)^2} = \sqrt{(-6-2)^2 + (-6-4)^2} = \sqrt{164} = 2\sqrt{41}$$
By the midpoint formula, we get.
$$\left(\frac{x_1 + x_2}{2}, \frac{y_1 + y_2}{2}\right) = \left(\frac{2-6}{2}, \frac{4-6}{2}\right) = (-2, -1)$$

5. By the distance formula, we get
$$d = \sqrt{(\Delta x)^2 + (\Delta y)^2} = \sqrt{\left(\frac{2}{\sqrt{3}} - \sqrt{3}\right)^2 + \left(\sqrt{2} + \sqrt{2}\right)^2}$$
$$= \sqrt{\left(-\frac{1}{\sqrt{3}}\right)^2 + \left(2\sqrt{2}\right)^2} = \sqrt{\frac{1}{3} + 8} = \sqrt{\frac{25}{3}} = \frac{5}{\sqrt{3}} \text{ or } \frac{5\sqrt{3}}{3}$$
By the midpoint formula, we get.
$$\left(\frac{x_1 + x_2}{2}, \frac{y_1 + y_2}{2}\right) = \left(\frac{\frac{2}{\sqrt{3}} + \sqrt{3}}{2}, \frac{\sqrt{2} - \sqrt{2}}{2}\right) = \left(\frac{\frac{2}{\sqrt{3}} + \sqrt{3}}{2}, 0\right)$$
The first component can be simplified by multiplying numerator and denominator by $\sqrt{3}$,

and then rationalizing the denominator: $\dfrac{\frac{2}{\sqrt{3}} + \sqrt{3}}{2} \cdot \dfrac{\sqrt{3}}{\sqrt{3}} = \dfrac{2+3}{2\sqrt{3}} = \dfrac{5}{2\sqrt{3}} = \dfrac{5}{2\sqrt{3}} \cdot \dfrac{\sqrt{3}}{\sqrt{3}} = \dfrac{5\sqrt{3}}{6}$

The midpoint is $\left(\dfrac{5\sqrt{3}}{6}, 0\right)$.

9. We use the tests on page 84 to check symmetry for the graph of $y = |x+5|$

y-axis test:	$y = |(-x) + 5|$ or $y = |5 - x| \Rightarrow$ No
x-axis test:	$(-y) = |x+5|$ or $y = -|x+5| \Rightarrow$ No
origin test:	$(-y) = |(-x) + 5|$ or $y = -|5 - x| \Rightarrow$ No
$y = x$ test:	$x = |y+5| \Rightarrow$ No

Next, we compute and plot a sufficient number of ordered pairs to discover the graph.

x	y
-7	$|-7+5| = 2$
-5	$|-5+5| = 0$
-2	$|-2+5| = 3$
0	$|0+5| = 5$
2	$|2+5| = 7$

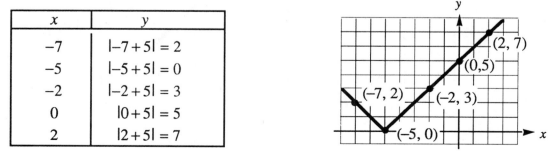

13. We use the tests on page 84 to check symmetry for the graph of $xy = 4$

y-axis test: $(-x)y = 4$ or $-xy = 4 \Rightarrow$ No

x-axis test: $x(-y) = 4$ or $-xy = 4 \Rightarrow$ No

origin test: $(-x)(-y) = 4$ or $xy = 4 \Rightarrow$ Yes

$y = x$ test: $yx = 4$ or $xy = 4 \Rightarrow$ Yes

Next, we compute and plot a sufficient number of ordered pairs in which x and y are positive and then use the symmetry with respect to the origin to discover the graph.

x	y
1	4
2	2
4	1

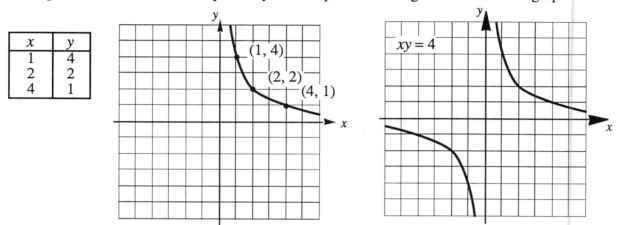

17. We use the tests on page 84 to check symmetry for the graph of $|x| + |y| = 6$

y-axis test: $|-x| + |y| = 6$ or $|x| + |y| = 6 \Rightarrow$ Yes

x-axis test: $|x| + |-y| = 6$ or $|x| + |y| = 6 \Rightarrow$ Yes

origin test: $|-x| + |-y| = 6$ or $|x| + |y| = 6 \Rightarrow$ Yes

$y = x$ test: $|y| + |x| = 6$ or $|x| + |y| = 6 \Rightarrow$ Yes

Next, we compute and plot a sufficient number of ordered pairs in which x and y are positive and then use the symmetries discovered above to construct the graph.

x	y
0	6
2	4
4	2
6	0

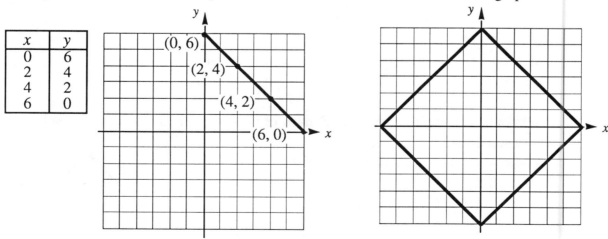

21. The equation is equivalent to $(x-0)^2+(y-(-6))^2 = 6^2$.
Comparing the equation with the general form of a circle
$(x-h)^2+(y-k)^2 = r^2$, we find that the center is
$(h, k) = (0, -6)$ and the radius is $r = 6$

25. The graph can be completed by reflecting a few points
of the existing graph through the y-axis and drawing a
line.

29. The graph can be completed by reflecting a few points
of the existing graph through the y-axis and drawing a
curve.

33. We multiply each side of the equation by $\frac{1}{3}$ and then complete
the square for both the x and y terms:

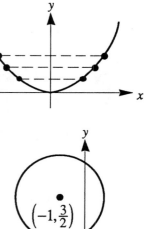

$$\tfrac{1}{3}\left(3x^2 + 3y^2 + 6x - 9y + 3\right) = \tfrac{1}{3}(0)$$

$$x^2 + y^2 + 2x - 3y + 1 = 0$$

$$\left(x^2 + 2x \quad\right) + \left(y^2 - 3y \quad\right) = -1$$

$$\left(x^2 + 2x + 1\right) + \left(y^2 - 3y + \tfrac{9}{4}\right) = -1 + 1 + \tfrac{9}{4}$$

$$(x+1)^2 + \left(y - \tfrac{3}{2}\right)^2 = \tfrac{9}{4}$$

The circle has center $(-1, \frac{3}{2})$ and radius $\frac{3}{2}$.

37. From the figure at right, it should seem reasonable that the center
of the circle is $(-3, 3)$. The radius is given as 3. In the form
$(x-h)^2+(y-k)^2 = r^2$, the equation of the circle is
$(x+3)^2+(y-3)^2 = 9$

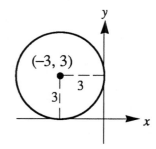

41. From the figure at right, it should seem clear that the center of the circle is $(0, -2)$. The radius is 6. In the form $(x-h)^2 + (y-k)^2 = r^2$, the equation of the circle is
$$x^2 + (y+2)^2 = 36$$

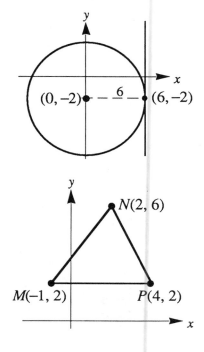

45. We compute the lengths of the three sides of the triangle:
$$d(M,N) = \sqrt{(2+1)^2 + (6-2)^2} = 5$$
$$d(N,P) = \sqrt{(4-2)^2 + (2-6)^2} = 2\sqrt{5}$$
$$d(P,M) = \sqrt{(-1-4)^2 + (2-2)^2} = 5$$
Because $MN = PM$, triangle MNP is an isosceles triangle

49. We compute the lengths of the three sides of the triangle:
$$d(A,B) = \sqrt{(-10+12)^2 + (14-4)^2} = \sqrt{104} = 2\sqrt{26}$$
$$d(B,C) = \sqrt{(8+10)^2 + (0-14)^2} = \sqrt{520} = 2\sqrt{130}$$
$$d(C,A) = \sqrt{(-12-8)^2 + (4-0)^2} = \sqrt{416} = 4\sqrt{26}$$
We can verify that this triangle is a right triangle by the Pythagorean theorem (see Problem 47):
$$AB^2 + CA^2 \overset{?}{=} BC^2$$
$$\left(2\sqrt{26}\right)^2 + \left(4\sqrt{26}\right)^2 \overset{?}{=} \left(2\sqrt{130}\right)^2$$
$$104 + 416 = 520$$
This is a right triangle with hypotenuse BC. By the midpoint formula, the midpoint M of BC is $\left(\dfrac{-10+8}{2}, \dfrac{14+0}{2}\right) = (-1, 7)$. The statement in the problem is verified by the distance formula:
$$d(A,M) = \sqrt{(-12+1)^2 + (4-7)^2} = \sqrt{130}$$
$$d(B,M) = \sqrt{(-10+1)^2 + (14-7)^2} = \sqrt{130}$$
$$d(C,M) = \sqrt{(8+1)^2 + (0-7)^2} = \sqrt{130}$$

53. We compute the lengths of the four sides of the quadrilateral :

$$d(J,K) = \sqrt{(10-3)^2 + (8-4)^2} = \sqrt{65}$$

$$d(K,G) = \sqrt{(16-10)^2 + (4-8)^2} = \sqrt{52}$$

$$d(G,H) = \sqrt{(10-16)^2 + (0-4)^2} = \sqrt{52}$$

$$d(H,J) = \sqrt{(3-10)^2 + (4-0)^2} = \sqrt{65}$$

Because the four sides are not all of equal length, the quadrilateral is not a rhombus.

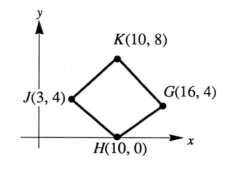

Section 2.2

1. The domain of the function is the set of all real numbers. The range is $[-5, +\infty)$ because x^2 is a nonnegative number and therefore $x^2 - 5$ can be no smaller than -5.

5. The domain of the function is the set of all real numbers. The value of the function for any x is the value of x less the greatest integer in x. (For example, $h(4.68) = 4.68 - 4 = 0.68$.) The range therefore is $[0, 1)$.

9. In order for the function to be defined, the radicand must be nonnegative. This means that the domain must be the set of all t such that $5 - 11t \geq 0$. Solving this inequality yields:

$5 - 11t \geq 0 \Rightarrow -11t \geq -5 \Rightarrow t \leq \frac{5}{11}$. The domain is $(-\infty, \frac{5}{11}]$.

13. a) $f(-2) = (-2) - 2|-2| = (-2) - 2(2) = (-2) - 4 = -6$

   b) $-f(2) = -(2 - 2|2|) = -(2 - 2(2)) = -(2 - 4) = 2$

17. a) First, note that $f(1) = 1 - 2|1| = 1 - 2 = -1$. It follows that $f(1) + 3 = -1 + 3 = 2$.

   b) From a), $f(1) = -1$. Also, $f(3) = 3 - 2|3| = 3 - 6 = -3$. It follows that
   $$f(1) + f(3) = (-1) + (-3) = -4$$

21. a) $g(-a) = (-a)^2 + 3(-a) - 1 = a^2 - 3a - 1$

   b) $-g(a) = -(a^2 + 3a - 1) = -a^2 - 3a + 1$

25. $\dfrac{f(3+h) - f(3)}{h} = \dfrac{\left[2(3+h)^2 - 9\right] - \left[2(3)^2 - 9\right]}{h} = \dfrac{\left[2\left(9 + 6h + h^2\right) - 9\right] - \left[2(9) - 9\right]}{h}$

$$= \dfrac{\left[9 + 12h + 2h^2\right] - [9]}{h} = \dfrac{12h + 2h^2}{h} = \dfrac{h(12 + 2h)}{h} = 12 + 2h$$

29. $\dfrac{f(x) - f(a)}{x - a} = \dfrac{\left[2x^2 - 9\right] - \left[2a^2 - 9\right]}{x - a} = \dfrac{2x^2 - 9 - 2a^2 + 9}{x - a} = \dfrac{2x^2 - 2a^2}{x - a}$

This rational expression can be reduced by factoring:

$\dfrac{2x^2 - 2a^2}{x - a} = \dfrac{2(x - a)(x + a)}{x - a} = 2(x + a)$ or $2x + 2a$

33. Let $y = \dfrac{2x}{x - 3}$. This equation can be solved for x in terms of y:

$y = \dfrac{2x}{x - 3} \Rightarrow y(x - 3) = 2x \Rightarrow xy - 3y = 2x \Rightarrow xy - 2x = 3y \Rightarrow x(y - 2) = 3y \Rightarrow x = \dfrac{3y}{y - 2}$

For a given range element y, one can use this last equation to determine the domain element x such that $h(x) = y$. Examining this equation shows that it is possible to determine such an x for any value y except for $y = 2$. Thus the range of h is all real numbers except $y = 2$

37. Let A be the name of the function we seek and p be the perimeter of the square. The input of this function is the p and the output is $A(p)$, the area of the square. We start with basic facts from geometry about squares:

$$\begin{pmatrix} \text{area of} \\ \text{square} \end{pmatrix} = \begin{pmatrix} \text{length of} \\ \text{square} \end{pmatrix}^2$$

and $\begin{pmatrix} \text{perimeter} \\ \text{of square} \end{pmatrix} = 4\begin{pmatrix} \text{length of} \\ \text{square} \end{pmatrix}$

From these relations it follows that $\begin{pmatrix} \text{area of} \\ \text{square} \end{pmatrix} = \left[\dfrac{1}{4}\begin{pmatrix} \text{perimeter} \\ \text{of square} \end{pmatrix}\right]^2$

Writing this as a function, we get $A(p) = \left(\dfrac{p}{4}\right)^2$ or $A(p) = \dfrac{p^2}{16}$. The domain of the function is $p > 0$, since the perimeter must be positive.

41. In the figure at right, the sides of the triangle are denoted as s and the height of the triangle is denoted as h. The triangle PQM is a right triangle, so by the Pythagorean theorem, $h^2 + \left(\frac{1}{2}s\right)^2 = s^2$, or

$$h = \sqrt{s^2 - \left(\tfrac{1}{2}s\right)^2} = \sqrt{s^2 - \tfrac{1}{4}s^2} = \sqrt{\tfrac{3}{4}s^2} = \dfrac{\sqrt{3}}{2}s$$

Let A be the name of the function we seek. The input of this function is the s and the output is $A(s)$, the area of the triangle.

From geometry, we know that $\begin{pmatrix} \text{area of} \\ \text{triangle} \end{pmatrix} = \dfrac{1}{2}\begin{pmatrix} \text{base of} \\ \text{triangle} \end{pmatrix}\begin{pmatrix} \text{height of} \\ \text{triangle} \end{pmatrix}$

So, $A(s) = \frac{1}{2} \cdot s \cdot \frac{\sqrt{3}}{2}s$ or $A(s) = \frac{\sqrt{3}}{4}s^2$. The domain of the function is $s > 0$, since the length of the side must be positive.

45. From the figure at right, we get
$$\begin{pmatrix} \text{Length} \\ \text{of rope} \end{pmatrix} = AB + BC.$$

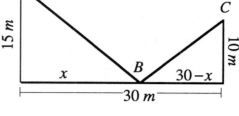

The line segment AB is the hypotenuse of a right triangle with legs 15 and x, so
$$AB = \sqrt{15^2 + x^2} = \sqrt{225 + x^2}$$
In a similar fashion, the line segment BC is the hypotenuse of a right triangle with legs 10 and $30 - x$, so
$$BC = \sqrt{10^2 + (30 - x)^2} = \sqrt{100 + (30 - x)^2}.$$
If we let d be the function that represents the length of the rope, then
$$d(x) = \sqrt{225 + x^2} + \sqrt{100 + (30 - x)^2}.$$
Because the peg is between the trees, it follows that the domain is $0 \le x \le 30$.

49. From the figure at right, we get
$$\begin{pmatrix} \text{total} \\ \text{area} \end{pmatrix} = \begin{pmatrix} \text{total} \\ \text{length} \end{pmatrix}\begin{pmatrix} \text{length} \\ \text{of pen} \end{pmatrix}$$

total length

There is 80 feet of fencing total, so
$$2\begin{pmatrix} \text{total} \\ \text{length} \end{pmatrix} + 4\begin{pmatrix} \text{length} \\ \text{of pen} \end{pmatrix} = 80 \text{ or } \begin{pmatrix} \text{total} \\ \text{length} \end{pmatrix} = 40 - 2\begin{pmatrix} \text{length} \\ \text{of pen} \end{pmatrix}$$

Let x be the length of each pen. Then $\begin{pmatrix} \text{total} \\ \text{length} \end{pmatrix} = 40 - 2x$.

Let A be the function we seek. The input of this function is the x and the output is $A(x)$, the total area of the pens. Thus,
$$A(x) = 40x - 2x^2$$
The length of each pen must be positive, and because there are four lengths of pen that in total cannot exceed 80 meters, the length of each pen cannot exceed 20 meters. The domain of the function is $0 \le x \le 20$.

53. Let x be the number of meters of phone line along the bank and c be the function we seek. The input of this function is the x, the distance of phone line strung along the bank, and the output is $c(x)$, the total cost of the phone line. The distance that the line is strung over the water is (by the Pythagorean theorem)
$$BC = \sqrt{20^2 + (30 - x)^2} = \sqrt{400 + (30 - x)^2}.$$

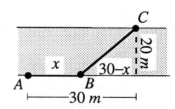

It should seem reasonable that
$$\begin{pmatrix} \text{total cost of} \\ \text{phone line} \end{pmatrix} = \begin{pmatrix} \text{cost of line} \\ \text{on bank} \end{pmatrix} + \begin{pmatrix} \text{cost of line} \\ \text{on water} \end{pmatrix}$$
$$= \begin{pmatrix} \text{cost / meter of} \\ \text{line on bank} \end{pmatrix}\begin{pmatrix} \text{length of line} \\ \text{on bank} \end{pmatrix} + \begin{pmatrix} \text{cost / meter of} \\ \text{line on water} \end{pmatrix}\begin{pmatrix} \text{length of line} \\ \text{on water} \end{pmatrix}$$

So, $\begin{pmatrix} \text{total cost of} \\ \text{phone line} \end{pmatrix} = 2 \cdot AB + 5 \cdot BC$ or $c(x) = 2x + 5\sqrt{400 + (30 - x)^2}.$

Because point B is between 0 and 30 meters of point A, the domain is $0 \le x \le 30$.

Section 2.3

1. The graph fails the vertical line test as shown in the figure at right. It is possible to intercept the graph at more than one point with a vertical line.

5. The graph passes the vertical line test. A few vertical lines are shown in the figure at right. None crosses the graph at more than one point.

9. a) The y-intercepts are found by setting $x = 0$ and solving the equation for y:

   $$y = \sqrt{9 - 0^2} = \sqrt{9} = 3 \Rightarrow \text{The } y\text{-intercept is } (0, 3)$$

   The x-intercepts are found by setting $y = 0$ and solving the equation for x:

   $$\text{for } x < 0: \ 0 = -\sqrt{9 - x^2} \ \Rightarrow \ 0 = 9 - x^2 \ \Rightarrow \ x^2 = 9 \ \Rightarrow \ x = -3$$

   $$\Rightarrow \text{An } x\text{-intercept is } (-3, 0)$$

   $$\text{for } x \geq 0: \ 0 = \sqrt{9 - x^2} \ \Rightarrow \ 0 = 9 - x^2 \ \Rightarrow \ x^2 = 9 \ \Rightarrow \ x = 3$$

   $$\Rightarrow \text{An } x\text{-intercept is } (3, 0)$$

   From the graph, the function is positive over $0 \leq x < 3$ and negative over $-3 < x < 0$.

   b) From the graph, the function is increasing nowhere and decreasing over $-3 < x < 0$ and $0 < x < 3$. There are no turning points.

13. The graph in question is the graph of $y = |x|$ translated -3 units horizontally (3 units left) and -2 units vertically (2 units down). The equation is
   $$y - (-2) = |x - (-3)|, \text{ or } y + 2 = |x + 3|.$$

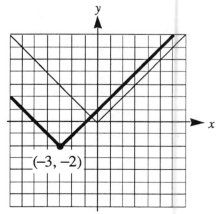
$(-3, -2)$

17. The graph in question is the graph of $y = x^3$ translated 2 units horizontally (2 units right) and −1 unit vertically (1 unit up). The equation is $y - 1 = (x - 2)^3$.

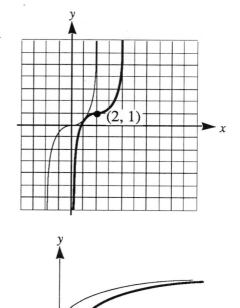

21. The graph of $y = \sqrt{x}$ is in the catalog on page 102.

The graph of $y = \sqrt{x - 1}$ is the graph of $y = \sqrt{x}$ translated 1 unit horizontally (1 unit right)

To find the y-intercept of $y = \sqrt{x - 1}$, we set $x = 0$ and solve for y:

$y = \sqrt{0 - 1}$ (this is not a real number)

$\Rightarrow$ no y-intercept

To find x-intercepts, we set $y = 0$ and solve for x:

$0 = \sqrt{x - 1} \Rightarrow x - 1 = 0 \Rightarrow x = 1 \Rightarrow$ x-intercept is $(1, 0)$.

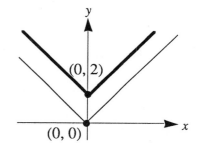

25. The graph of $y = |x|$ is in the catalog on page 102.

The graph of $y - 2 = |x|$ is the graph of $y = |x|$ translated 2 units vertically (2 units up)

To find the y-intercept of $y - 2 = |x|$, we set $x = 0$ and solve for y:

$y - 2 = |0| \Rightarrow y = 2 \Rightarrow$ y-intercept is $(0, 2)$

To find x-intercepts, we set $y = 0$ and solve for x:

$0 - 2 = |x| \Rightarrow |x| = -2 \Rightarrow$ no x-intercepts

29. The graph of $y = \sqrt{9 - x^2}$ is in the catalog on page 102 in which $a = 3$. The graph of $y - 1 = \sqrt{9 - (x - 3)^2}$ is the graph of $y = \sqrt{x}$ translated 3 units right and 1 unit up. To find the y-intercept of $y - 1 = \sqrt{9 - (x - 3)^2}$, we set $x = 0$ and solve for y:

$y - 1 = \sqrt{9 - (0 - 3)^2} \Rightarrow y - 1 = 0 \Rightarrow y = 1$

$\Rightarrow$ y-intercept is $(0, 1)$.

To find x-intercepts, we set $y = 0$ and solve for x:

$0 - 1 = \sqrt{9 - (x - 3)^2} \Rightarrow -1 = \sqrt{9 - (x - 3)^2}$

$\Rightarrow$ no x-intercepts.

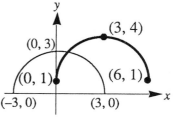

33. The graph is the graph of $y = \sqrt{x}$ translated vertically -2 units (2 units down). To find the y-intercept of $y = \sqrt{x} - 2$, we set $x = 0$ and solve for y:
$$y = \sqrt{0} - 2 \implies y = -2 \implies y\text{-intercept is } (0, -2).$$
To find x-intercepts, we set $y = 0$ and solve for x:
$$0 = \sqrt{x} - 2 \implies \sqrt{x} = 2$$
$$\implies x = 4 \implies x\text{-intercept is } (4, 0).$$

37. The graph is the graph of $y = x^3$ translated horizontally -2 units (2 units left) and vertically 7 units (7 units up). To find the y-intercept of $y = (x+2)^3 + 7$, we set $x = 0$ and solve for y:
$$y = (0+2)^3 + 7 \implies y = 8 + 7$$
$$\implies y = 15 \implies y\text{-intercept is } (0, 15).$$
To find x-intercepts, we set $y = 0$ and solve for x:
$$0 = (x+2)^3 + 7 \implies (x+2)^3 = -7 \implies x+2 = -\sqrt[3]{7}$$
$$\implies x = -2 - \sqrt[3]{7} \implies x\text{-intercept is } (-2 - \sqrt[3]{7},\, 0).$$

41. First, we complete the square on the radicand $3 - x^2 - 2x$:
$$3 - x^2 - 2x = -(x^2 + 2x \quad) + 3 = -(x^2 + 2x + 1) + 3 + 1 = -(x+1)^2 + 4 \; .$$
The equation in question becomes $y = \sqrt{4 - (x+1)^2}$; its graph is the graph of $y = \sqrt{4 - x^2}$ translated -1 unit horizontally (1 unit to the left). (The graph of $y = \sqrt{4 - x^2}$ is discussed on page 102). To find the y-intercept of $y = \sqrt{4 - (x+1)^2}$, we set $x = 0$ and solve for y:
$$y = \sqrt{4 - (0+1)^2} = \sqrt{3} \implies y\text{-intercept is } (0, \sqrt{3}).$$
To find x-intercepts, we set $y = 0$ and solve for x:
$$0 = \sqrt{4 - (x+1)^2} \implies 4 - (x+1)^2 = 0$$
$$\implies (x+1)^2 = 4 \implies x + 1 = \pm 2$$
$$\implies x = 1 \text{ or } -3 \implies x\text{-intercepts are } (1, 0),\ (-3, 0)$$

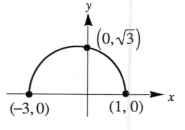

45. First, we sketch $y = \sqrt{25 - x^2} + 2$ and $y = 6$ on the same coordinate plane. The graph of $y = \sqrt{25 - x^2} + 2$ is the graph of $y = \sqrt{25 - x^2}$ translated 2 units vertically (2 units up). (The graph of $y = \sqrt{25 - x^2}$ is discussed on page 102.)

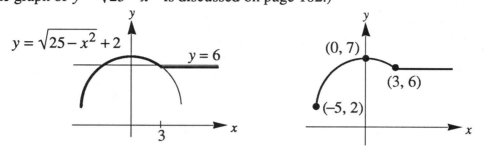

49. We test f to determine if it is even or odd by simplifying $f(-x)$:
$$f(-x) = -(-x)^3 + 4(-x) = x^3 - 4x = -[x^3 - 4x] = -f(x)$$
This means that the function is odd and its graph is symmetric with respect to the origin. We complete the graph using this symmetry in the figure below.
To find the y-intercept of the graph, we evaluate $f(0)$:
$$f(0) = -(0)^3 + 4(0) = 0 \Rightarrow y\text{-intercept is } (0, 0).$$
To find x-intercepts, we set $f(x) = 0$ and solve for x by factoring.
$$0 = x^3 - 4x \Rightarrow x(x^2 - 4) = 0 \Rightarrow x = 0 \text{ or } x^2 - 4 = 0$$
$$x^2 - 4 = 0 \Rightarrow x^2 = 4 \Rightarrow x = \pm 2$$
$$\Rightarrow x\text{-intercepts are } (-2, 0), (0, 0), \text{ and } (2, 0).$$

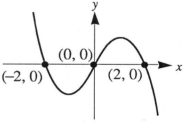

53. We test f to determine if it is even or odd by simplifying $f(-x)$:
$$f(-x) = \frac{1}{|-x|} = \frac{1}{|x|} = f(x)$$
This means that the function is even and its graph is symmetric with respect to the y-axis. We complete the graph using this symmetry in the figure at right. There are no intercepts

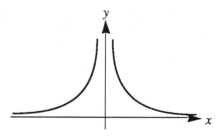

57. We test f to determine if it is even or odd by simplifying $f(-x)$:
$$f(-x) = ((-x)^2 - 3(-x))^2 = (x^2 + 3x)^2$$
This is neither $f(x)$ nor $-f(x)$. The function is neither even nor odd.

61. Consider $f(-x) = (-x)^n$. If n is an even integer, then $f(-x) = (-x)^n = x^n$. If n is an odd integer, then $f(-x) = (-x)^n = -x^n$. This implies that f is an even function if n is an even integer, and that f is an odd function if n is an odd integer.

Section 2.4

1. a) The graph of $y = \frac{3}{4}x$ is a line with slope $\frac{3}{4}$ and y-intercept 0.

b) The graph of $y = \frac{3}{4}x - 5$ is a line with slope $\frac{3}{4}$ and y-intercept -5.

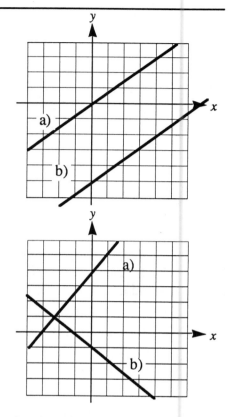

5. a) The graph of $y = \frac{5}{4}x + 4$ is a line with slope $\frac{5}{4}$ and y-intercept 4.

b) The graph of $y = -\frac{4}{5}x - 1$ is a line with slope $-\frac{4}{5}$ and y-intercept -1.

9. Because the lines are parallel, the slope of the line in question is the same as the slope of $3x - 7y = 5$. The slope of this line can be determined by rewriting its equation:

$$3x - 7y = 5 \;\Rightarrow\; -7y = -3x + 5 \;\Rightarrow\; y = \frac{3}{7}x - \frac{5}{7} \;\Rightarrow\; \text{the slope is } \frac{3}{7}.$$

The point-slope form of the line in question is $y - 1 = \frac{3}{7}(x - 2)$. Solving this equation for y yields the desired form of the equation:

$$y - 1 = \frac{3}{7}(x - 2) \;\Rightarrow\; y - 1 = \frac{3}{7}x - \frac{6}{7} \;\Rightarrow\; y = \frac{3}{7}x + \frac{1}{7} \;.$$

13. The perpendicular bisector passes through the midpoint of the line segment between the given points. Its slope is the negative reciprocal of this line segment. The coordinates of the midpoint are found using the midpoint formula: $\left(\dfrac{5+1}{2}, \dfrac{10-2}{2}\right) = (3,4).$

Also, the slope of the line segment is $\dfrac{-2-10}{1-5} = 3$, so the slope of the perpendicular bisector is $-\dfrac{1}{3}$. The point-slope form of the perpendicular bisector is $y - 4 = -\frac{1}{3}(x - 3)$. Solving this equation for y yields the desired form:

$$y - 4 = -\frac{1}{3}(x - 3) \;\Rightarrow\; y - 4 = -\frac{1}{3}x + 1 \;\Rightarrow\; y = -\frac{1}{3}x + 5 \;.$$

17. Because $h(2) = -3$, the graph of h passes through the point $(2, -3)$. Likewise, $h(-9) = -3$, implies that the graph of h passes through the point $(-9, -3)$. From the figure at right, the graph $y = h(x)$ is the line $y = -3$. This implies that $h(x) = -3$.

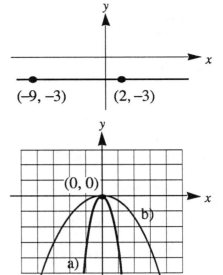

21. The graph of a) is discussed on page 117. The graph of b) opens down as well since the coefficient a is negative.

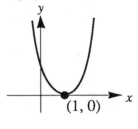

25. We follow the steps to complete the square described in Example 6 on page 119.

$$f(x) = \left(x^2 - 2x \qquad\right) + 1$$
$$= \left(x^2 - 2x + \mathbf{1}\right) + 1 - \mathbf{1}$$
$$= (x - 1)^2$$

The graph of $f(x) = (x-1)^2$ is the graph of $y = x^2$ translated 1 unit to the right. The vertex is $(1, 0)$.

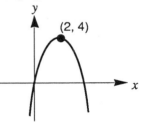

29. We follow the steps to complete the square described in Example 6 on page 119.

$$f(x) = 4x - x^2$$
$$= -\left(x^2 - 4x \qquad\right)$$
$$= -\left(x^2 - 4x + \mathbf{4}\right) + \mathbf{4}$$
$$= -(x - 2)^2 + 4$$

This means that the graph of f is the graph of $y = -x^2$ translated 2 units to the right and 4 units up. The vertex is $(2, 4)$.

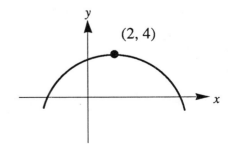

33. We follow the steps to complete the square described in Example 6 on page 119.

$$f(x) = -\tfrac{1}{4}x^2 + x + 3$$
$$= -\tfrac{1}{4}\left(x^2 - 4x \qquad\right) + 3$$
$$= -\tfrac{1}{4}\left(x^2 - 4x + \mathbf{4}\right) + 3 + \mathbf{1}$$
$$= -\tfrac{1}{4}(x - 2)^2 + 4$$

The graph of f is the graph of $y = -\tfrac{1}{4}x^2$ translated 2 units to the right and 4 units up. The vertex is $(2, 4)$.

37. The coefficient of the x^2 term is -1, so the graph of f opens down. This means that f has a maximum value. The x-coordinate for the vertex of its graph is $x = -\frac{b}{2a} = -\frac{(-4)}{2(-1)} = -2$.
We determine the maximum by computing $f(-2)$: $f(-2) = 2-4(-2)-(-2)^2 = 6$.
The maximum value of the function f is 6; it occurs for $x = -2$.

41. The coefficient of the x^2 term is -2, so the graph of f opens down. This means that f has a maximum value. The x-coordinate for the vertex of its graph is $x = -\frac{b}{2a} = -\frac{5}{2(-2)} = \frac{5}{4}$.
We determine the maximum by computing $f(\frac{5}{4})$: $f(\frac{5}{4}) = 2-4(\frac{5}{4})-(\frac{5}{4})^2 = -\frac{15}{8}$.
The maximum value of the function f is $-\frac{15}{8}$; it occurs for $x = \frac{5}{4}$.

45. Given that $v_0 = 1200$, the height of the projectile is $f(t) = 1200t - 16t^2$. The maximum value of this function occurs at $t = -\frac{b}{2a} = -\frac{1200}{2(-16)} = 37.5$ seconds. The greatest height attained by the projectile therefore is $f(37.5) = 1200(37.5) - 16(37.5)^2 = 22{,}500$ feet.

49. The area of the rectangle is a function of the width w of the rectangle. Because the perimeter is 300,

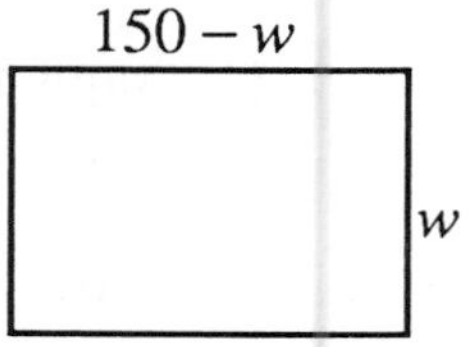

$$2\begin{pmatrix}\text{width of}\\\text{rectangle}\end{pmatrix} + 2\begin{pmatrix}\text{length of}\\\text{rectangle}\end{pmatrix} = 300$$

$$\text{or}\quad \begin{pmatrix}\text{length of}\\\text{rectangle}\end{pmatrix} = 150 - \begin{pmatrix}\text{width of}\\\text{rectangle}\end{pmatrix} = 150 - w$$

Since $\begin{pmatrix}\text{length of}\\\text{rectangle}\end{pmatrix}\begin{pmatrix}\text{width of}\\\text{rectangle}\end{pmatrix} = (150-w)w$, the function we seek is $A(w) = (150-w)w$, or $A(w) = -w^2 + 150w$. The maximum value of the function occurs when the width is $w = -\frac{b}{2a} = -\frac{150}{2(-1)} = 75$ meters. This implies that the length is $150 - 75 = 75$ meters (The rectangle is a square.) The maximum value is (75 meters)(75 meters) = 5625 sq. meters.

53. a) The three smaller triangles are similar to the large triangle, so they are also right isosceles triangles. From this, it follows that the base of each of the two smallest triangles is of length h. The base of the rectangle therefore is $12-2h$. The area of the rectangle is given by $A(h) = h(12-2h) = 12h-2h^2$. The maximum value of h is 6, the height of the triangle, so the domain of the function is $[0, 6]$.

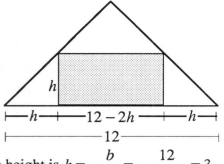

b) The maximum value of the function occurs when the height is $h = -\dfrac{b}{2a} = -\dfrac{12}{2(-2)} = 3$

The maximum area is $A(3) = 12(3)-2(3)^2 = 18$.

57. The values of x that satisfy the inequality are those for which the graph of the function $f(x) = 2x^2+9x+4$ is above the x-axis (see page 119). Because the coefficient of the x^2 term is 2, the graph of f opens up. Its x-intercepts are found by setting $f(x) = 0$ and solving the resulting equation for x:

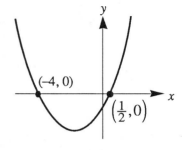

$$2x^2+9x+4 = 0 \implies (2x+1)(x+4) = 0$$

$$\implies 2x+1 = 0 \text{ or } x+4 = 0 \implies x = -\frac{1}{2} \text{ or } x = -4$$

$$\implies \text{ the } x\text{-intercepts are } (-\tfrac{1}{2}, 0) \text{ and } (-4, 0).$$

From the graph, we get the solution for the inequality: $(-\infty, -4)$ or $(-\frac{1}{2}, +\infty)$.

61. The values of x that satisfy the inequality are those for which the graph of the function $f(x) = 2x^2+2x+1$ is below the x-axis (see page 119). Because the coefficient of the x^2 term is 2, the graph of f opens up. Its x-intercepts are found by setting $f(x) = 0$ and solving the resulting equation for x:

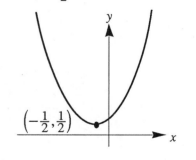

$$2x^2+2x+1 = 0 \implies x = \frac{-(2)\pm\sqrt{(2)^2-4(2)(1)}}{2(2)} = \frac{-2\pm\sqrt{-4}}{4}$$

Because this equation has no real roots, the graph has no x-intercepts. The graph opens up and has no x-intercepts, so the entire graph is above the x-axis. (Its vertex, which can be found by completing the square, is $(-\frac{1}{2}, \frac{1}{2})$.) The solution set is empty.

Section 2.5

1. a) The graph of $y = 3f(x)$ is the graph of $y = f(x)$ expanded vertically by a factor of 3.

b) The graph of $y = f(3x)$ is the graph of $y = f(x)$ compressed horizontally by a factor of $\frac{1}{3}$.

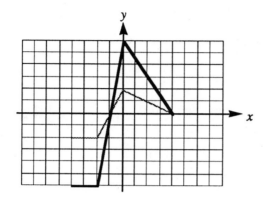

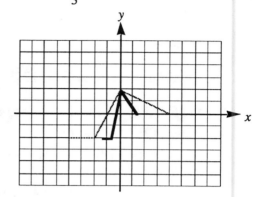

5. a) The graph of $y = -g(x)$ is the graph of $y = g(x)$ reflected through the x-axis.

b) The graph of $y = g(-x)$ is the graph of $y = g(x)$ reflected through the y-axis.

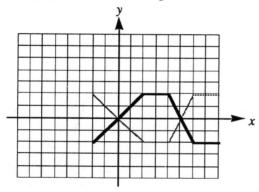

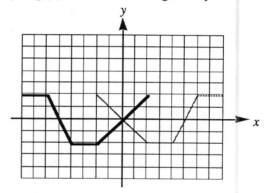

9. a) The graph of $y = f(-x)$ is the graph of $y = f(x)$ reflected through the y-axis.

b) The equation can be rewritten as
$$y - 3 = f(-(x + 1)).$$

This is the graph found in a) translated -1 units horizontally and 3 units vertically.

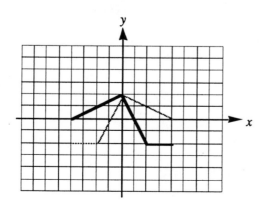

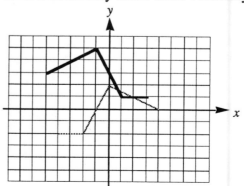

13. The graph in question is a vertical expansion (by a factor of 2) and reflection through the

y-axis of the graph of $y = \sqrt{x}$ (see catalog on page 102 and summary on page 131).
Both graphs are shown in the figure at right.

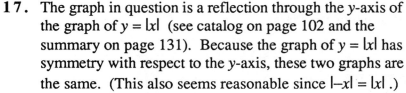

17. The graph in question is a reflection through the y-axis of the graph of $y = |x|$ (see catalog on page 102 and the summary on page 131). Because the graph of $y = |x|$ has symmetry with respect to the y-axis, these two graphs are the same. (This also seems reasonable since $|-x| = |x|$.)

21. Because $|-u| = |u|$, the equation in question can be rewritten as $y - 2 = -|x - 4|$. The graph of $y = -|x|$ (shown in the figure at right) is a reflection through the x-axis of the graph of $y = |x|$ (see catalog on page 102 and the summary on page 131). The graph we seek is a horizontal translation (4 units right) and a vertical translation (2 units up).

(4, 2)

25. The equation in question can be rewritten as

$y = \sqrt{36 - \left(\frac{1}{2}x\right)^2}$. The graph of $y = \sqrt{36 - x^2}$ is shown in the figure at right (see catalog on page 102). The graph we seek is a horizontal expansion of this graph by a factor of 2 (see the summary on page 131).

(0, 6)

(−12, 0)   (12, 0)

29. The graph of a) is in the catalog on page 102. The graph of b) is a translation of the graph of a) horizontally (4 units left) and a vertically (2 units up). (The equation in b) is also in the point-slope form of Section 2.4)

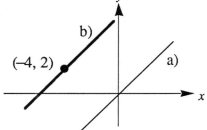

b)

(−4, 2)   a)

33. a) The greatest integer function is described on page 90. From its definition, it follows that

$$-2 \leq x < -1 \Rightarrow [\![x]\!] = -2$$
$$-1 \leq x < 0 \Rightarrow [\![x]\!] = -1$$
$$0 \leq x < 1 \Rightarrow [\![x]\!] = 0$$
$$1 \leq x < 2 \Rightarrow [\![x]\!] = 1$$
$$2 \leq x < 3 \Rightarrow [\![x]\!] = 2$$

and so on.
This is used to construct the graph of $y = [\![x]\!]$ at right.

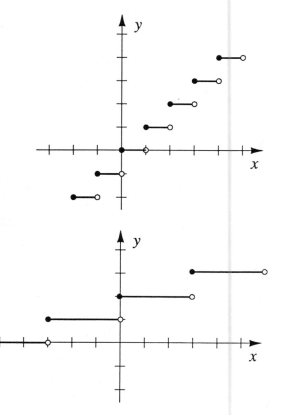

b) The graph of this equation is the graph of a) expanded horizontally by a factor of 3 and then translated 2 units up.

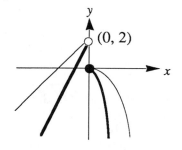

37. a) The graph of $y = f(x)$ for $x \geq 0$ is the graph of $y = x + 2$. For $x < 0$, the graph is the graph of $y = -x^2$.

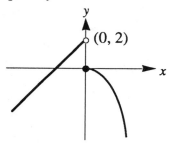

b) The graph of $y = f(2x)$ is the graph of $y = f(x)$ compressed horizontally by a factor of 2.

41. First, we complete the square on the radicand $32 - 4x - x^2$:

$$32 - 4x - x^2 = -(x^2 + 4x \quad) + 32 = -(x^2 + 4x + 4) + 32 + 4 = -(x+2)^2 + 36.$$

The equation in question becomes $y = \sqrt{36 - (x+2)^2}$; its graph is the graph of

$y = \sqrt{36 - x^2}$ translated -2 units horizontally (2 units to the left). (The graph of

$y = \sqrt{36 - x^2}$ is discussed on page 102). To find the y-intercept of $y = \sqrt{36 - (x+2)^2}$,

we set $x = 0$ and solve for y:

$$y = \sqrt{36 - (0+2)^2} = 4\sqrt{2} \quad \Rightarrow \quad y\text{-intercept is } (0, 4\sqrt{2}).$$

To find x-intercepts, we set $y = 0$ and solve for x.

$$0 = \sqrt{36 - (x+2)^2} \quad \Rightarrow \quad 36 - (x+2)^2 = 0$$

$$\Rightarrow \quad (x+2)^2 = 36 \quad \Rightarrow \quad x + 2 = \pm 6$$

$$\Rightarrow \quad x = 4 \text{ or } -8 \quad \Rightarrow \quad x\text{-intercepts are } (4, 0), (-8, 0)$$

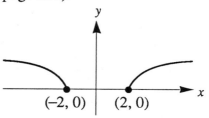

45. a) The graph of $y = f(|x|)$ is the graph of $y = f(x)$ for $x \geq 0$. The graph is completed for $x < 0$ by reflecting across the y-axis. (See summary on page 131).

b) The graph of $y = -2f(|x|)$ is a vertical expansion (by a factor of 2) and reflection through the y-axis of the graph determined in a). (See summary on page 131)

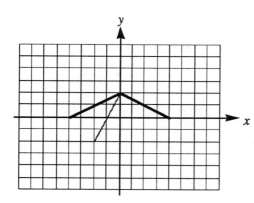

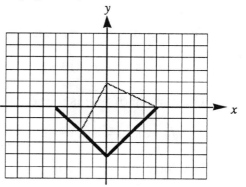

49. a) The graph of $y = \sqrt{|x|}$ is the graph of $y = \sqrt{x}$ for $x \geq 0$. The graph is completed for $x < 0$ by reflecting across the y-axis. (See summary on page 131).

b) The graph of $y = \sqrt{x - 2}$ is a horizontal translation (2 units right) of the graph of $y = \sqrt{x}$. The graph of $y = \sqrt{|x| - 2}$ is completed for $x < 0$ by reflecting across the y-axis. (See summary on page 131).

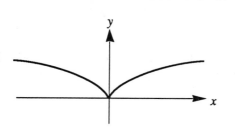

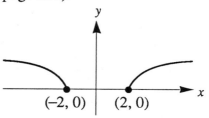

Section 2.6

1. $(f+g)(x) = f(x) + g(x) = \left(2x^4 - x^2\right) + \left(x^2 + 3\right)^2$
$$= \left(2x^4 - x^2\right) + \left(x^4 + 6x^2 + 9\right) = 3x^4 + 5x^2 + 9$$

The domain of $f+g$ is the set of all real numbers.
$$(f-g)(x) = f(x) - g(x) = \left(2x^4 - x^2\right) - \left(x^2 + 3\right)^2$$
$$= \left(2x^4 - x^2\right) - \left(x^4 + 6x^2 + 9\right) = x^4 - 7x^2 - 9 \;.$$

The domain of $f-g$ is the set of all real numbers.
$$(fg)(x) = f(x) \cdot g(x) = \left(2x^4 - x^2\right)\left(x^2 + 3\right)^2 = 2x^8 + 11x^6 + 12x^4 - 9x^2 \;.$$

The domain of fg is is the set of all real numbers.
$$\left(\frac{f}{g}\right)(x) = \frac{f(x)}{g(x)} = \frac{\left(2x^4 - x^2\right)}{\left(x^2 + 3\right)^2} \;.$$

The divisor $g(x)$ has no zeros, so the domain of $\dfrac{f}{g}$ is the set of all real numbers.

5. The function f is undefined for $x = -4$, and the function g is undefined for $x = 0$.
$$(f+g)(x) = f(x) + g(x) = \frac{x}{x+4} + \frac{x+2}{x} = \frac{x^2 + (x+2)(x+4)}{x(x+4)} = \frac{2x^2 + 6x + 8}{x^2 + 4x} \;.$$

The domain of $f+g$ is the set of all real numbers except for 0 and -4.
$$(f-g)(x) = f(x) - g(x) = \frac{x}{x+4} - \frac{x+2}{x} = \frac{x^2 - (x+2)(x+4)}{x(x+4)} = \frac{-6x - 8}{x^2 + 4x}$$

The domain of $f-g$ is the set of all real numbers except for 0 and -4.
$$(fg)(x) = f(x) \cdot g(x) = \frac{x}{x+4} \frac{x+2}{x} = \frac{x+2}{x+4}$$

The domain of fg is the set of all real numbers except for 0 and -4.
$$\left(\frac{f}{g}\right)(x) = \frac{f(x)}{g(x)} = \frac{\left(\dfrac{x}{x+4}\right)}{\left(\dfrac{x+2}{x}\right)} = \frac{x}{x+4} \cdot \frac{x}{x+2} = \frac{x^2}{(x+4)(x+2)} = \frac{x^2}{x^2 + 6x + 8} \;.$$

Since $g(-2) = 0$, the domain of $\dfrac{f}{g}$ is the set of all real numbers except for 0, -2, and -4.

9. $(f \circ g)(x) = f[g(x)] = f\left[\sqrt{x-1}\right] = \left(\sqrt{x-1}\right)^2 - \dfrac{2}{\sqrt{x-1}} = x - 1 - \dfrac{2}{\sqrt{x-1}}$

$(g \circ f)(x) = g[f(x)] = g\left[x^2 - \dfrac{2}{x}\right] = \sqrt{\left(x^2 - \dfrac{2}{x}\right) - 1} = \sqrt{\dfrac{x^3 - 2 - x}{x}} = \dfrac{\sqrt{x^4 - 2x - x^2}}{x}$

13. In determining both $(f \circ g)$ and $(g \circ f)$ below, a compound fraction is simplified in the manner described on page 24.

$$(f \circ g)(x) = f[g(x)] = f\left[\sqrt[3]{\frac{1-5x}{x}}\right] = \frac{1}{\left(\sqrt[3]{\frac{1-5x}{x}}\right)^3 + 5} = \frac{1}{\frac{1-5x}{x} + 5}$$

$$= \frac{1}{\frac{1-5x}{x} + 5} \cdot \frac{x}{x} = \frac{x}{1-5x+5x} = \frac{x}{1} = x$$

$$(g \circ f)(x) = g[f(x)] = g\left[\frac{1}{x^3+5}\right] = \sqrt[3]{\frac{1-5\left(\frac{1}{x^3+5}\right)}{\left(\frac{1}{x^3+5}\right)}}$$

$$= \sqrt[3]{\frac{1-\frac{5}{x^3+5}}{\left(\frac{1}{x^3+5}\right)} \cdot \frac{x^3+5}{x^3+5}} = \sqrt[3]{\frac{x^3+5-5}{1}} = \sqrt[3]{x^3} = x$$

17. $(f \circ g)(x) = f[g(x)] = f\left[\frac{1}{x^2}\right] = 4$

$(g \circ f)(x) = g[f(x)] = g[4] = \frac{1}{4^2} = \frac{1}{16}$

21. $(f \circ f)(x) = f[f(x)] = f\left[-\sqrt{9-x^2}\right]$

$$= -\sqrt{9-\left(-\sqrt{9-x^2}\right)^2} = -\sqrt{9-\left(9-x^2\right)} = -\sqrt{x^2} = -|x|$$

25. The graph is constructed by the technique of adding y-coordinates.

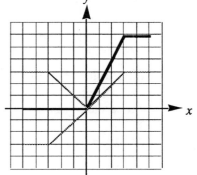

29. The graph of $y = 2f(x)$ is a vertical expansion (by a factor of 2) of the graph of $y = f(x)$. The graph of $y = h(x) + 2f(x)$ is constructed by the technique of adding y-coordinates.

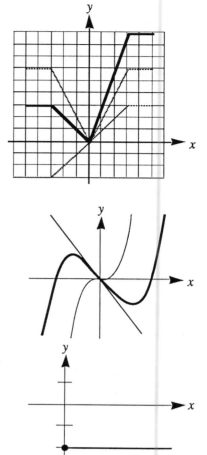

33. The graph of $f(x) = \frac{1}{2}x^3$ is a vertical compression of the graph (by a factor of 2) of graph of $y = x^3$ (see page 102). The graph of $g(x) = -2x$ is line passing through the origin with a slope of -2. The graph of $y = f(x) + g(x)$ is constructed by the technique of adding y-coordinates.

37. The graph we seek is a vertical expansion (by a factor of 2) and reflection through the y-axis of the graph of $y = u(x)$ (This graph is shown on page 137.)

41. Below are the graphs of $y = 2u(x)$, $y = -u(x-3)$, and $y = 2u(x-5)$

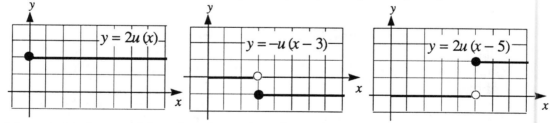

The graph of $y = 2u(x) - u(x-3) + 2u(x-5)$ is constructed by the technique of adding y-coordinates.
You can also consider the graph in the following way.
For $0 \le x < 3$,
$$2u(x) - u(x-3) + 2u(x-5) = 2(1) - 0 + 2(0) = 2$$
For $3 \le x < 5$,
$$2u(x) - u(x-3) + 2u(x-5) = 2(1) - 1 + 2(0) = 1$$
For $x \ge 3$,
$$2u(x) - u(x-3) + 2u(x-5) = 2(1) - 1 + 2(1) = 3$$

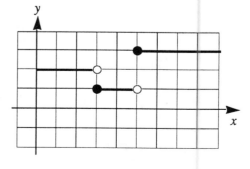

45. There are many reasonable possible answers. Here are a few examples:

$$f(x) = \frac{1}{x} \quad \text{and} \quad g(x) = (x-1)^2 .$$

$$f(x) = \frac{1}{x^2} \quad \text{and} \quad g(x) = x-1 .$$

$$f(x) = \frac{1}{(x+1)^2} \quad \text{and} \quad g(x) = x-2 .$$

49. There are many reasonable possible answers. Here are a few examples:

$$f(x) = \sqrt{25-x} \quad \text{and} \quad g(x) = (2x+1)^2 .$$

$$f(x) = \sqrt{25-x^2} \quad \text{and} \quad g(x) = 2x+1$$

$$f(x) = \sqrt{25-(x+1)^2} \quad \text{and} \quad g(x) = 2x .$$

49. There are many reasonable possible answers. Note that $|x-2| = |2-x|$. Here is one example:

$$f(x) = |x| + \frac{3}{x^3} \quad \text{and} \quad g(x) = 2-x .$$

Section 2.7

1. These functions are inverses:

$$g[f(a)] = g[4a+2] = \tfrac{1}{4}((4a+2)-2) = \tfrac{1}{4}(4a) = a$$

$$f[g(b)] = f\left[\tfrac{1}{4}(b-2)\right] = 4\left(\tfrac{1}{4}(b-2)\right) + 2 = (b-2)+2 = b$$

5. These functions are not inverses:

$$g[f(a)] = g\left[\tfrac{1}{2}\sqrt{25-4a^2}\right] = \tfrac{1}{2}\sqrt{25 - 4\left(\tfrac{1}{2}\sqrt{25-4a^2}\right)^2} = \tfrac{1}{2}\sqrt{25 - 4\left(\tfrac{1}{4}\left(25-4a^2\right)\right)}$$

$$= \tfrac{1}{2}\sqrt{25 - \left(25-4a^2\right)} = \tfrac{1}{2}\sqrt{4a^2} = \tfrac{1}{2}\sqrt{4a^2} = \tfrac{1}{2}2|a| = |a|$$

$$f[g(b)] = g\left[\tfrac{1}{2}\sqrt{25-4b^2}\right] = \tfrac{1}{2}\sqrt{25 - 4\left(\tfrac{1}{2}\sqrt{25-4b^2}\right)^2} = \tfrac{1}{2}\sqrt{25 - 4\left(\tfrac{1}{4}\left(25-4b^2\right)\right)}$$

$$= \tfrac{1}{2}\sqrt{25 - \left(25-4b^2\right)} = \tfrac{1}{2}\sqrt{4b^2} = \tfrac{1}{2}\sqrt{4b^2} = \tfrac{1}{2}2|b| = |b|$$

For example, $g[f(-2)] \neq -2$:

$$g[f(-2)] = g\left[\tfrac{1}{2}\sqrt{25-4(-2)^2}\right]$$

$$= g\left[\tfrac{1}{2}\sqrt{25-16}\right] = g\left[\tfrac{3}{2}\right] = \tfrac{1}{2}\sqrt{25-4\left(\tfrac{3}{2}\right)^2} = \tfrac{1}{2}\sqrt{25-9} = 2 \neq -2$$

9. The graph of h is a vertical expansion (by a factor of two) and reflection (about the x-axis) of the graph of $y = \dfrac{1}{x-2}$ which is a translation of the graph of $y = \dfrac{1}{x}$ (this graph is in the catalog on page 102). The function appears to be one-to-one from its graph. Suppose that for some real numbers a and b, $h(a) = h(b)$. So, in the manner of Example 4 on page 146,

$$h(a) = h(b) \implies \frac{-2}{a-2} = \frac{-2}{b-2} \implies -2(b-2) = -2(a-2)$$

$$\implies -2b + 4 = -2a + 4 \implies -2b = -2a \implies b = a$$

This means that the function is indeed one-to-one

13. The graph appears to be one-to-one by the horizontal line test (some horizontal lines are shown below left). The inverse is sketched by first reflecting a few points of the graph of the function through the line $y = x$ and then drawing a curve through them (below right).

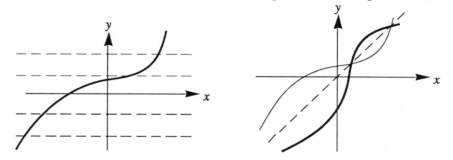

17. The graph appears to be one-to-one by the horizontal line test (some horizontal lines are shown below left). The inverse is sketched by first reflecting a few points of the graph of the function through the line $y = x$ and then drawing a curve through them.

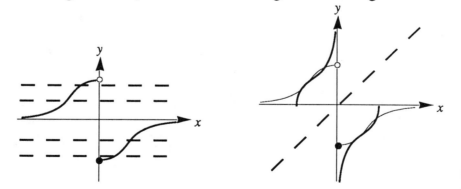

21. We follow the steps outlined on page 148.

$$G(y) = x \implies -2(y-3) = x \implies -2y + 6 = x \implies -2y = x - 6 \implies y = 3 - \frac{x}{2}.$$

The inverse of G is $G^{-1}(x) = 3 - \dfrac{x}{2}$.

25. We follow the steps outlined on page 148.
$$p(y) = x \implies \sqrt[3]{y+4} = x \implies y+4 = x^3 \implies y = x^3 - 4$$
The inverse of p is $p^{-1}(x) = x^3 - 4$.

29. We follow the steps outlined on page 148.
$$S(y) = x \implies \frac{y}{y+4} = x \implies y = x(y+4) \implies y = xy + 4x$$
$$\implies y - xy = 4x \implies y(1-x) = 4x \implies y = \frac{4x}{1-x}$$
The inverse of S is $S^{-1}(x) = \frac{4x}{1-x}$.

33. The graph of $g(x) = |x-1|$ is the graph of $y = |x|$ translated 1 unit to the right. With the restriction $x \leq 1$, then the function becomes $g(x) = 1-x$. Following the steps outlined on page 148, we obtain
$$g(y) = x \implies 1-y = x \implies y = 1-x.$$
The inverse of g is $g^{-1}(x) = 1-x$. From the figure below right, the inverse is defined for $x > 0$ only.

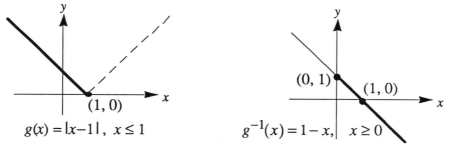

$g(x) = |x-1|, \; x \leq 1$ $g^{-1}(x) = 1-x, \; x \geq 0$

37. Following the steps outlined on page 148, we obtain
$$G(y) = x \implies \sqrt{36 - y^2} = x \implies 36 - y^2 = x^2$$
$$\implies y^2 = 36 - x^2 \implies y = \pm\sqrt{36 - x^2}$$
The restriction on the domain of G requires that $x \leq 0$. It follows that the range of $G^{-1}(x)$ is such that $G^{-1}(x) \leq 0$, so we choose $G^{-1}(x) = -\sqrt{36 - x^2}$. From the figure at right, the inverse is defined for $x > 0$ only.

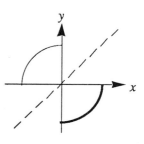

41. Following the steps outlined on page 148, we obtain

$$f(y) = x \Rightarrow \frac{12}{y^2 + 4} = x \Rightarrow 12 = x(y^2 + 4) \Rightarrow 12 = xy^2 + 4x$$

$$\Rightarrow xy^2 = 12 - 4x \Rightarrow y^2 = \frac{12 - 4x}{x} \Rightarrow y = \pm\sqrt{\frac{12 - 4x}{x}}$$

The restriction on the domain of f requires that $x \leq 0$. It follows that the range of $f^{-1}(x)$ is such that $f^{-1}(x) \leq 0$, so we choose $f^{-1}(x) = -\sqrt{\frac{12 - 4x}{x}}$. From the figure, the inverse is defined for $x > 0$ only.

45. We follow the steps outlined on page 148.

$$F(y) = x \Rightarrow \tfrac{9}{5}y + 32 = x \Rightarrow \tfrac{9}{5}y = x - 32 \Rightarrow y = \tfrac{5}{9}(x - 32).$$

The inverse of F is $F^{-1}(x) = \tfrac{5}{9}(x - 32)$. This function can be used to determine the temperature in degress Celsius given the temperature in degrees Fahrenheit.

Miscellaneous Exercises for Chapter 2

1. By the distance formula, we get

$$d = \sqrt{(\Delta x)^2 + (\Delta y)^2} = \sqrt{(6+1)^2 + (-12-12)^2} = 25$$

By the midpoint formula, we get.

$$\left(\frac{x_1 + x_2}{2}, \frac{y_1 + y_2}{2}\right) = \left(\frac{6-1}{2}, \frac{-12+12}{2}\right) = \left(\tfrac{5}{2}, 0\right)$$

5. We are given $(h, k) = (2, 6)$ and $r = 2$, so the equation of the circle in the form $(x - h)^2 + (y - k)^2 = r^2$ is

$(x - 2)^2 + (y - 6)^2 = 4.$

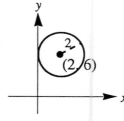

9. Because the circle is tangent to the x-axis and its radius is 5, the y-coordinate of the center of the circle is 5. To determine the x-coordinate of the center we let $y = 5$ in the equation $y = \tfrac{1}{2}x + 2$:

$$5 = \tfrac{1}{2}x + 2 \Rightarrow \tfrac{1}{2}x = 3 \Rightarrow x = 6 \Rightarrow \text{The center}$$

is $(6, 5)$

The equation of the circle is $(x - 5)^2 + (y - 6)^2 = 25.$

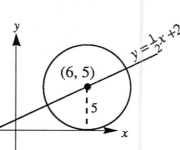

13. The slope of the line $3x + 4y = 16$ can be determined by solving the equation for y:

$$3x + 4y = 16 \implies 4y = -3x + 16 \implies y = -\frac{3}{4}x + 4 \implies \text{the slope is } -\frac{3}{4}.$$

Because the line we seek is perpendicular to this line, its slope m is the negative reciprocal of $\frac{3}{4}$. Thus, $m = -\dfrac{1}{\left(-\frac{3}{4}\right)} = \frac{4}{3}$. Because the line passes the point $(0, -2)$, its y-intercept is $b = -2$. Using the slope-intercept form $y = mx + b$, we get $y = \frac{4}{3}x - 2$.

17. We use the tests on page 84 to check symmetry for the graph of $x^2 + xy + 2y^2 = 7$

y-axis test:	$(-x)^2 + (-x)y + 2y^2 = 7$ or $x^2 - xy + 2y^2 = 7 \implies$ No	
x-axis test:	$x^2 + x(-y) + 2(-y)^2 = 7$ or $x^2 - xy + 2y^2 = 7 \implies$ No	
origin test:	$(-x)^2 + (-x)(-y) + 2(-y)^2 = 7$ or $x^2 + xy + 2y^2 = 7 \implies$ Yes	
$y = x$ test:	$y^2 + yx + 2x^2 = 7$ or $2x^2 - xy + y^2 = 7 \implies$ No	

The graph of the equation has symmetry with respect to the origin only.

21. We use the tests on page 84 to check symmetry for the graph of $y^2 = x^3 - \dfrac{1}{x}$

y-axis test: $\quad y^2 = (-x)^3 - \dfrac{1}{(-x)}$ or $y^2 = -x^3 + \dfrac{1}{x} \implies$ No

x-axis test: $\quad (-y)^2 = x^3 - \dfrac{1}{x}$ or $y^2 = x^3 - \dfrac{1}{x} \implies$ Yes

origin test: $\quad (-y)^2 = (-x)^3 - \dfrac{1}{(-x)}$ or $y^2 = -x^3 + \dfrac{1}{x} \implies$ No

$y = x$ test: $\quad x^2 = y^3 - \dfrac{1}{y} \implies$ No

The graph of the equation has symmetry with respect to the x-axis only.

25. The function f is defined for all values x except for those that make its denominator $2x - 6$ equal to zero. Because $2x - 6 = 0 \implies 2x = 6 \implies x = 3$, the domain is the set of all real numbers except 3.

29. By the definition of principal square root (see page 10), $\sqrt{x - 3}$ is nonnegative. This implies that $3 + \sqrt{x - 3} \geq 3$. Therefore the range for the function is $[3, +\infty)$.

33.
$$\frac{g(x + h) - g(x)}{h} = \frac{\left[12 - (x + h)^2\right] - \left[12 - x^2\right]}{h} = \frac{\left[12 - \left(x^2 + 2xh + h^2\right)\right] - \left[12 - x^2\right]}{h}$$

$$= \frac{12 - x^2 - 2xh - h^2 - 12 + x^2}{h} = \frac{-2xh - h^2}{h} = \frac{h(-2x - h)}{h} = -2x - h$$

37. In determining $\dfrac{g(x+h)-g(x)}{h}$ below, a compound fraction is simplified (see page 24).

$$\frac{g(x+h)-g(x)}{h} = \frac{\left[-\dfrac{1}{x+h}\right]-\left[-\dfrac{1}{x}\right]}{h} = \frac{-\dfrac{1}{x+h}+\dfrac{1}{x}}{h} = \frac{-\dfrac{1}{x+h}+\dfrac{1}{x}}{h}\cdot\frac{x(x+h)}{x(x+h)}$$

$$= \frac{-x+(x+h)}{hx(x+h)} = \frac{h}{hx(x+h)} = \frac{1}{x(x+h)}$$

41. In determining $\dfrac{f(x+h)-f(x)}{h}$ below, we use the conjugate of the numerator (see page 26).

$$\frac{f(x+h)-f(x)}{h} = \frac{\sqrt{2(x+h)}-\sqrt{2x}}{h} = \frac{\sqrt{2(x+h)}-\sqrt{2x}}{h}\cdot\frac{\sqrt{2(x+h)}+\sqrt{2x}}{\sqrt{2(x+h)}+\sqrt{2x}}$$

$$= \frac{2(x+h)-2x}{h\left(\sqrt{2(x+h)}+\sqrt{2x}\right)} = \frac{2h}{h\left(\sqrt{2(x+h)}+\sqrt{2x}\right)} = \frac{2}{\sqrt{2(x+h)}+\sqrt{2x}}$$

45. The graph is the graph of $y = x^2$ translated vertically -16 units (16 units down). To find the y-intercept of $y = x^2 - 16$ we set $x = 0$ and solve for y:

$y = 0^2 - 16 = -16 \Rightarrow$ y-intercept is $(0, -16)$.

To find x-intercepts, we set $y = 0$ and solve for x.

$0 = x^2 - 16 \Rightarrow x^2 = 16 \Rightarrow x = \pm 4$

$\Rightarrow$ x-intercept is $(-4, 0)$ and $(4, 0)$.

49. The graph is the graph of $y = |x|$ translated vertically -9 units (9 units down). To find the y-intercept of $y = |x| - 16$ we set $x = 0$ and solve for y:

$y = |0| - 9 = -9 \Rightarrow$ y-intercept is $(0, -9)$.

To find x-intercepts, we set $y = 0$ and solve for x.

$0 = |x| - 9 \Rightarrow |x| = 9 \Rightarrow x = \pm 9$

$\Rightarrow$ x-intercept is $(-9, 0)$ and $(9, 0)$.

53. The graph opens down as well since the coefficient of x^2 is $-\dfrac{2}{5}$, a negative number. The only intercept is the origin.

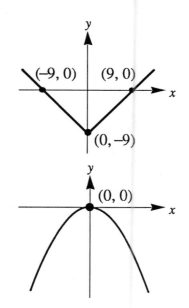

5 7. The equation can be rewritten $y+2 = |x+4|$. The graph of this equation is the graph of $y = |x|$ translated -4 units horizontally (4 units left) and -2 units vertically (2 units down). To find its y-intercept we set $x = 0$ and solve for y:

$$y+2 = |0+4| \implies y+2 = 4 \implies y = 2$$
$$\implies y\text{-intercept is } (0, 2).$$

To find x-intercepts, we set $y = 0$ and solve for x.
$$0+2 = |x+4| \implies |x+4| = 2 \implies x+4 = \pm 2$$
$$\implies x = -6 \text{ or } -2 \implies x\text{-intercept are } (-6, 0), (-2, 0).$$

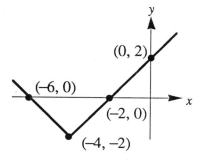

6 1. The graph of this equation is the graph of $y = \sqrt{x}$ translated 5 units left and translated 3 units up. To find its y-intercept we set $x = 0$ and solve for y:

$$y-3 = \sqrt{0+5} \implies y-3 = \sqrt{5} \implies y = 3+\sqrt{5}$$
$$\implies y\text{-intercept is } (0, 3+\sqrt{5}).$$

To find x-intercepts, we set $y = 0$ and solve for x.
$$0-3 = \sqrt{x+5} \implies -3 = \sqrt{x+5}$$
$$\implies \text{no real roots} \implies \text{no } x\text{-intercepts}$$

6 5. The graph of $y = \sqrt{9-x^2}$ is in the catalog on page 102 in which $a = 3$. The graph of $y+3 = \sqrt{9-(x-3)^2}$ is the graph of $y = \sqrt{9-x^2}$ translated 3 units right and 3 units down. To find its y-intercept, we set $x = 0$ and solve for y:

$$y+3 = \sqrt{9-(0-3)^2} \implies y+3 = 0 \implies y = -3$$
$$\implies y\text{-intercept is } (0, -3).$$

To find x-intercepts, we set $y = 0$ and solve for x.
$$0+3 = \sqrt{9-(x-3)^2} \implies 9 = 9-(x-3)^2$$
$$\implies 0 = -(x-3)^2 \implies x-3 = 0 \implies x = 3$$
$$\implies x\text{-intercept is } (3, 0).$$

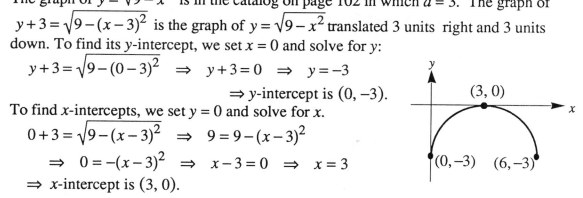

6 9. First, we sketch $y = 4-x^2$, $y = 4$, and $y = x-5$ on the same coordinate plane. The graph we seek is constructed in the manner of Example 10 on page 106.

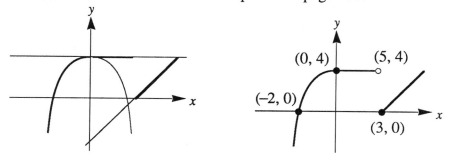

73. First, we sketch $y = |x-6|$ and $y = -|x+2|$ on the same coordinate plane. The graph we seek is constructed by the technique of adding y-coordinates.

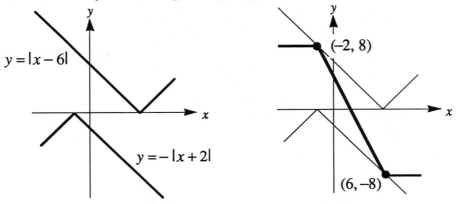

77. We test f to determine if it is even or odd by simplifying $f(-x)$:
$$f(-x) = ((-x)-8)^3 = -(x+8)^3$$
This is not equal to either $f(x)$ or $-f(x)$. This means that the function f is neither even nor odd.

81. We test f to determine if it is even or odd by simplifying $f(-x)$:
$$f(-x) = \frac{(-x)^{3/5}}{\left((-x)^{4/3}+3\right)\left((-x)^2-1\right)} = \frac{-x^{3/5}}{\left(x^{4/3}+3\right)\left(x^2-1\right)} = -\frac{x^{3/5}}{\left(x^{4/3}+3\right)\left(x^2-1\right)} = -f(x)$$
This means that the function f is an odd function.

85. $(u \circ r)(x) = u[r(x)] = u[\sqrt{x}] = \sqrt{x}+2 = h(x)$

89. $(q \circ r)(x) = q[r(x)] = q[\sqrt{x}] = \dfrac{4}{\sqrt{x}} = h(x)$

93. $(u \circ u)(x) = u[u(x)] = u[x+2] = (x+2)+2 = x+4 = h(x)$

97. The values of x vary from -6 to 5, so the domain of the function is $[-6, 5]$. The values of y vary from -1 to 3, so the domain of the function is $[-1, 3]$.

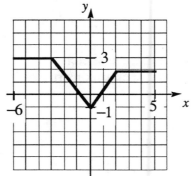

101. The graph of g passes through the point $(-2, -1)$, so
$$g(-2) = -1.$$
The graph of g passes through the point $(2, 1)$, so
$$g(2) = 1.$$
The graph of g passes through the point $(4, -1)$, so
$$g(4) = -1.$$

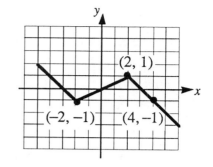

105. a) The graph of $y = -g(x)$ is the graph of $y = g(x)$ reflected through the x-axis (see Summary on page 131).

b) The graph of $y = -g(x)$ is the graph of $y = g(x)$ reflected through the y-axis (see Summary on page 131).

c) The graph of $y = -g(-x)$ is the graph of $y = g(x)$ reflected through the x-axis and the y-axis   (see Summary on page 131).

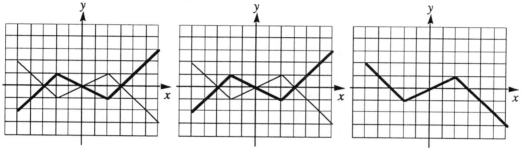

109. a) If $f(x) \geq 0$, then $|f(x)| = f(x)$. If $f(x) < 0$, then $|f(x)| = -f(x)$. It follows then that
$$y = \tfrac{1}{2}(f(x) + |f(x)|) = \tfrac{1}{2}(f(x) + f(x)) = f(x) \text{ for } f(x) \geq 0,$$
and
$$y = \tfrac{1}{2}(f(x) + |f(x)|) = \tfrac{1}{2}(f(x) - f(x)) = 0 \text{ for } f(x) \leq 0.$$
This implies that the the value of this function is the same as $f(x)$ when the graph of $f(x)$ is on or above the x-axis, and its value is zero when the graph of $f(x)$ is below the x-axis.

b) In the manner of a),
$$y = \tfrac{1}{2}(f(x) - |f(x)|) = \tfrac{1}{2}(f(x) - f(x)) = 0 \text{ for } f(x) \geq 0,$$
and
$$y = \tfrac{1}{2}(f(x) - |f(x)|) = \tfrac{1}{2}(f(x) + f(x)) = f(x) \text{ for } f(x) \leq 0.$$
This implies that the value of this function is zero when the graph of $f(x)$ is on or above the x-axis, and its value is the same as of $y = f(x)$ when the graph of $f(x)$ is below the x-axis.

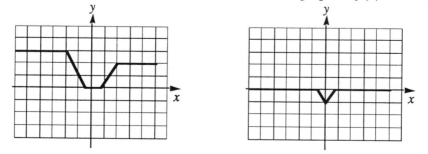

113. Suppose that there exist two real numbers x_1 and x_2 such that $f(x_1) = f(x_2)$. In the manner of Example 4 on page 146, we find that

$$f(x_1) = f(x_2) \Rightarrow 7(x_1 - 4) = 7(x_2 - 4) \Rightarrow 7x_1 - 28 = 7x_2 - 28 \Rightarrow 7x_1 = 7x_2 \Rightarrow x_1 = x_2.$$

This shows that no two distinct numbers exist. Thus f is one-to-one. To determine f^{-1}, we set $f(y) = x$ and solve for x:

$$f(y) = x \Rightarrow 7(y - 4) = x \Rightarrow 7y - 28 = x \Rightarrow 7y = x + 28 \Rightarrow y = \tfrac{1}{7}x + 4.$$

The inverse function of f is $f^{-1}(x) = \tfrac{1}{7}x + 4$.

117. Suppose that there exist two real numbers x_1 and x_2 such that $f(x_1) = f(x_2)$. In the manner of Example 4 on page 146, we get

$$f(x_1) = f(x_2) \Rightarrow |5 - x_1| = |5 - x_2|$$

If two numbers have the same absolute value then the numbers themselves are either equal or opposites. So,

$$|5 - x_1| = |5 - x_2| \Rightarrow 5 - x_1 = 5 - x_2 \text{ or } -(5 - x_1) = 5 - x_2.$$

From this first equation we obtain

$$5 - x_1 = 5 - x_2 \Rightarrow -x_1 = -x_2 \Rightarrow x_1 = x_2$$

However, the second equation yields different results:

$$-(5 - x_1) = 5 - x_2 \Rightarrow -5 + x_1 = 5 - x_2 \Rightarrow x_1 = 10 - x_2$$

This means that any pair of numbers x_1 and x_2 that satisfies this last equation $x_1 = 10 - x_2$ are such that $f(x_1) = f(x_2)$. For example, for $x_1 = 3$ and $x_2 = 7$, we find that

$$f(3) = |5 - 3| = |2| = 2 \text{ and } f(7) = |5 - 7| = |-2| = 2.$$

Thus f is not one-to-one.

121. The coefficient of the x^2 term is 1, so the graph of f opens up. This means that f has a minimum value. The x-coordinate for the vertex of the graph of f is $x = -\dfrac{b}{2a} = -\dfrac{(0)}{2(1)} = 0$.

We determine the minimum by computing $f(0)$: $f(0) = (0)^2 - 2 = -2$.
The minimum value of the function f is -2; it occurs for $x = 0$.

125. The function F is a rational expression with a constant numerator, so it has its maximum value for the value of x that causes the denominator $x^2 + 6$ to be as small as possible. Consider the graph of $y = x^2 + 6$. The coefficient of the x^2 term is 1, so this graph opens up. The x-coordinate for the vertex of $y = x^2 + 6$ is $x = -\dfrac{b}{2a} = -\dfrac{(0)}{2(1)} = 0$.

We determine the maximum value of F by computing $F(0)$: $F(0) = \dfrac{4}{(0)^2 + 6} = \dfrac{2}{3}$.

The maximum value of the function F is $\tfrac{2}{3}$; it occurs for $x = 0$.

129. The midpoints of the four sides are determined using the midpoint formula (see page 81):

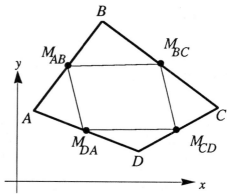

$$M_{AB} = \left(\frac{x_1 + x_2}{2}, \frac{y_1 + y_2}{2}\right) \quad M_{BC} = \left(\frac{x_2 + x_3}{2}, \frac{y_2 + y_3}{2}\right)$$

$$M_{CD} = \left(\frac{x_3 + x_4}{2}, \frac{y_3 + y_4}{2}\right) \quad M_{DA} = \left(\frac{x_4 + x_1}{2}, \frac{y_4 + y_1}{2}\right)$$

We use the distance formula to compute the lengths of the sides:

$$d\left(M_{AB}, M_{BC}\right) = \sqrt{\left(\frac{x_2 + x_3}{2} - \frac{x_1 + x_2}{2}\right)^2 + \left(\frac{y_2 + y_3}{2} - \frac{y_1 + y_2}{2}\right)^2} = \sqrt{\left(\frac{x_3 - x_1}{2}\right)^2 + \left(\frac{y_3 - y_1}{2}\right)^2}$$

$$d\left(M_{DA}, M_{CD}\right) = \sqrt{\left(\frac{x_3 + x_4}{2} - \frac{x_4 + x_1}{2}\right)^2 + \left(\frac{y_3 + y_4}{2} - \frac{y_4 + y_1}{2}\right)^2} = \sqrt{\left(\frac{x_3 - x_1}{2}\right)^2 + \left(\frac{y_3 - y_1}{2}\right)^2}$$

$$d\left(M_{AB}, M_{DA}\right) = \sqrt{\left(\frac{x_4 + x_1}{2} - \frac{x_1 + x_2}{2}\right)^2 + \left(\frac{y_4 + y_1}{2} - \frac{y_1 + y_2}{2}\right)^2} = \sqrt{\left(\frac{x_4 - x_2}{2}\right)^2 + \left(\frac{y_4 - y_2}{2}\right)^2}$$

$$d\left(M_{BC}, M_{CD}\right) = \sqrt{\left(\frac{x_3 + x_4}{2} - \frac{x_2 + x_3}{2}\right)^2 + \left(\frac{y_3 + y_4}{2} - \frac{y_2 + y_3}{2}\right)^2} = \sqrt{\left(\frac{x_4 - x_2}{2}\right)^2 + \left(\frac{y_4 - y_2}{2}\right)^2}$$

Thus $d\left(M_{AB}, M_{BC}\right) = d\left(M_{CD}, M_{DA}\right)$ and $d\left(M_{BC}, M_{CD}\right) = d\left(M_{DA}, M_{AB}\right)$. This completes the proof.

133. The area of the trapezoid is given on the front inside cover of the text:

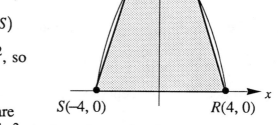

$$\left(\begin{array}{c}\text{area of}\\\text{trapezoid}\end{array}\right) = \frac{1}{2}\left(\begin{array}{c}\text{height of}\\\text{trapezoid}\end{array}\right)\left(\begin{array}{c}\text{sum of bases}\\\text{of trapezoid}\end{array}\right)$$

$$= \frac{1}{2}\left(\begin{array}{c}y\text{-coordinate}\\\text{of }P\end{array}\right)(PQ + RS)$$

The point P is on the graph of $y = 8 - \frac{1}{2}x^2$, so the coordinates of P are $(x, 8 - \frac{1}{2}x^2)$. By the symmetry of the figure, the coordinates of Q are $(x, -(8 - \frac{1}{2}x^2))$. The x-intercepts of $y = 8 - \frac{1}{2}x^2$ are $(-4, 0)$ and $(4, 0)$. From this we determine that $PQ = 2x$, $RS = 8$, and the y-coordinate of P is $8 - \frac{1}{2}x^2$. If we let $A(x)$ represent the area of the trapezoid in terms of x, we get

$$A(x) = \frac{1}{2}(8 - \frac{1}{2}x^2)(2x + 8) = (8 - \frac{1}{2}x^2)(x + 4).$$

The x-coordinate of the point P must be between 0 and 4, so domain of A is $[0, 4]$

Chapter 3

Section 3.1

---

1. For this graph, it appears that $y \to 0$ as $x \to \pm\infty$. From the discussion on page 163, the behavior of the graph of a polynomial function is such that $y \to \pm\infty$ as $x \to \pm\infty$. This graph is not the graph of a polynomial function.

5. For this graph, there is a corner just to the left of the y-axis. From the discussion on page 163, the graph of a polynomial function is smooth without corners. This graph is not the graph of a polynomial function.

9. For this graph, $y \to +\infty$ as $x \to \pm\infty$, so the polynomial function is of even degree and its leading coefficient is positive (see the discussion on page 163). Possible candidates for the function we seek are I, IV, and VI. The y-intercept is positive, so we can eliminate VI (its y-intercept is zero). The graph is not symmetric with respect to the y-axis, so the function we seek is not an even function. But IV is an even function, since $P_4(-x) = (-x)^4 - 16(-x)^2 - 2 = x^4 - 16x^2 - 2 = P_4(x)$. It remains that the function whose graph is shown is I.

13. The graph of a) is discussed in Example 3.

The graph of b) is a vertical translation of -8 units (8 units down) of the graph in a).
To find the y-intercept, we set $x = 0$ and solve for y:
$$y = 0^3 - 8 = -8 \Rightarrow y\text{-intercept is } (0, -8)$$
To find x-intercepts, we set $y = 0$ and solve for x:
$$0 = x^3 - 8 \Rightarrow x^3 = 8 \Rightarrow x = 2 \Rightarrow x\text{-intercept is } (2, 0).$$

17. The graph of a) is a vertical compression by a factor of $\frac{1}{2}$ of the graph of $y = x^5$ (the graph of $y = x^5$ is discussed in Example 3). The graph of b) is a vertical translation of 16 units (16 units up) of the graph in a).
To find the y-intercept of $y = \frac{1}{2}x^5 + 16$, we set $x = 0$ and solve

for y: $\qquad y = \frac{1}{2}0^5 + 16 = 16 \Rightarrow y\text{-intercept is } (0, 16)$
To find x-intercepts, we set $y = 0$ and solve for x:
$$0 = \frac{1}{2}x^5 + 16 \Rightarrow x^5 = -32 \Rightarrow x = -2$$
$$\Rightarrow x\text{-intercept is } (-2, 0).$$

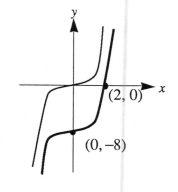

21. The graph of a) is a vertical reflection of the graph of $y = x^6$ (this graph is discussed in Example 2).

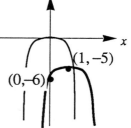

The graph of b) is a translation of −5 units vertically (5 units down) and 1 unit horizontally (1 unit right) of the graph in a).

To find its y-intercept, we set $x = 0$ and solve for y:
$$y = -(0-1)^2 - 5 = - \quad \Rightarrow \quad y\text{-intercept is } (0, -6)$$
To find its x-intercepts, we set $y = 0$ and solve for x:
$$0 = -(x-1)^6 - 5 \quad \Rightarrow \quad (x-1)^6 = -5$$
$$\Rightarrow \quad x - 1 = \sqrt{-5} \quad \Rightarrow \quad \text{no } x\text{-intercepts.}$$

25. If we were to simplify $f(x)$ by multiplying, the leading term would be x^3. This implies (by the discussion on page 163) that the behavior of the graph is such that

$y \to -\infty$ as $x \to -\infty$ and $y \to +\infty$ as $x \to +\infty$.

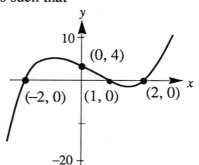

To find the y-intercept, we evaluate $f(0)$:
$$f(0) = (0-1)(0+2)(0-2) = (-1)(2)(-2) = 4$$
$$\Rightarrow \quad y\text{-intercept is } (0, 4)$$
To find x-intercepts, we set $f(x) = 0$ and solve for x.
$$0 = (x-1)(x+2)(x-2) \quad \Rightarrow x - 1 = 0 \text{ or } x + 2 = 0$$
or $x - 2 = 0 \Rightarrow x = 1$ or $x = -2$ or $x = 2$

$\Rightarrow$ x-intercepts are $(1, 0)$, $(-2, 0)$ and $(2, 0)$

29. If we were to simplify $w(x)$ by multiplying, the leading term would be x^4. This implies (by the discussion on page 163) that the behavior of the graph is such that

$y \to +\infty$ as $x \to -\infty$ and $y \to +\infty$ as $x \to +\infty$. To find the y-intercept, we evaluate $w(0)$:
$$w(0) = ((0)^2 + 5(0) + 4)((0)^2 - 10(0)) = 0 \quad \Rightarrow \quad y\text{-intercept is } (0, 0)$$
To find x-intercepts, we set $w(x) = 0$ and solve for x.

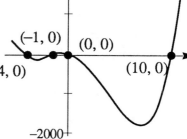

$$0 = (x^2 + 5x + 4)(x^2 - 10x)$$
$$\Rightarrow x^2 + 5x + 4 = 0 \text{ or } x^2 - 10x = 0$$
$$x^2 + 5x + 4 = 0 \quad \Rightarrow \quad (x+4)(x+1) = 0$$
$$\Rightarrow x + 4 = 0 \text{ or } x + 1 = 0 \Rightarrow x = -1 \text{ or } x = -4$$
$$x^2 - 10x = 0 \quad \Rightarrow \quad x(x-10) = 0 \quad \Rightarrow \quad x = 0 \text{ or}$$
$$x - 10 = 0 \quad \Rightarrow \quad x = 0 \text{ or } x = 10$$
$\Rightarrow$ x-intercepts are $(-4, 0)$, $(-1, 0)$, $(0, 0)$, and $(10, 0)$

33. If we were to simplify $D(x)$ by multiplying, the leading term would be $-x^3$. This implies (by the discussion on page 163) that the behavior of the graph is such that

$y \to +\infty$ as $x \to -\infty$ and $y \to -\infty$ as $x \to +\infty$. To find the y-intercept, we evaluate $D(0)$:

$$D(0) = -\tfrac{1}{2}(2(0)-15)((0)^2 - 9) = -\tfrac{1}{2}(-15)(-9) = -\tfrac{135}{2}$$

$$\Rightarrow y\text{-intercept is } (0, -\tfrac{135}{2})$$

To find x-intercepts, we set $D(x) = 0$ and solve for x.

$$0 = -\tfrac{1}{2}(2x-15)(x^2-9)$$

$$\Rightarrow 2x-15 = 0 \text{ or } x^2-9 = 0$$

$$2x-15 = 0 \Rightarrow x = \tfrac{15}{2}$$

$$x^2-9 = 0 \Rightarrow x^2 = 9 \Rightarrow x = \pm 3$$

$\Rightarrow x$-intercepts are $(\tfrac{15}{2}, 0)$, $(-3, 0)$ and $(3, 0)$

37. Because the leading term is $-x^4$, the graph is such that $y \to -\infty$ as $x \to -\infty$ and $y \to -\infty$ as $x \to +\infty$. To find the y-intercept, we evaluate $Q(0)$:

$$Q(0) = -(0)^4 + 4(0)^3 + 21(0)^2 = 0 \Rightarrow y\text{-intercept is } (0, 0).$$

To find x-intercepts, we set $Q(x) = 0$ and solve for x.

$$0 = -x^4 + 4x^3 + 21x^2 \Rightarrow 0 = -x^2(x^2 - 4x - 21)$$

$$\Rightarrow 0 = -x^2 \text{ or } 0 = x^2 - 4x - 21$$

$$0 = -x^2 \Rightarrow x = 0$$

$$0 = x^2 - 4x - 21 \Rightarrow (x+3)(x-7) = 0$$

$$\Rightarrow x+3 = 0 \text{ or } x-7 = 0 \Rightarrow x = -3 \text{ or } x = 7$$

$$\Rightarrow x\text{-intercepts are } (-3, 0), (0, 0) \text{ and } (7, 0)$$

41. The function can be rewritten in descending powers as $y(x) = -\tfrac{1}{2}x^4 + 8x^2$ Because the leading term is $-\tfrac{1}{2}x^4$, the graph is such that $y \to -\infty$ as $x \to \pm\infty$. To find the y-intercept, we evaluate $y(0)$: $y(0) = -\tfrac{1}{2}(0)^4 + 8(0)^2 = 0 \Rightarrow y$-intercept is $(0, 0)$

To find x-intercepts, we set $y(x) = 0$ and solve for x.

$$0 = -\tfrac{1}{2}x^4 + 8x^2 \Rightarrow 0 = -\tfrac{1}{2}x^2(x^2 - 16)$$

$$\Rightarrow 0 = -\tfrac{1}{2}x^2 \text{ or } 0 = x^2 - 16$$

$$0 = -\tfrac{1}{2}x^2 \Rightarrow x = 0$$

$$0 = x^2 - 16 \Rightarrow x^2 - 16 \Rightarrow x = \pm 4$$

$$\Rightarrow x\text{-intercepts are } (-4, 0), (0, 0), (4, 0)$$

45. Because the leading term is $-x^3$, the graph is such that $y \to +\infty$ as $x \to -\infty$ and $y \to -\infty$ as $x \to +\infty$. To find the y-intercept, we evaluate $g(0)$:
$$g(0) = -(0)^3 - 3(0)^2 + 7(0) + 21 = 21 \implies y\text{-intercept is } (0, 21)$$
To find x-intercepts, we set $g(x) = 0$ and solve for x.
(The solution involves factoring by grouping.)
$$0 = -x^3 - 3x^2 + 7x + 21 \implies 0 = -x^2(x+3) + 7(x+3)$$
$$\implies 0 = (-x^2 + 7)(x+3) \implies 0 = -x^2 + 7 \text{ or } x + 3 = 0$$
$$0 = -x^2 + 7 \implies x^2 = 7 \implies x = \pm\sqrt{7}$$
$$0 = x + 3 \implies x = -3$$
$$\implies x\text{-intercepts are } (-\sqrt{7}, 0), (-3, 0) \text{ and } (\sqrt{7}, 0)$$

49. First, note that $Q(-x) = -\frac{1}{2}(-x)^4 + 4(-x)^2 = -\frac{1}{2}x^4 + 4x^2 = Q(x)$

This means that Q is an even function and its graph is symmetric with respect to the y-axis. Because its leading term is $-\frac{1}{2}x^4$, the graph is such that $y \to -\infty$ as $x \to \pm\infty$. To find the y-intercept, we evaluate $Q(0)$: $Q(0) = -\frac{1}{2}(0)^4 + 4(0)^2 = 0 \implies y\text{-intercept is } (0, 0)$

To find x-intercepts, we set $Q(x) = 0$ and solve for x.
$$0 = -\frac{1}{2}x^4 + 4x^2 \implies 0 = -\frac{1}{2}x^2(x^2 - 8)$$
$$\implies 0 = -\frac{1}{2}x^2 \text{ or } x^2 - 8 = 0$$
$$0 = -\frac{1}{2}x^2 \implies x = 0$$
$$0 = x^2 - 8 \implies x^2 = 8 \implies x = \pm 2\sqrt{2}$$
$$\implies x\text{-intercepts are } (-2\sqrt{2}, 0), (0, 0) \text{ and } (2\sqrt{2},$$
$0)$

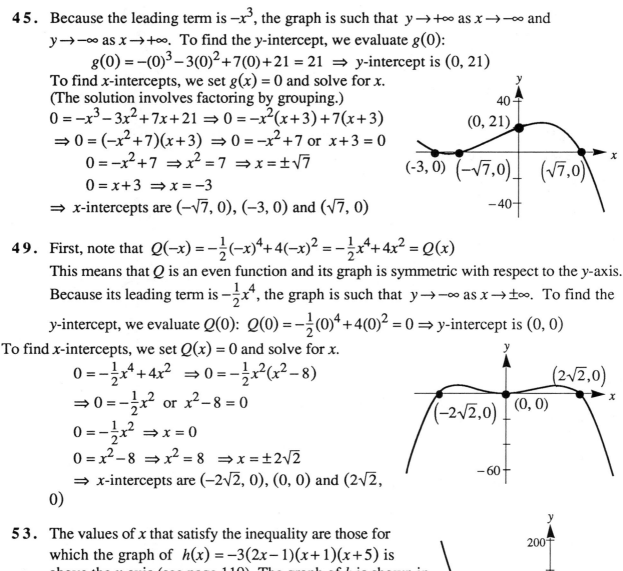

53. The values of x that satisfy the inequality are those for which the graph of $h(x) = -3(2x-1)(x+1)(x+5)$ is above the x-axis (see page 119). The graph of h is shown in the solution to Problem 31. Its x-intercepts are $(-5, 0)$, $(-1, 0)$, and $(\frac{1}{2}, 0)$ (these can be determined in the manner of Problem 33). From the graph, we get the solution for the inequality: $(-\infty, -5]$ or $[-1, \frac{1}{2}]$.

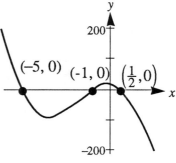

57. The inequality is equivalent to $x^4 - 9x^2 + 20 \geq 0$. The values of x that satisfy the inequality are those for which the graph of the function $n(x) = x^4 - 9x^2 + 20$ is above or on the x-axis (see page 119). The graph of n is shown in the solution to Problem 51. Its x-intercepts are $(\pm\sqrt{5}, 0)$ and

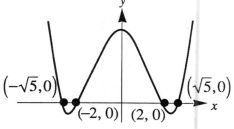

$(\pm 2, 0)$ (these can be determined in the manner of Problem 33). From the graph, we get the solution for the inequality: $(-\infty, -\sqrt{5}]$ or $[-2, 2]$ or $[\sqrt{5}, +\infty)$.

61. The sketch of f is a vertical compression of the graph of $y = x^3$ (by a factor of $\frac{1}{6}$). The graph of g is a line with slope -2 and y-intercept 1. The graph of h is constructed using the technique of adding y-coordinates (See Section 2.6).

65. Let $f(x) = ax^n$. First, suppose that n is an even positive integer. Then
$$f(-x) = a(-x)^n = ax^n = f(x),$$
which implies that f is an even function. Next, suppose n is an odd positive integer. Then
$$f(-x) = a(-x)^n = a(-x^n) = a(-x^n) = f(x),$$
which implies that f is an odd function.

Section 3.2

---

1. Below is the division tableau

$$
\begin{array}{r}
x^2 \qquad\; -5 \\
2x+1\overline{)2x^3 + x^2 - 10x + 2} \\
\underline{2x^3 + x^2} \qquad\qquad \\
-10x + 2 \\
\underline{-10x - 5} \\
7
\end{array}
$$

$P(x) = (2x+1)\left(x^2 - 5\right) + 7$

5. Below is the division tableau

$$\begin{array}{r} x^3 +2x-1 \\ x^2-3\overline{\smash{\big)}\,x^5+0x^4-x^3-x^2+8x+4} \\ \underline{x^5 -3x^3} \\ 2x^3-x^2+8x \\ \underline{2x^3 -6x} \\ -x^2+14x+4 \\ \underline{-x^2 +3} \\ 14x+1 \end{array}$$

$$P(x)=\left(x^2-3\right)\left(x^3+2x-1\right)+(14x+1)$$

9. The divisor is $x-6$, so we enter 6 in the upper left corner of the division tableau. Also, note that we insert a zero for the x-term.

$$\begin{array}{r|rrrrr} 6 & 1 & -4 & -17 & 0 & 2 \\ & & 6 & 12 & -30 & -180 \\ \hline & 1 & 2 & -5 & -30 & \big| -178 \end{array}$$

Thus $P(6)=-178$ $P(x)=(x-6)(x^3+2x^2-5x-30)-178$

13. The divisor is $x-4$, so we enter 4 in the upper left corner of the division tableau.

$$\begin{array}{r|rrrrrr} 4 & \frac{1}{2} & 7 & -2 & 1 & -5 & 9 \\ & & 2 & 36 & 136 & 548 & 2172 \\ \hline & \frac{1}{2} & 9 & 34 & 137 & 543 & \big| 2181 \end{array}$$

Thus $P(4)=2181$ $P(x)=(x-4)(\frac{1}{2}x^3+9x^3+34x^2+137x+543)+2181$

17. The divisor is $x-3$, so we enter 3 in the upper left corner of the division tableau.

$$\begin{array}{r|rrrr} 3 & 2 & -11 & 17 & -6 \\ & & 6 & -15 & 6 \\ \hline & 2 & -5 & 2 & \big| 0 \end{array}$$

$$P(x)=(x-3)(2x^2-5x+2)$$

21. The divisor is $x+\frac{2}{3}$, so we enter $-\frac{2}{3}$ in the upper left corner of the division tableau.

$$\begin{array}{r|rrrrr} -\frac{2}{3} & 6 & 4 & -18 & 9 & 14 \\ & & -4 & 0 & 12 & -14 \\ \hline & 6 & 0 & -18 & 21 & \big| 0 \end{array}$$

$$P(x)=(x+\frac{2}{3})(6x^3-18x+21)$$

25. A real number is a root of the equation if and only if it is a zero for $P(x) = x^3 - 7x + 6$. Using synthetic division, we investigate each of the potential roots.   By the factor theorem, a remainder of zero indicates the potential root is a actual root.

$$
\begin{array}{r|rrrr}
-2 & 1 & 0 & -7 & 6 \\
   &   & -2 & 4 & 6 \\
\hline
   & 1 & -2 & -3 & 12
\end{array}
\qquad
\begin{array}{r|rrrr}
1 & 1 & 0 & -7 & 6 \\
  &   & 1 & 1 & -6 \\
\hline
  & 1 & 1 & -6 & 0
\end{array}
$$

$$
\begin{array}{r|rrrr}
2 & 1 & 0 & -7 & 6 \\
  &   & 2 & 4 & -6 \\
\hline
  & 1 & 2 & -3 & 0
\end{array}
\qquad
\begin{array}{r|rrrr}
3 & 1 & 0 & -7 & 6 \\
  &   & 3 & 9 & 6 \\
\hline
  & 1 & 3 & 2 & 12
\end{array}
$$

The roots of the equation are 1 and 2.

29. The equation is equivalent to $2x^4 - 7x^3 - 18x^3 + 49x + 28 = 0$. A real number x is a root of this equation if and only if it is a zero for the function $P(x) = 2x^4 - 7x^3 - 18x^3 + 49x - 28$. Using synthetic division, we investigate each of the potential roots. By the factor theorem, a remainder of zero indicates the potential root is a actual root.

$$
\begin{array}{r|rrrrr}
-\frac{1}{2} & 2 & -7 & -18 & 49 & 28 \\
 &   & -1 & 4 & 7 & -28 \\
\hline
 & 2 & -8 & -14 & 56 & 0
\end{array}
$$

$$
\begin{array}{r|rrrrr}
\sqrt{7} & 2 & -7 & -18 & 49 & 28 \\
 &   & 2\sqrt{7} & 14-7\sqrt{7} & -49-4\sqrt{7} & -28 \\
\hline
 & 2 & 2\sqrt{7}-7 & -4-7\sqrt{7} & 4\sqrt{7} & 0
\end{array}
$$

$$
\begin{array}{r|rrrrr}
1 & 2 & -7 & -18 & 49 & 28 \\
 &   & 2 & -5 & -23 & 26 \\
\hline
 & 2 & -5 & -23 & 26 & 54
\end{array}
\qquad
\begin{array}{r|rrrrr}
4 & 2 & -7 & -18 & 49 & 28 \\
 &   & 8 & 4 & -56 & -28 \\
\hline
 & 2 & 1 & -14 & -7 & 0
\end{array}
$$

The roots of the equation are $-\frac{1}{2}$, $\sqrt{7}$, and 4.

33. Because $(2, 0)$ is an intercept of the graph, 2 is a zero of f. By the factor theorem, $f(x) = (x-2)Q(x)$, where is $Q(x)$ found by dividing $f(x)$ by $(x-2)$.

$$
\begin{array}{r|rrrr}
2 & 1 & 12 & 12 & -80 \\
  &   & 2 & 28 & 80 \\
\hline
  & 1 & 14 & 40 & 0
\end{array}
\qquad \Rightarrow f(x) = (x-2)(x^2 + 14x + 40)
$$

To find the x-intercepts, we set $f(x) = 0$ and solve the resulting equation.

$(x-2)(x^2 + 14x + 40) = 0 \Rightarrow x-2 = 0 \quad \text{or} \quad x^2 + 14x + 40 = 0$.

$x-2 = 0 \Rightarrow x = 2 \qquad x^2 + 14x + 40 = 0 \Rightarrow (x+10)(x+4) = 0 \Rightarrow x = -4 \text{ or } -10$

$\Rightarrow x$-intercepts are $(2, 0)$, $(-4, 0)$, $(-10, 0)$

Because the leading coefficient is positive and the degree of the function is odd, the table on page 162 indicates that $y \rightarrow +\infty$ as $x \rightarrow +\infty$ and $y \rightarrow -\infty$ as $x \rightarrow -\infty$. Computing and plotting a few more points gives us the graph.

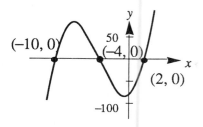

37. The values of x that satisfy the inequality are those for which the graph of $f(x) = x^3 - 7x + 6$ is below the x-axis (see page 119). The graph of f is shown in the solution to Problem 31. Its x-intercepts are $(1, 0)$, $(2, 0)$, and $(-3, 0)$ (these can be determined in the manner of Problem 33). From the graph, we get the solution for the inequality: $(-\infty, -3)$ or $(1, 2)$.

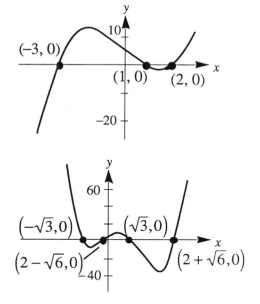

41. The inequality is equivalent to
$$x^4 - 4x^3 - 5x^2 + 12x + 6 < 0.$$
The values of x satisfying this inequality are those for which the graph of $f(x) = x^4 - 4x^3 - 5x^2 + 12x + 6$ is below the x-axis (see page 119). The graph of f is shown in the solution to Problem 35. Its x-intercepts are $(\pm\sqrt{3}, 0)$ (these were given in Problem 31) and $(2\pm\sqrt{6}, 0)$ (these can be determined in the manner of Problem 33). From the graph, we get the solution for the inequality:
$$(-\sqrt{3}, 2-\sqrt{6}) \text{ or } (\sqrt{3}, 2+\sqrt{6}).$$

45. Let $g(x) = x^n - a^n$. Because $g(a) = a^n - a^n = 0$, a is a zero of the function g. By the factor theorem, this implies that $(x - a)$ is a factor of $x^n - a^n$.

49. $f(-4) = -(-4)^3 - 3(-4)^2 + 2(-4) + 2 = 10 \;\Rightarrow\; f(-4) > 0$
$f(-3) = -(-3)^3 - 3(-3)^2 + 2(-3) + 2 = -4 \;\Rightarrow\; f(-3) < 0$
Because $f(-4)$ and $f(-3)$ have different signs, it follows from the location theorem of Problem 47 that there exists a zero in the interval $[-4, -3]$.

53. First, note that $f(1) = 1 - 11 = -10$ and $f(2) = 16 - 11 = 5$. By the location theorem, a zero for the f exists in the interval $[1, 2]$. Using the location theorem with the first table at right, we further determine that there is zero in the interval $[1.8, 1.9]$. Using the location theorem again with the second table at right, we determine that the zero is in the interval

x	$f(x)$	x	$f(x)$
1.1	−9.5359	1.81	−0.2672
1.2	−8.9264	**1.82**	**−0.0280**
1.3	−8.1439	**1.83**	**0.2151**
1.4	−7.1584	1.84	0.4623
1.5	−5.9375	1.85	0.7135
1.6	−4.4464	1.86	0.9688
1.7	−2.6479	1.87	1.2283
1.8	**−0.5024**	1.88	1.4920
1.9	**2.0321**	1.89	1.7599

$[1.82, 1.83]$. The value of $f(1.825) = 0.9306$, so the location theorem implies that the zero is between 1.82 and 1.825. Thus, to the nearest 0.01, the zero of the function is

1.82. This is the approximation of $\sqrt[3]{11}$ required.

Section 3.3

---

1. Setting $P(x) = 0$ and solving the resulting equation yields the zeros of P

$$P(x) = 0 \implies (x^2 - 7x + 12)(x^2 - 4x - 8) = 0 \implies x^2 - 7x + 12 = 0 \text{ or } x^2 - 4x - 8 = 0$$

$$x^2 - 7x + 12 = 0 \implies (x - 3)(x - 4) = 0 \implies x - 3 = 0 \text{ or } x - 4 = 0 \implies x = 3 \text{ or } x = 4$$

$$x^2 - 4x - 8 = 0 \implies x = \frac{-(-4) \pm \sqrt{(-4)^2 - 4(1)(-8)}}{2} = \frac{4 \pm \sqrt{48}}{2} = \frac{4 \pm 4\sqrt{3}}{2} = 2 \pm 2\sqrt{3}$$

Rational zeros: $3, 4$ Real zeros: $3, 4, 2 + 2\sqrt{3}, 2 - 2\sqrt{3}$

Complex zeros: $3, 4, 2 + 2\sqrt{3}, 2 - 2\sqrt{3}$

5. Setting $f(x) = 0$ equal to zero and solving the resulting equation yields the zeros of f.

$$f(x) = 0 \implies (x^2 - 4)(x^4 - 7x^2 + 12) = 0 \implies x^2 - 4 = 0 \text{ or } x^4 - 7x^2 + 12 = 0$$

$$x^2 - 4 = 0 \implies x^2 = 4 \implies x = \pm 2$$

$$x^4 - 7x^2 + 12 = 0 \implies (x^2 - 3)(x^2 - 4) = 0 \implies x^2 - 3 = 0 \text{ or } x^2 - 4 = 0.$$

$$x^2 - 3 = 0 \implies x^2 = 3 \implies x = \pm\sqrt{3}$$

(the other factor repeats the zero already found).

Rational zeros: $-2, 2$ Real zeros: $-2, 2, -\sqrt{3}, \sqrt{3}$ Complex zeros: $-2, 2, -\sqrt{3}, \sqrt{3}$

9. Because $1 + i\sqrt{3}$ is a zero of P, the conjugate root theorem on page 188 insures that its conjugate $1 - i\sqrt{3}$ is also a zero of P. So,

$$P(x) = 1(x + 4)^2 \left(x - \left(1 + i\sqrt{3}\right)\right)\left(x - \left(1 - i\sqrt{3}\right)\right) = \left(x^2 + 8x + 16\right)\left(x^2 - 2x + 4\right)$$

or $P(x) = x^4 + 6x^3 + 4x^2 + 64$.

13. $(2x^2 - 3)(x - 4)(3x - 8) = 0 \implies 2x^2 - 3 = 0 \text{ or } x - 4 = 0 \text{ or } 3x - 8 = 0$

$$2x^2 - 3 = 0 \implies x^2 = \frac{3}{2} \implies x = \pm\sqrt{\frac{3}{2}} \text{ or, simplified, } \pm\frac{\sqrt{6}}{2}$$

$$x - 4 = 0 \implies x = 4 \quad \text{and} \quad 3x - 8 = 0 \implies x = \frac{8}{3}$$

Rational zeros: $4, \frac{8}{3}$ Real zeros: $4, \frac{8}{3}, -\frac{\sqrt{6}}{2}, \frac{\sqrt{6}}{2}$ Complex zeros: $4, \frac{8}{3}, -\frac{\sqrt{6}}{2}, \frac{\sqrt{6}}{2}$

17. $(x^2+12x+20)(x^2+2x-6)(x+8) = 0$

$$\Rightarrow x^2+12x+20 = 0 \ \text{ or } \ x^2+2x-6 = 0 \ \text{ or } \ x+8 = 0$$

$x^2+12x+20 = 0 \Rightarrow (x+10)(x+2) = 0 \Rightarrow x+10 = 0 \text{ or } x+2 = 0 \Rightarrow x = -2 \text{ or } x = -10$

$x^2+2x-6 = 0 \Rightarrow x = \dfrac{-2\pm\sqrt{2^2-4(1)(-6)}}{2} = \dfrac{-2\pm\sqrt{28}}{2} = \dfrac{-2\pm2\sqrt{7}}{2} = -1\pm\sqrt{7}$

$x+8 = 0 \Rightarrow x = -8$

Rational roots: $-10, -8, -2$ Real roots: $-10, -8, -2, -1+\sqrt{7}, -1-\sqrt{7}$

Complex roots: $-10, -8, -2, -1+\sqrt{7}, -1-\sqrt{7}$

21. If $\dfrac{p}{q}$ is a zero of P, then q is a factor of 2 (the leading coefficient of P) and p is a factor of

-6 (the constant term of P). This implies that the potential zeros of P are

$$\frac{p}{q} = \pm\frac{1, 2, 3, 6}{1, 2} = \pm 1, \pm 2, \pm 3, \pm 6, \pm\frac{1}{2}, \pm\frac{3}{2}$$

In the manner of Example 5 on page 185, we check each of these potential zeros using synthetic division. We find that 2 is an actual zero.

$$\begin{array}{r|rrrr}
2 & 2 & -5 & 5 & -6 \\
  &   & 4 & -2 & 6 \\
\hline
  & 2 & -1 & 3 & 0
\end{array}$$

It follows that $P(x) = (x-2)(2x^2-x+3)$. The zeros of the reduced function $2x^2-x+3$ has no rational zeros since the equation $2x^2-x+3 = 0$ has nonreal roots. Thus 2 is the only rational zero of the function.

25. According to the rational root theorem, the potential rational zeros of P are

$$\frac{p}{q} = \pm\frac{1, 2, 3, 5, 6, 10, 15, 30}{1} = \pm 1, \pm 2, \pm 3, \pm 5 \pm 6, \pm 10, \pm 15, \pm 30$$

We check each of these potential zeros using synthetic division. We find first that 2 is a zero.

$$\begin{array}{r|rrrr}
2 & 1 & 6 & -1 & -30 \\
  &   & 2 & 16 & 30 \\
\hline
  & 1 & 8 & 15 & 0
\end{array}$$

It follows that $P(x) = (x-2)(x^2+8x+15)$. The zeros of the reduced function $x^2+8x+15$ can be found by factoring:

$$x^2+8x+15 = 0 \Rightarrow (x+3)(x+5) = 0$$
$$\Rightarrow x+3 = 0 \ \text{ or } \ x+5 = 0 \Rightarrow x = -3 \ \text{ or } \ x = -5.$$

The zeros of P are $2, -3,$ and -5.

In lineared factored form, the function is $P(x) = (x-2)(x+3)(x+5)$

29. According to the rational root theorem, the potential rational zeros of P are

$$\frac{p}{q} = \pm\frac{1, 2, 3, 4, 6, 9, 11, 12, 18, 22, 33, 36, 44, 66, 99, 132, 198, 396}{1, 2}$$

$$= \pm 1, \pm 2, \pm 3, \pm 4, \pm 6, \pm 9, \pm 11, \pm 12, \pm 18, \pm 22, \pm 33, \pm 36,$$
$$\pm 44, \pm 66, \pm 99, \pm 132, \pm 198, \pm 396, \pm\tfrac{1}{2}, \pm\tfrac{3}{2}, \pm\tfrac{9}{2}, \pm\tfrac{11}{2}, \pm\tfrac{33}{2}, \pm\tfrac{99}{2}$$

In the manner of Example 5 on page 185, we check each of these potential zeros using synthetic division. We find first that 2 is a zero.

2	2	−7	−44	97	204	−396
		4	−6	−100	−6	−396
	2	−3	−50	−3	198	0

It follows that $R(x) = (x-2)(2x^4 - 3x^3 - 50x^2 - 3x + 198)$. Checking 2 again with the reduced function , we get

2	2	−3	−50	−3	198
		4	2	−96	−198
	2	1	−48	−99	0

This implies that

$R(x) = (x-2)(x-2)(2x^3 + x^2 - 48x - 99) = (x-2)^2(2x^3 + x^2 - 48x - 99)$

Working with the reduced equation, $2x^3 + x^2 - 48x - 99$, we find that -3 is also a zero.

−3	2	1	−48	−99
		−6	15	99
	2	−5	−33	0

Thus, $R(x) = (x-2)^2(x+3)(2x^2 - 5x - 33)$. The remaining zeros are found by solving the equation $2x^2 - 5x - 33 = 0$. $2x^2 - 5x - 33 = 0 \Rightarrow (2x-11)(x-3) = 0 \Rightarrow 2x - 11 = 0$ or $x - 3 = 0 \Rightarrow x = \frac{11}{2}$ or $x = 3$. The zeros of R are 2 (multiplicity 2), -3, and $\frac{11}{2}$. In lineared factored form, the function is $R(x) = 2(x-2)^2(x+3)(x-\frac{11}{2})$.

33. The leading term is x^4, so it follows from the discussion on page 163 that $y \to +\infty$ as $x \to \pm\infty$. The y-intercept is found by evaluating $g(0)$:

$$g(0) = (0)^4 - 2(0)^3 - 38(0)^2 - 6(0) + 105 = 105$$
$$\Rightarrow y\text{-intercept is } (0, 105).$$

From Problem 27, the zeros of g are -5, 7, and $\pm\sqrt{3}$. The x-intercepts are $(-5, 0)$, $(-\sqrt{3}, 0)$, $(\sqrt{3}, 0)$, $(7, 0)$.

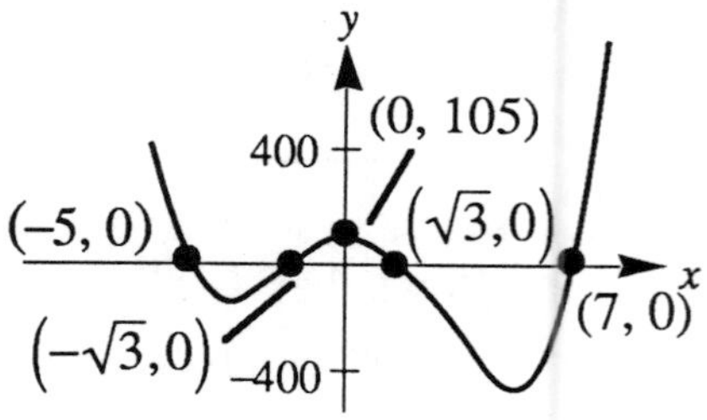

37. We are seeking four zeros for the function, not necessarily distinct. Because one of the zeros of this real function is $-2-2i$, the conjugate root theorem indicates that the conjugate of $-2-2i$, namely $-2+2i$, is also a zero.

$$
\begin{array}{r|rrrrr}
-2-2i & 1 & 0 & -16 & -64 & -64 \\
& & -2-2i & 8i & 48+16i & 64 \\
\hline
-2+2i & 1 & -2-2i & -16+8i & -16+16i & \Big|\ 0 \\
& & -2+2i & 8-8i & 16-16i & \\
\hline
& 1 & -4 & -8 & \Big|\ 0 &
\end{array}
$$

From the last line of the division tableau, the reduced function is x^2-4x-8. Solving the equation $x^2-4x-8=0$, we get

$$x = \frac{-(-4)\pm\sqrt{4^2-4(1)(-8)}}{2} = \frac{4\pm\sqrt{48}}{2} = \frac{4\pm4\sqrt{3}}{2} = 2\pm2\sqrt{3}$$

The zeros of the function are $-2-2i$, $-2+2i$, $2-2\sqrt{3}$, and $2+2\sqrt{3}$.

41. We are seeking six zeros for the function, not necessarily distinct. Because one of the zeros of this real function is $-i$, the conjugate root theorem indicates that i is also a zero.

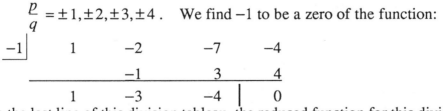

From the last line of the division tableau, the reduced function is x^3-2x^2-7x-4. The rational root theorem suggests that the potential rational zeros of the reduced function are

$$\frac{p}{q} = \pm1,\pm2,\pm3,\pm4 .$$ We find -1 to be a zero of the function:

$$
\begin{array}{r|rrrr}
-1 & 1 & -2 & -7 & -4 \\
& & -1 & 3 & 4 \\
\hline
& 1 & -3 & -4 & \Big|\ 0
\end{array}
$$

From the last line of this division tableau, the reduced function for this division is x^2-3x-4. The zeros of this reduced function are found by setting it equal to zero and solving the equation:

$$x^2-3x-4=0 \Rightarrow (x-4)(x+1)=0 \Rightarrow x-4=0 \text{ or } x+1=0 \Rightarrow x=4 \text{ or } x=-1$$

The zeros of the function are 1, $-i$, i, 4, and -1 .

45. Let $f(x) = x^3 + 27$. The roots of -27 are the zeros of f. Setting $f(x) = 0$, we get
$x^3 + 27 = 0 \Rightarrow (x+3)(x^2 - 3x + 9) \Rightarrow x + 3 = 0$ or $x^2 - 3x + 9 = 0$
$x - 3 = 0 \Rightarrow x = 3$

$$x^2 - 3x + 9 = 0 \Rightarrow x = \frac{-3 \pm \sqrt{3^2 - 4(1)(9)}}{2} = \frac{-3 \pm \sqrt{-27}}{2} = -\frac{3}{2} \pm \frac{3\sqrt{3}}{2} i$$

The complex roots of -27 are -3, $-\frac{3}{2} - \frac{3\sqrt{3}}{2} i$, and $-\frac{3}{2} + \frac{3\sqrt{3}}{2} i$.

Section 3.4

1. The zeros of the denominator are 3, -6, and -1, so the domain is the set of all real numbers except 3, -6, or -1. The zeros of the numerator are 8 and -3; these are the zeros of the function P.

5. The zeros of the denominator are the solutions of $x^2 - 2x - 8 = 0$.
$x^2 - 2x - 8 = 0 \Rightarrow (x-4)(x+2) = 0 \Rightarrow x - 4 = 0$ or $x + 2 = 0 \Rightarrow x = 4$ or $x = -2$
The domain is the set of all real numbers except -2 or 4.
The zeros of the numerator are the solutions of $x^3 - 1 = 0$.
$x^3 - 1 = 0 \Rightarrow x^3 = 1 \Rightarrow x = 1$.
This is the zero of the function f.

9. The domain of r is the set of all real numbers except -5.
Factoring and reducing, we find that

$$r(x) = \frac{x^2 - 25}{x + 5} = \frac{(x+5)(x-5)}{x+5} = x - 5, \quad x \neq -5$$

The graph of r is identical to the graph of $y = x - 5$, except
that the graph of r is undefined for $x = -5$. This is
indicated on the graph by an open circle.

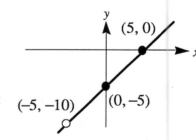

13. The graph of a) is discussed in Example 4 on
page 193.

The graph of b) is a reflection (about the y-axis)
and a vertical expansion (by a factor of 2). There
are no intercepts. The vertical asymptote is $x = 0$
and the horizontal asymptote is $y = 0$.

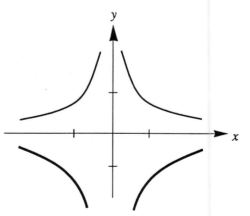

17. The graph of a) is discussed in Example 3 on page 193.

The function in b) is an improper rational function. By long division we get

$$x-1\overline{\smash{\big)}2x-3} \quad \Rightarrow \quad y = 2 - \frac{1}{x-1}$$
$$\underline{2x-2}$$
$$-1$$

This is a translation (1 unit to the right, 2 units up) of the graph of $y = -\frac{1}{x}$, which is a reflection of the graph in a). The y-intercept is determined

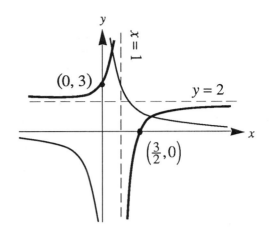

by setting $x = 0$ and solving the equation for y: $y = \frac{2(0)-3}{(0)-1} = 3 \Rightarrow y$-intercept is $(0, 3)$.

To find x-intercepts, we set $y = 0$ and solve for x:

$$0 = \frac{2x-3}{x-1} \quad \Rightarrow 0 = 2x-3 \quad \Rightarrow x = \frac{3}{2} \quad \Rightarrow x\text{-intercept is } (\tfrac{3}{2}, 0).$$

The vertical asymptote is $x = 1$ and the horizontal asymptote is $y = 2$.

21. The graph has vertical asymptotes $x = -2$ and $x = 2$, so the function we seek has factors $x-2$ and $x+2$ in the denominator. Because $(x-2)(x+2) = x^2 - 4$, possible answers are III and V. The graph appears to have symmetry with respect to the y-axis, so the function we seek is an even function. Of the two possibilities, V is the even function (III is an odd function). V is the solution.

25. We rewrite the function as $g(x) = \dfrac{x}{(x-4)(x+4)}$

This implies vertical asymptotes at $x = -4$ and $x = 4$. Because the degree of the numerator is less than the degree of the denominator, the graph has a horizontal asymptote $y = 0$. Because $g(0) = 0$, the y-intercept is $(0, 0)$. The zero of the numerator is 0, so the x-intercept is also $(0, 0)$. We construct the sign graph and use this to sketch the graph

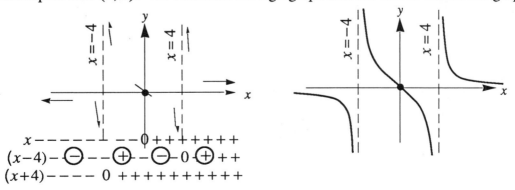

29. We rewrite the function as $P(x) = \dfrac{2x^2}{(x-3)(x+3)}$.

This implies vertical asymptotes at $x = -3$ and $x = 3$. Because the degree of the numerator is equal to the degree of the denominator, the horizontal asymptote is determined by the leading coefficients of the numerator and denominator, so the horizontal asymptote is $y = \dfrac{2}{1}$ or $y = 2$. Because $P(0) = 0$, the y-intercept is $(0, 0)$. The zero of the numerator is 0, so the x-intercept is also $(0, 0)$. We construct the sign graph and use this to sketch the graph.

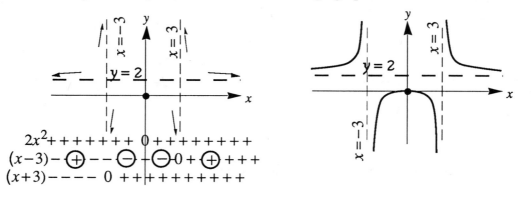

33. The zeros of the denominator determine that the vertical asymptotes are $x = -3$ and $x = 7$. Because the degree of the numerator is equal to the degree of the denominator, the horizontal asymptote is determined by the leading coefficients of the numerator and denominator, so the horizontal asymptote is $y = \dfrac{1}{1}$ or $y = 1$. Because $Q(0) = 0$, the y-intercept is $(0, 0)$. The zeros of the numerator are 0 and 4, so the x-intercepts are $(0, 0)$ and $(4, 0)$. We construct the sign graph and use this to sketch the graph.

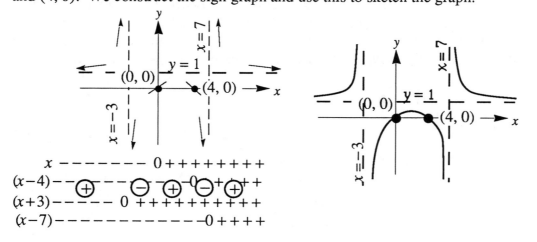

37. The zeros of the denominator determine that the vertical asymptotes are $x = -3$ and $x = 2$. Because the degree of the numerator is equal to the degree of the denominator, the horizontal asymptote is determined by the leading coefficients of the numerator and denominator, so the horizontal asymptote is $y = \frac{1}{3}$. Because $f(0) = \frac{4}{9}$, the y-intercept is $(0, \frac{4}{9})$. The zeros of the numerator are 4 and -1, so the x-intercepts are $(4, 0)$ and $(-1, 0)$. We construct the sign graph and use this to sketch the graph.

41. We rewrite the function as $F(x) = \dfrac{(x-3)(x+3)}{(x-6)}$. This implies vertical asymptotes at $x = 6$. Because the degree of the numerator is greater than that of the denominator, the graph has an oblique asymptote. By long division, we find that $y = x + 6 + \dfrac{27}{x-6}$, so the oblique asymptote is $y = x + 6$. Because $f(0) = \frac{3}{2}$, the y-intercept is $(0, \frac{3}{2})$. The zeros of the numerator are -3 and 3, so the x-intercepts are $(-3, 0)$ and $(3, 0)$. We construct the sign graph and use this to sketch the graph.

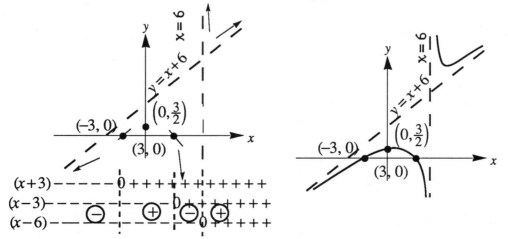

45. We rewrite the function as $W(x) = \dfrac{x(x-3)(x+3)}{(x-4)(x+4)}$. This implies vertical asymptotes at

$x = 4$ and $x = -4$. Because the degree of the numerator is greater than that of the denominator, the graph has an oblique asymptote. By long division, we find that

$y = x + \dfrac{7x}{x^2 - 16}$, so the oblique asymptote is $y = x$. Because $W(0) = 0$, the y-intercept is

$(0, 0)$. The zeros of the numerator are 0, -3 and 3, so the x-intercepts are $(0, 0)$, $(-3, 0)$ and $(3, 0)$. We construct the sign graph and use this to sketch the graph.

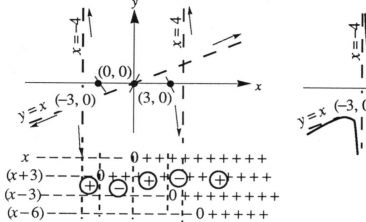

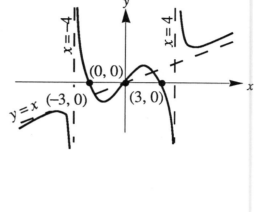

49. The values of x that satisfy the inequality are those for which the graph of the function $g(x) = \dfrac{x}{x^2 - 16}$ is below the x-axis (see page 119). The graph of g is shown in the solution to Problem 25. Its x-intercept is $(0, 0)$, and its vertical asymptotes are $x = -4$ and $x = 4$. From the graph, we get the solution for the inequality: $(-\infty, -4)$ or $(0, 4)$.

53. The values of x that satisfy the inequality are those for which the graph of the function $F(x) = \dfrac{x^2 - 9}{x - 6}$ is above or on the x-axis (see page 119). The graph of F is shown in the solution to Problem 41. Its x-intercepts are $(-3, 0)$ and $(3, 0)$, and its vertical asymptote is $x = 6$. From the graph, we get the solution for the inequality: $[-3, 3]$ or $[6, +\infty)$.

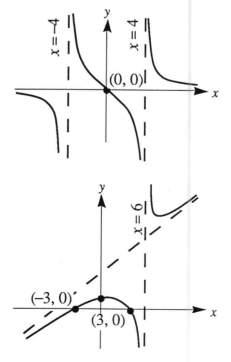

57. We rewrite the function as

$$f(x) = \frac{x^4+1}{2x^2} = \frac{x^4}{2x^2} + \frac{1}{2x^2} = \tfrac{1}{2}x^2 + \frac{1}{2x^2}$$

The graph is constructed by adding y-coordinates of the graphs of $y = \tfrac{1}{2}x^2$ and $y = \frac{1}{2x^2}$. At the extremes, the graph behaves as

$y = \tfrac{1}{2}x^2$. Near $x=0$, the graph behaves as $y = \frac{1}{2x^2}$.

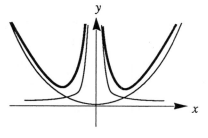

Section 3.5

1. Because the radicand x^2-1 must be positive in order for the function to be defined, the domain of the function is the set of real numbers such that $x^2-1 > 0$. This implies that the domain is $(-\infty, -1)$ or $(1, \infty)$.

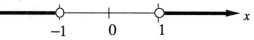

The function has no zeros because the numerator is 1.

5. Because the radicand $8-x^2$ must be positive in order for the function to be defined, the domain of the function is the set of real numbers such that $8-x^2 > 0$, or $x^2-8 < 0$. The solution of this inequality is $(-2\sqrt{2}, 2\sqrt{2})$. Additionally, the value of x cannot be zero; otherwise the denominator would be zero. The domain therefore is $(-2\sqrt{2}, 0)$ or $(0, 2\sqrt{2})$.

The zeros of the function are those values of x that make the numerator zero. Solving the equation $2x-5 = 0$, we obtain $x = \frac{5}{2}$. The zero of the function is $\frac{5}{2}$.

9. From the discussion on page 206, we determine that the domain of the function in a) is $x \geq 0$ and the graph is concave down.

The graph of $y = (x-1)^{3/4}$ is a translation of 1 unit horizontally (1 unit left) of the graph of $y = x^{3/4}$ in a). Its y-intercept is determined by setting $x = 0$ and solving the equation:

$$y = (0-1)^{3/4} \Rightarrow \text{no } y\text{-intercept}$$

To find x-intercepts, we set $y = 0$ and solve for x.

$$0 = (x-1)^{3/4} \Rightarrow 0 = x-1 \Rightarrow x = 1$$

$$\Rightarrow x\text{-intercept is } (1, 0).$$

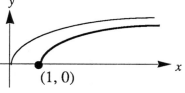

13. First, consider the graph of $y = x^{1/4}$. From the discussion on page 206, we determine that the domain of this function is $x \geq 0$ and the graph is concave down. The graph in a) is a reflection (through the x-axis) and a vertical compression (by a factor of $\frac{1}{2}$) of the graph of $y = x^{1/4}$. The graph in b) is a translation of 1 unit horizontally (1 unit left) of the graph in a). The y-intercept is determined by setting $x = 0$ and solving the equation.

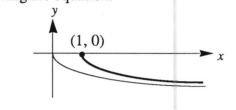

$$y = -\tfrac{1}{2}(0-1)^{1/4} \Rightarrow \text{no } y\text{-intercept}$$

To find x-intercepts, we set $y = 0$ and solve for x.

$$0 = -\tfrac{1}{2}(x-1)^{1/4} \Rightarrow 0 = x-1$$

$$\Rightarrow x = 1 \Rightarrow x\text{-intercept is } (1, 0).$$

17. The graph of $y = x^{2/3}$ is discussed in Example 2 on page 204. The graph of a) is a vertical compression (by a factor of $\frac{1}{2}$) of the graph of $y = x^{2/3}$. The graph in b) is a translation of 4 units horizontally (4 units right) of the graph in a). The y-intercept is determined by setting $x = 0$ and solving the equation.

$$y = \tfrac{1}{2}(0-4)^{2/3} = \tfrac{1}{2}(-4)^{2/3} = 2^{1/3} \Rightarrow y\text{-intercept is } (0, 2^{1/3})$$

To find x-intercepts, we set $y = 0$ and solve for x.

$$0 = \tfrac{1}{2}(x-4)^{2/3} \Rightarrow 0 = x-4 \Rightarrow x = 4$$

$$\Rightarrow x\text{-intercept is } (4, 0).$$

21. The radicand of H is the function $y = x^3 - 16x$. This can be rewritten in factored form as $y = x(x-4)(x+4)$. Using the techniques of Section 3.1 for sketching the graphs of polynomial functions, we get the graph below left.

The domain of H is $[-4, 0]$ or $[4, +\infty)$. The graph of H is sketched below right using the steps on page 209.

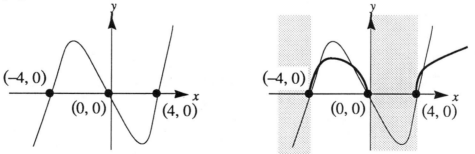

2 5. The radicand of f is the function $y = \dfrac{1}{x-4}$. This is a translation of the graph $y = \dfrac{1}{x}$ by 4 units horizontally (4 units right). Using the techniques of Section 3.4 for graphing rational functions, we get the graph at below left.

The domain of f is $(4, +\infty)$. The graph of f is sketched below right using the steps on page 209.

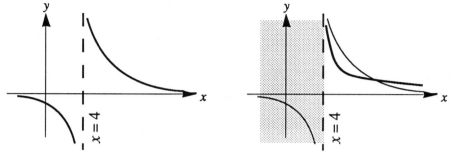

2 9. The radicand of G is the function $y = \dfrac{4x}{x^2+2}$. Using the techniques of Section 3.4 for graphing rational functions, we get the graph at below left.

The domain of G is $[0, +\infty)$. The graph of G is sketched below right using the steps on page 209.

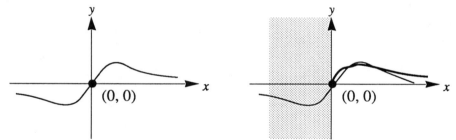

3 3. The graph of $y = \frac{1}{2}x$ is a line passing through the origin with slope $\frac{1}{2}$. The graph of $= x^{2/3}$ is sketched in Example 2 on page 204. The graph of T is constructed by adding y-coordinates of the graphs of $y = \frac{1}{2}x$ and $y = x^{2/3}$. The y-intercept is determined by evaluating $T(0)$ and solving the equation.

$T(0) = \frac{1}{2}(0) + (0)^{2/3} = 0 \implies y$-intercept is $(0, 0)$.

To find x-intercepts, we set $T(x) = 0$ and solve for x:

$$0 = \frac{1}{2}x + x^{2/3} \implies 0 = x + 2x^{2/3}$$
$$\implies 0 = x^{2/3}(x^{1/3} + 2)$$
$$\implies x^{2/3} = 0 \text{ or } x^{1/3} + 2 = 0 \implies x = 0 \text{ or } x = -8$$
$$\implies x\text{-intercepts are } (0, 0) \text{ and } (-8, 0).$$

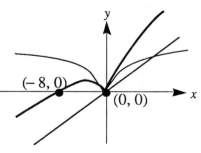

37. The radicand $x - 1$ must be positive, so the domain of the function is $(1, +\infty)$. The vertical asymptote of the graph is $x = 1$ and horizontal asymptote is $y = 0$. The graph is at right.

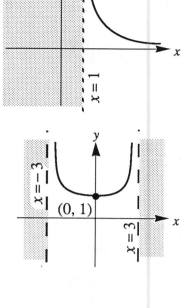

41. The radicand $9 - x^2$ must be positive, so the domain of the function is $(-3, 3)$. The vertical asymptotes of the graph are $x = -3$ and $x = 3$. The horizontal asymptote is $y = 0$. The graph is at right.

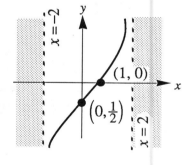

45. The radicand $4 - x^2$ must be positive, so the domain of the function is $(-2, 2)$. The zero of the numerator is 1, so the x-intercept is $(1, 0)$. The vertical asymptotes of the graph are $x = -2$ and $x = 2$. The graph at right shows these results.

49. The values of x that satisfy the inequality are those for which the graph of $f(x) = x^{2/3} - \dfrac{1}{x}$ is above the x-axis (see page 119). The graph of f is shown in the solution to Problem 35. Its x-intercept is $(1, 0)$, and its vertical asymptote is $x = 0$. From the graph, we get the solution for the inequality: $(-\infty, 0)$ or $(1, +\infty)$.

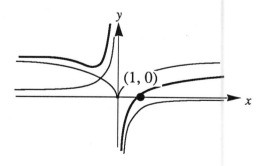

Miscellaneous Exercises for Chapter 3

1. This graph is discussed in Example 2 on page 160.

5. This graph is discussed in part a) of Example 2 on page 204.

9. This graph is a horizontal translation of the graph (by −8 units) discussed in part a) of Example 2 on page 204.

13. The graph of the equation is a translation of −15 units (1 unit down) and 2 units horizontally (2 units right) of the graph in Problem 1.

17. First, consider the graph of $y = x^{5/4}$. From the discussion on page 206, we determine that the domain for the function is $x \geq 0$ and the graph is concave up. The graph of $y = -2x^{5/4}$ is a vertical expansion (by a factor of 2) and reflection (through the x-axis) of this graph. The graph of the function in question then is a translation of 2 units vertically (2 units up) and −1 unit horizontally (1 unit left) of the graph of $y = -2x^{5/4}$.

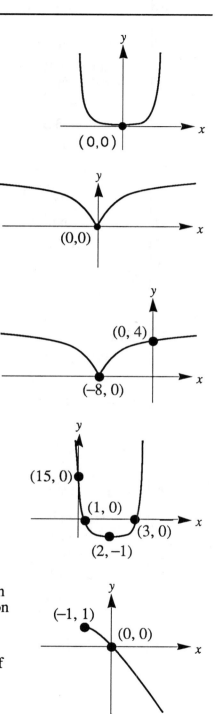

21. The function can be simplified and written as $y = x^4 + 5x^3 - 14x^2$. This is a fourth degree polynomial function with leading coefficient 1, so it follows from the discussion on page 163 that $y \to +\infty$ as $x \to \pm\infty$.

We let $x = 0$ to find the y-intercept:

$y = (0)^4 + 5(0)^3 - 14(0)^2 \Rightarrow$ y-intercept is $(0, 0)$

We let $y = 0$ to find the x-intercepts:

$0 = x^4 + 5x^3 - 14x^2 \Rightarrow x^2(x-2)(x+7) = 0$

$\Rightarrow x^2 = 0$ or $x - 2 = 0$ or $x + 7 = 0$

$\Rightarrow x = 0$ or $x = 2$ or $x = -7$

$\Rightarrow$ x-intercepts are $(0, 0)$, $(2, 0)$, and $(-7, 0)$.

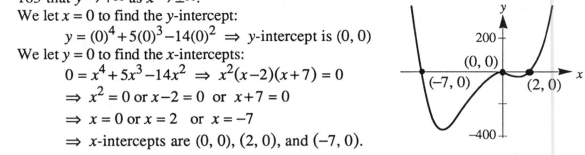

25. The leading term of the function is x^3. This implies (by the discussion on page 163) that the behavior of the graph is such that $y \to -\infty$ as $x \to -\infty$ and $y \to +\infty$ as $x \to +\infty$.

To find the y-intercept, we set $x = 0$ and solve the resulting equation:

$y = (0)^3 - 36(0) = 0$

$\Rightarrow$ y-intercept is $(0, 0)$

To find x-intercepts, we set $y = 0$ and solve for x.

$0 = x^3 - 36x \Rightarrow x(x-6)(x+6) = 0$

$\Rightarrow x = 0, x - 6 = 0,$ or $x + 6 = 0$

$\Rightarrow x = 0, x = 6,$ or $x = -6$

$\Rightarrow$ x-intercepts are $(-6, 0)$, $(0, 0)$ and $(6, 0)$.

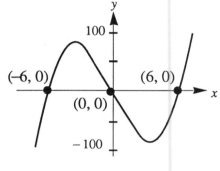

29. The leading term of the function (when simplified) is x^5. This implies (by the discussion on page 163) that the behavior of the graph is such that $y \to -\infty$ as $x \to -\infty$ and $y \to +\infty$ as $x \to +\infty$. To find the y-intercept, we set $x = 0$ and solve the resulting equation:

$y = 0(0-2)^2(0+4)^2 = 0 \Rightarrow$ y-intercept is $(0, 0)$

To find x-intercepts, we set $y = 0$ and solve for x.

$0 = x(x-2)^2(x+4)^2$

$\Rightarrow x = 0$ or $x - 2 = 0$ or $x + 4 = 0$

$\Rightarrow x = 0, x = 2,$ or $x = -4$

$\Rightarrow$ x-intercepts are $(-4, 0)$, $(0, 0)$ and $(2, 0)$.

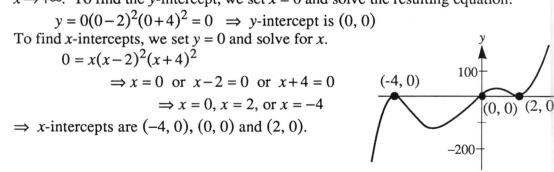

33. The function can be rewritten as $P(x) = \dfrac{-5(x-3)(x+3)}{(x+5)(x-2)}$.

The zeros of the denominator determine that the vertical asymptotes are $x = -5$ and $x = 2$. Because the degree of the numerator is equal to the degree of the denominator, the horizontal asymptote is determined by the leading coefficients of the numerator and denominator, so the horizontal asymptote is $y = \dfrac{-5}{1} = -5$. Since $P(0) = -\dfrac{9}{2}$ the y-intercept is $(0, -\dfrac{9}{2})$. The zeros of the numerator are 3 and -3, so the x-intercepts are $(3, 0)$ and $(-3, 0)$. We construct the sign graph and use this to sketch the graph.

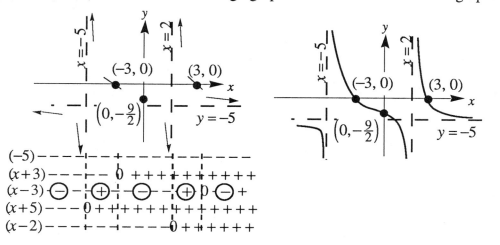

37. The denominator has no real zeros, so there are no vertical asymptotes. Because the degree of the numerator is equal to the degree of the denominator, the horizontal asymptote is determined by the leading coefficients of the numerator and denominator, so the horizontal asymptote is $y = \dfrac{1}{1} = 1$. Since $P(0) = 0$ the y-intercept is $(0, \dfrac{9}{2})$. The zeros of the numerator are 0 and 6, so the x-intercepts are $(0, 0)$ and $(6, 0)$. We construct the sign graph and use this to sketch the graph.

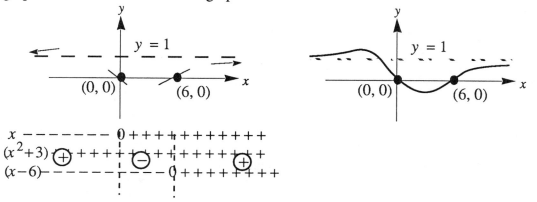

41. The function can be rewritten as $P(x) = \dfrac{(x^2-1)(x^2-16)}{2x(x^2-4)} = \dfrac{(x-1)(x+1)(x-4)(x+4)}{2x(x-2)(x+2)}$

Because the degree of the numerator is greater than the degree of the denominator, an oblique asymptote is determined by long division. We find that $P(x) = \frac{1}{2}x - \dfrac{13x^2-16}{2x^3-8x}$; this implies that the oblique asymptote is $y = \frac{1}{2}x$. We construct the sign graph and use this to sketch the graph.

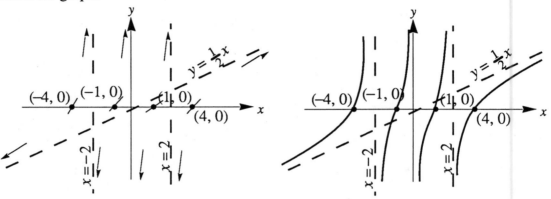

```
x - - - - - - - - - - 0 + + + + + + + + +
(x+2) - - - - - -0+ + + + + + + + + + + +
(x - 2) - - - - - - - - - - - - -0 + + + + +
        -    +   - + - +    -       +
(x+1) - - - - - - - 0 + + + + + + + + + + +
(x-1)  - - - - - - - - - - - 0+ + + + + + +
(x-4) - - - - - - - - - - - - - - - -0 + + +
(x+4) - - -0 + + + + + + + + + + + + + + + +
```

45. The function can be rewritten as $P(x) = \left[\dfrac{x+1}{(x-4)(x+3)}\right]^2 = \dfrac{(x+1)^2}{(x-4)^2(x+3)^2}$.

Because the degree of the numerator is less than the degree of the denominator, the horizontal asymptote is $y = 0$.

Since $P(0) = \dfrac{(0+1)^2}{(0-4)^2(0+3)^2} = \frac{1}{144}$ the y-intercept is

$(0, \frac{1}{144})$. The zeros of the numerator are 3 and −3, so

the x-intercepts are $(3, 0)$ and $(-3, 0)$. A sign graph is unnecessary for this function since the square of a real number is nonnegative.

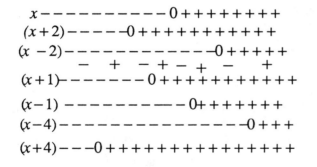

49. The radicand of f is the function $y = 16 - x^4$. The graph of this function is a reflection and vertical translation of the graph of $y = x^4$ (this is discussed in Example 2 on page 160); its graph is shown below left. The domain of H is $[-4, 0]$ or $[4, +\infty)$. The graph of H is sketched below right using the steps on page 209.

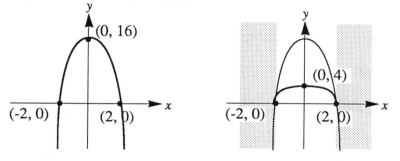

53. The radicand $x^2 - 1$ must be nonnegative and $x \neq 4$, so the domain of the function is $(-\infty, -1]$ or $[1, 4)$ or $(4, +\infty)$ The zeros of the numerator are ± 1, so the x-intercept s are $(-1, 0)$ and $(1, 0)$. The vertical asymptote of the graph is $x = 4$. As $x \to -\infty$, $y \to \dfrac{x}{|x|}$ or -1, and as $x \to +\infty$,

$y \to \dfrac{x}{|x|}$ or 1.

57. The graph of $y = x^{2/3}$ is sketched in Example 2 on page 204. The graph of $y = -\frac{1}{2}x$ is a line passing through the origin with slope $-\frac{1}{2}$. The graph of the function in question is constructed by adding y-coordinates of the graphs of $y = -\frac{1}{2}x$ and $y = x^{2/3}$. The y-intercept is determined by setting $x = 0$ and solving the equation.

$$y = (0)^{2/3} - \frac{1}{2}(0) = 0 \implies y\text{-intercept is } (0, 0)$$

To find x-intercepts, we set $y = 0$ and solve for x:

$$0 = x^{2/3} - \frac{1}{2}x \implies 0 = 2x^{2/3} - x$$
$$\implies 0 = x^{2/3}(2 - x^{1/3})$$
$$\implies x^{2/3} = 0 \text{ or } 2 - x^{1/3} = 0 \implies x = 0 \text{ or } x = 8$$
$$\implies x\text{-intercepts are } (0, 0) \text{ and } (8, 0).$$

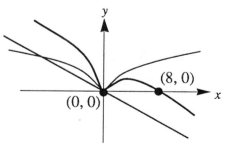

61. The divisor is $x - 4$ so we enter 4 in the upper left corner of the division tableau.

```
4|   2   -3    1    -4
           8   20    84
     2    5   21  |  80
```

$P(4) = 80$ $P(x) = (x - 4)(2x^2 + 5x + 21) + 80$

65. The divisor is $x - \frac{2}{3}$, so we enter $-\frac{2}{3}$ in the upper left corner of the division tableau.

$$
\begin{array}{r|rrrrr}
\frac{2}{3} & 12 & -2 & -7 & -1 & 2 \\
 & & 8 & 4 & -2 & -2 \\
\hline
 & 12 & 6 & -3 & -3 & \bigm| \ 0
\end{array}
$$

$P(\frac{2}{3}) = -3$ $\qquad\qquad P(x) = (x - \frac{2}{3})(12x^3 + 6x^2 - 3x - 3)$

69. The equation is equivalent to $2x^4 + 5x^3 - 15x^2 + 46x - 20 = 0$. If $\frac{p}{q}$ is a root of this equation, then q is a factor of 2 and p is a factor of -20. This implies that the potential roots are

$$\frac{p}{q} = \frac{1, 2, 4, 5, 10, 20}{1, 2} = \pm 1, \pm 2, \pm 4, \pm 5, \pm 10, \pm 20, \pm \frac{1}{2}, \pm \frac{5}{2}$$

We begin to check each of these potential zeros using synthetic division. We find that -5 is an actual zero.

$$
\begin{array}{r|rrrrr}
-5 & 2 & 5 & -15 & 46 & -20 \\
 & & -10 & 25 & -50 & 20 \\
\hline
 & 2 & -5 & 10 & -4 & \bigm| \ 0
\end{array}
$$

It follows that the equation can be factored as $(x + 5)(2x^3 - 5x^2 + 10x - 4) = 0$.

The rational zeros $\frac{p}{q}$ of the equation $2x^3 - 5x^2 + 10x - 4 = 0$ are such that q is a factor of 2 and p is a factor of -4. Potential rational roots are

$$\frac{p}{q} = \frac{1, 2, 4}{1, 2} = \pm 1, \pm 2, \pm 4, \pm \frac{1}{2}$$

We find that $\frac{1}{2}$ is an actual zero:

$$
\begin{array}{r|rrrr}
\frac{1}{2} & 2 & -5 & 10 & -4 \\
 & & 1 & -2 & 4 \\
\hline
 & 2 & -4 & 8 & \bigm| \ 0
\end{array}
$$

It follows that the equation can be factored as $(x + 5)(x - \frac{1}{2})(2x^2 - 4x + 8)$. The equation $2x^2 - 4x + 8 = 0$ has no real roots (the discriminant of this quadratic equation is negative). The real roots to the equation are -5 and $\frac{1}{2}$.

73. a) The *x*-intercepts of the function are found by setting *y* = 0 and solving for *x*:

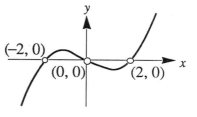

$$0 = x^3 - 4x \implies 0 = x(x-2)(x+2)$$

$$\implies x = 0 \text{ or } x - 2 = 0 \text{ or } x + 2 = 0$$

$$\implies x = 0 \text{ or } x = 2 \text{ or } x = -2$$

$$\implies x\text{-intercepts are } (-2, 0), (0, 0), \text{ and } (2, 0).$$

Using the techniques of Section 3.1, we sketch the graph.

b) The values of *x* that satisfy the inequality are those for which the graph of part a) is below the *x*-axis (see page 119). From its graph, we get the solution for the inequality: $(-\infty, -2)$ or $(0, 2)$.

77. a) The *x*-intercepts of the function are found by setting *y* = 0 and solving for *x*:

$$0 = x^4 - 20x^2 \implies 0 = x^2(x^2 - 20)$$

$$\implies x^2 = 0 \text{ or } x^2 - 20 = 0$$

$$x^2 = 0 \implies x = 0$$

$$x^2 - 20 = 0 \implies x^2 = 20 \text{ or } x = \pm\sqrt{20} = \pm 2\sqrt{5}$$

$$\implies x\text{-intercepts are } (-2\sqrt{5}, 0), (0, 0), (2\sqrt{5}, 0).$$

Using the techniques of Section 3.1, we sketch the graph.

b) The values of *x* that satisfy the inequality are those for which the graph of part a) is above or on the *x*-axis (see page 119). From its graph, we get the solution for the inequality: $(-\infty, -2\sqrt{5}]$ or 0 or $[2\sqrt{5}, +\infty)$.

81. a) The equation can be rewritten as $y = \dfrac{x-2}{x(x-5)}$

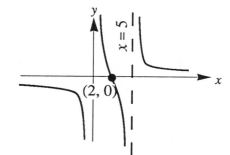

The *x*-intercepts of the function are found by setting *y* = 0 and solving for *x*:

$$y = \frac{0-2}{0(0-5)} = \frac{-2}{0} \implies \text{no } y\text{-intercept}$$

The only zero of the numerator is 2, so the *x*-intercept is (2, 0). The horizontal asymptote is *y* = 0 and the vertical asymptotes are *x* = 0 and *x* = 5.

Using the techniques of Section 3.4, we sketch the graph.

b) The values of *x* that satisfy the inequality are those for which the graph of part a) is above or on the *x*-axis (see page 119). From its graph, we get the solution for the inequality: $(0, 2]$ or $(5, +\infty)$.

85. First, consider $y = \dfrac{x}{x-5}$; its graph has horizontal

asymptote $y = 1$, vertical asymptote $x = 5$, and intercept $(0, 0)$. Recall that $|x| = -x$ for $x < 0$ and $|x| = x$ for $x > 0$. The graph of the equation in a) then is

$y = -\dfrac{x}{x-5}$ for $x < 0$ and $y = \dfrac{x}{x-5}$ for $x \geq 0$. This

means that for $x < 0$ the graph of $y = \dfrac{|x|}{x-5}$ is the

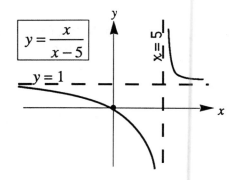

reflection though the x-axis of $y = \dfrac{x}{x-5}$ for $x < 0$ and is the same as the graph of $y = \dfrac{x}{x-5}$

for $x > 0$. The graph of b) follows from the discussion of pages 130-31.

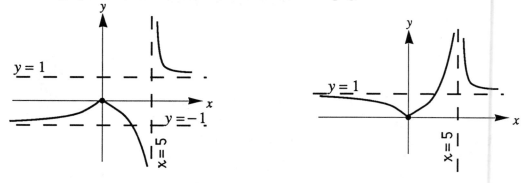

89. Because one of the zeros of this real function is i, the conjugate root theorem indicates that the conjugate of i, namely $-i$, is also a zero. The linear factored form of the function is P

$$P(x) = a(x+i)(x-i)(x-2)(x+2) \text{ or}$$
$$P(x) = a\,[(x+i)(x-i)][(x-2)(x+2)] = a\,[x^2+1][x^2-4] = a\,[x^4-3x^2-4]$$

It remains to find the value of a. We evaluate $P(\sqrt{5})$ in terms of a.

$$P(\sqrt{5}) = a\,[(\sqrt{5})^4-3(\sqrt{5})^2-4] = a\,[25-15-4] = 6a$$

Since $P(\sqrt{5}) = 6a = 12$, it follows that $a = 2$. The function is $P(x) = 2[x^4-3x^2-4]$, or $P(x) = 2x^4-6x^2-8$.

Chapter 4

Section 4.1

1. Keystrokes: 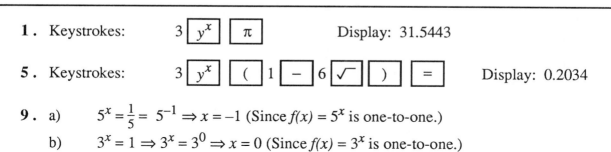 3 y^x π Display: 31.5443

5. Keystrokes: 3 y^x (1 − 6 $\sqrt{\ }$) =   Display: 0.2034

9. a) $5^x = \dfrac{1}{5} = 5^{-1} \Rightarrow x = -1$ (Since $f(x) = 5^x$ is one-to-one.)

     b) $3^x = 1 \Rightarrow 3^x = 3^0 \Rightarrow x = 0$ (Since $f(x) = 3^x$ is one-to-one.)

13. $2^x(4x + 1) = 0 \Rightarrow 2^x = 0$ or $4x + 1 = 0$. Since $2^x > 0$ for all real values of x, the only

     solution is $x = -\dfrac{1}{4}$.

17. a) Refer to page 220 of the text for the general shape of the graph of $f(x) = b^x$. A few coordinates on the graph are $(-1, \frac{1}{4})$, $(\frac{1}{2}, 2)$, and $(1, 4)$.

b) Refer to the graph of $f(x) = 4^x$ in part a). Recall from Section 2.5 that to graph $y = -f(-x) = -4^{-x}$, reflect $y = f(x)$ through the x-axis and the y-axis.

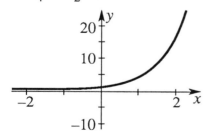

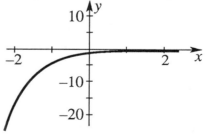

21. a) Refer to page 220 of the text for the general shape of the graph of $f(x) = b^x$. A few coordinates on the graph are $(-1, \frac{1}{2})$, $(1, 2)$, and $(2, 4)$.

b) Refer to the graph of $f(x) = 2^x$ in part a). Recall from Section 2.3 that to graph $y - 3 = f(x)$, shift the graph of $y = f(x)$ three units vertically. Ordered pairs corresponding to the three given in part a) are $(-1, \frac{7}{2})$, $(1, 5)$, and $(2, 7)$. The line $y = 3$ is a horizontal asymptote.

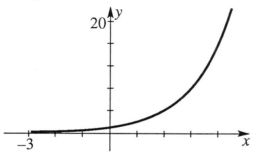

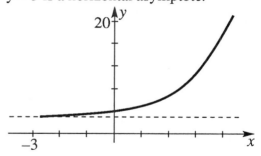

25. a) Refer to page 220 of the text for the general shape of the graph of $f(x) = b^x$, $0 < b < 1$. A few coordinates on the graph are $(-1, 4)$, $(\frac{1}{2}, \frac{1}{2})$, and $(1, \frac{1}{4})$.

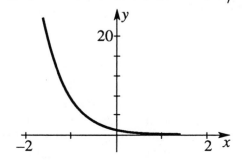

b) Refer to the graph of $f(x) = (\frac{1}{4})^x$ in part a). Recall from Section 2.3 that to graph $y - 1 = f(x - 3)$, shift $y = f(x)$ three units right and one unit up. Ordered pairs corresponding to the three given in part a) are $(2, 5)$, $(\frac{7}{2}, \frac{3}{2})$, and $(4, \frac{5}{4})$. The line $y = 1$ is a horizontal asymptote.

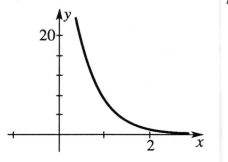

29. Answers vary. One possibility is $f(x) = (\frac{1}{2})^x$ and $g(x) = |x|$. Then $f[g(x)] = f(|x|) = (\frac{1}{2})^{|x|}$.

33. a) Refer to page 220 of the text for the general shape of the graph of $f(x) = 7^x$. Recall from Section 2.5 that to graph $y = f(-x) = 7^{-x}$, reflect the graph of $y = f(x)$ through the y-axis. A few points on the graph are $(-1, 7)$, $(0, 1)$, and $(1, \frac{1}{7})$.

b) Refer to the graph of $f(x) = 7^{-x}$ in part a). Since $7^{2-x} = 7^{-(x-2)}$, shift $y = 7^{-x}$ two units to the right. Ordered pairs corresponding to the three given in part a) are $(1, 7)$, $(2, 1)$, and $(3, \frac{1}{7})$.

37. a) Refer to Example 9 (page 222) of the text. A few points on the graph are $(-1, \frac{1}{2})$, $(1, \frac{1}{2})$, $(0, 1)$, $(-\sqrt{2}, \frac{1}{4})$, and $(\sqrt{2}, \frac{1}{4})$.

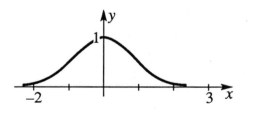

b) Refer to the graph in part a). If $f(x) = 2^{-x^2}$, then $f(x - 2) = 2^{-(x-2)^2}$. Thus we shift the graph in a) two units to the right. Points corresponding to the ordered pairs given in part a) are $(1, \frac{1}{2})$, $(3, \frac{1}{2})$, $(2, 1)$, $(-\sqrt{2} + 2, \frac{1}{4})$, and $(\sqrt{2} + 2, \frac{1}{4})$, respectively.

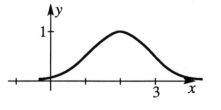

41. a) Refer to Example 2 (page 219) of the text for the general shape of the graph of $f(x) = 3^x$. Recall from Section 2.5 that to graph $y = f(|x|) = 3^{|x|}$, we keep all the points of $y = 3^x$ to the right of the y-axis. Complete the graph by noting that $y = 3^{|x|}$ is symmetric with respect to the y-axis.

b) Refer to Example 3 (page 219) of the text for the general shape of the graph of $f(x) = (\frac{1}{3})^x$. Recall from Section 2.5 that to graph $y = f(|x|) = (\frac{1}{3})^{|x|}$, we keep all the points of $y = (\frac{1}{3})^x$ to the right of the y-axis. Complete the graph by noting that $y = (\frac{1}{3})^{|x|}$ is symmetric with respect to the y-axis.

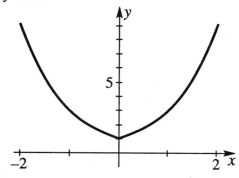

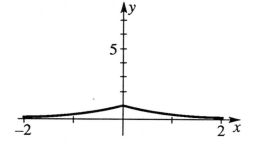

45. Refer to the formula on page 223: $A = P(1 + \frac{r}{n})^{nt} \Rightarrow A = 7000(1 + \frac{0.06}{2})^{2t}$.

After 3 years the amount will be $A = 7000(1 + \frac{0.06}{2})^{2(3)} = \$8358.37$.

After 4.5 years the amount will be $A = 7000(1 + \frac{0.06}{2})^{2(4.5)} = \$9133.41$.

49. If P dollars is invested at 7% compounded weekly, the accumulated amount after a year will be $A = P(1 + \frac{0.07}{52})^{52} = 1.07246P$ dollars. At 7.5% compounded annually, the accumulated amount will be $A = P(1 + \frac{0.075}{1})^1 = 1.075P$ dollars, so that is the better investment.

53. After $\frac{1}{2}$ an hour, there will be 200 animals; after 1 hour there will be 400 animals, and so on. The following table shows the progression:

Time	$\frac{1}{2}$	1	$\frac{3}{2}$	2	$\frac{5}{2}$	3	$\frac{7}{2}$	4
Number	200	400	800	1600	3200	6400	12800	25600

So after 3 hours there will be 6400 animals, and after 4 hours there will be 25600.

57. a) Substitute $Q_0 = 40$ and $t = 3$ into the given formula: $Q(3) = 40(0.9)^3 = 29$ (to the nearest mg).

b) Substituting various values of t into the given formula gives $Q(4) = 40(0.9)^4 = 26$, $Q(5) = 40(0.9)^5 = 24$, $Q(6) = 40(0.9)^6 = 21$, $Q(7) = 40(0.9)^7 = 19$, so it takes 7 hours.

61. a) False, because $f(x) = x^2$ is not one-to-one; for example $5^2 = (-5)^2$, but $5 \neq -5$.
b) True, because $g(x) = 2^x$ is one-to-one.

65. The sketch is shown on the right. The
graph of $f(x) = \frac{1}{2}x^2$ is below the graph of

$g(x) = (\frac{1}{2})^{x^2}$ for all values of x between -1
and 1. Therefore the solution to the
inequality

$\frac{1}{2}x^2 \leq (\frac{1}{2})^{x^2}$ is $-1 \leq x \leq 1$.

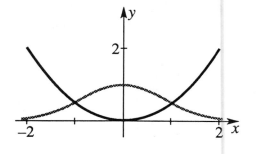

Section 4.2

1. Keystrokes: $1\ \boxed{+}\ .05\ \boxed{=}\ \boxed{y^x}\ .05\ \boxed{^1\!/_x}\ \boxed{=}$ Display: 2.6533

5. Keystrokes: $1.55\ \boxed{e^x}$ Display: 4.7115

9. Keystrokes: $2\ \boxed{+}\ 5\ \boxed{\sqrt{\ }}\ \boxed{=}\ \boxed{e^x}$ Display: 69.1355

13. Refer to page 229 (Figure 9) for the sketch
of $y = e^x = f(x)$. Recall from Section 2.3
that to graph
$y - 2 = f(x - 1)$, we shift the graph of
$y = f(x)$ one unit right and two units up.
Since $y = 0$ is a horizontal asymptote for
$y = e^x$, the line $y = 2$ is a horizontal
asymptote for $y - 2 = e^{x-1}$.

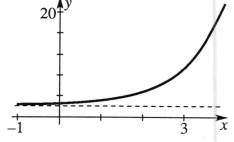

17. Answers vary. One possibility is $f(x) = e^x$ and $g(x) = 3 - x$. Then
$f[g(x)] = f(3 - x) = e^{3-x}$.

21. Answers vary. One possibility is $f(x) = e^x$ and $g(x) = \frac{x-5}{4}$. Then

$f[g(x)] = f(\frac{x-5}{4}) = e^{(x-5)/4}$.

25. Use the fact that as $x \to 0$, $(1 + x)^{1/x} \to e$: $\quad (1 + x)^{3/x} = ((1 + x)^{1/x})^3 \to (e)^3$ as $x \to 0$.

29. We will use the fact that as $a \to 0$, $(1 + a)^{1/a} \to e$. Let $a = x - 1$, which also means $x = 1 + a$. Also, $x \to 1$ if and only if $a \to 0$. Now, $x^{1/(x-1)} = (1 + a)^{1/a} \to e$ as $a \to 0$.

33. Refer to Example 3 (page 229) of the text for the general shape of the graph of $f(x) = e^x$. Recall from Section 2.5 that to graph $y = f(|x|) = e^{|x|}$, we keep all the points of $y = e^x$ to the right of the y-axis. Complete the graph by noting that $y = e^{|x|}$ is symmetric with respect to the y-axis.

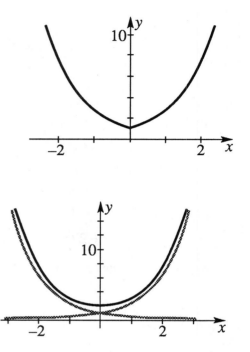

37. First, we graph $f(x) = e^x$, and $g(x) = e^{-x}$ (see figure at right). Then we use the technique of adding y-coordinates discussed in Section 2.6 . The graph of $y = e^x + e^{-x}$ is symmetric with respect to the y-axis, since for $h(x) = e^x + e^{-x}$,

$$h(-x) = e^{-x} + e^x = h(x).$$

41. If P dollars is invested for a year at 8% compounded quarterly, the accumulated amount will be $A = P(1 + \frac{0.08}{4})^4 = 1.0824P$ dollars. At 7.75% compounded continuously, the accumulated amount for one year will be $A = Pe^{0.0775(1)} = 1.0805P$ dollars, so the better investment is 8% compounded quarterly.

45. a) See Example 9 on page 234. Starting with the population growth model $P(t) = P_0 e^{kt}$, substitute $P_0 = 41666000$: $P(t) = 41666000e^{kt}$. The year 1940 corresponds to $t = 0$, and we are given that in 1960 ($t = 20$), the population was 54973000:

$$54973000 = P(20) = 41666000e^{k(20)} \Rightarrow \frac{54973000}{41666000} = e^{20k},$$

Raising both sides of the last equation to the $\frac{1}{20}$ power (and reducing the fraction on the left side) gives $(\frac{54973}{41666})^{1/20} = e^k$. Therefore the growth model is

$$P(t) = 41666000[e^k]^t = 41666000[(\frac{54973}{41666})^{1/20}]^{\,t} .$$

The year 2000 corresponds to $t = 60 \Rightarrow P(60) = 41666000[\ (\frac{54973}{41666})^{1/20}]^{60}$

$$= 41666000(\frac{54973}{41666})^3 \approx 95,694,000.$$

b) The year 1980 corresponds to $t = 40 \Rightarrow P(40) = 41666000[\ (\frac{54973}{41666})^{1/20}]^{40}$

$$= 41666000(\frac{54973}{41666})^2 \approx 72,530,000.$$

Compared to the actual population of 75,370,000 the error is

$$\frac{75370000 - 72530000}{75370000} \approx 0.038 \text{ or less than } 4\% \ .$$

49. a) See Example 10 on page 236. Starting with the formula for Newton's law of cooling $T(t) = A + (T_0 - A) \ e^{kt}$, substitute $T_0 = 350°$ and $A = 75°$:

$$T(t) = 75 + (350 - 75) \ e^{kt} \Rightarrow T(t) = 75 + 275 \ e^{kt}.$$

Twenty minutes later the cake is $210° \Rightarrow 210 = T(20) = 75 + 275 \ e^{k(20)}$. Solving for e^k,

$210 - 75 = 275e^{k(20)} \Rightarrow \frac{135}{275} = e^{k(20)} \Rightarrow (\frac{27}{55})^{1/20} = e^k$. Therefore the cooling function is

$$T(t) = 75 + 275[(\frac{27}{55})^{1/20}]^t \ , \text{ or } T(t) = 75 + 275(\frac{27}{55})^{t/20} \ .$$

b) Thirty minutes after removal, the temperature will be $T(30) = 75 + 275[(\frac{27}{55})^{1/20}]^{30}$

$$= 75 + 275(\frac{27}{55})^{3/2} \approx 170°.$$

53. Since the employee learned 8 numbers in 2 days, we have $8 = N(2) = 30 - 30e^{k(2)}$. Solving for e^k, we get $8 - 30 = -30e^{k(2)} \Rightarrow \frac{22}{30} = e^{k(2)} \Rightarrow (\frac{11}{15})^{1/2} = e^k$. Therefore

$$N(t) = 30 - 30[e^k]^t = 30 - 30[\ (\frac{11}{15})^{1/2}]^t \ , \text{ or } N(t) = 30 - 30(\frac{11}{15})^{t/2} \ .$$

After 5 days she will have learned

$$N(5) = 30 - 30[\ (\frac{11}{15})^{1/2}]^5 \approx 16 \text{ numbers.}$$

57. $\dfrac{e^x(2e^{-2x} + 1) - (e^{2x} + 1)e^{-x}}{e^{4x}} = \dfrac{2e^{-x} + e^x - e^x - e^{-x}}{e^{4x}} = \dfrac{e^{-x}}{e^{4x}} = e^{-x-4x} = e^{-5x} \text{ or } \dfrac{1}{e^{5x}}$

Section 4.3

1. $5^3 = 125 \Leftrightarrow \log_5 125 = 3$

5. $4^{-2} = 0.0625 \Leftrightarrow \log_4 0.0625 = -2$

9. $2.7^k = 5 \Leftrightarrow \log_{2.7} 5 = k$

13. $\log_{10}1000 = 3 \Leftrightarrow 10^3 = 1000$

17. $\log_{2.7}2 = 3t^2 \Leftrightarrow 2.7^{3t^2} = 2$

21. Since $10^3 < 3250 < 10^4$, $3 < \log_{10}3250 < 4$

25. See the box entitled Logarithmic Function on page 245 in the text. The inverse function of $f(x) = 10^x$ is $f^{-1}(x) = \log_{10}x$.

29. Following the steps outlined on page 148, we obtain $f(y) = x \Rightarrow \log_3(y + 5) = x$ $\Rightarrow y + 5 = 3^x \Rightarrow y = 3^x - 5$. The inverse of f is $f^{-1}(x) = 3^x - 5$.

33. The domain is all real values of x such that $x^2 > 0 \Leftrightarrow x \neq 0$.

37. Refer to page 248 in the text (Figure 18) for the sketch of $y = \ln x = f(x)$. A few points are (approx.) (1, 0), (2.72, 1), and (4, 1.39). Recall from Section 2.5 that the graph of $y = f(2x) = \ln 2x$ is a compression of the graph of $y = f(x)$ towards the y-axis. A few points corresponding to those mentioned above are $(\frac{1}{2}, 0)$, (1.36, 1), and (2, 1.39), respectively.

41. Refer to page 246 in the text for the general graph of $y = \log_b x$. A few points on the graph of $\log_4 x = f(x)$ are (4, 1), $(2, \frac{1}{2})$, and $(8, \frac{3}{2})$. Recall from Section 2.5 that the graph of $y = f(-x) = \log_4 (-x)$ is a reflection of the graph of $y = f(x)$ across the y-axis. This means that the points corresponding to those mentioned above are $(-4, 1)$, $(-2, \frac{1}{2})$, and $(-8, \frac{3}{2})$, respectively.

45. Using a property of logarithms, $\ln x^5 = 5\ln x$. Refer to page 248 in the text (Figure 18) for the sketch of $y = \ln x = f(x)$. A few points are (approx.) (1, 0), (2.7, 1), and (4, 1.4). Recall from Section 2.5 that the graph of $y = 5f(x) = 5 \ln x$ is an expansion of the graph of $y = f(x)$ away from the x-axis by a factor of 5. A few points corresponding to those mentioned above are (1, 0), (2.7, 5), and (4, 7), respectively.

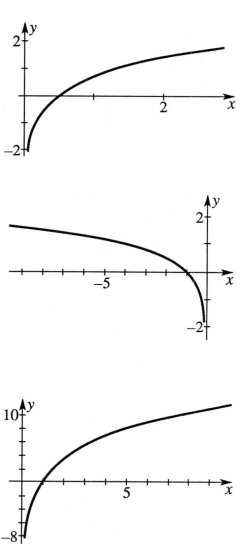

49. Using a property of logarithms,

$\ln \sqrt[4]{x} = \ln x^{1/4} = \frac{1}{4}\ln x$. Refer to page 248 in the text (Figure 18) for the sketch of $y = \ln x = f(x)$. A few points are (approx.) $(1, 0)$, $(2.7, 1)$, and $(4, 1.4)$. Recall from Section 2.5 that the graph of $y = \frac{1}{4}f(x)$

$= \frac{1}{4}\ln x$ is a compression of the graph of

$y = f(x)$ toward from the x-axis. A few points corresponding to those mentioned above are $(1, 0)$, $(2.7, 0.25)$, and $(4, 0.35)$, respectively.

53. Using a property of logarithms, $\log_8 x^4 = 4 \log_8 x$, for $x > 0$. First sketch $y = \log_8 x$ $= f(x)$. (See page 246 in the text for a general graph.) A few points on the graph are $(1, 0)$, $(2, \frac{1}{3})$, and $(8, 1)$. The graph of

$y = 4f(x) = 4 \log_8 x$ is an expansion of the graph of $y = f(x)$ away from the x-axis by a factor of 4. So points

corresponding to those mentioned above are $(1, 0)$, $(2, \frac{4}{3})$, and $(8, 4)$. Following the hint, the graph of $y = \log_8 x^4$ is symmetric with respect to the y- axis.

57. Refer to the properties of logarithms (in the text) on page 242. Applying the first property,

$$\log_b \frac{a^2}{x} + \log_b 5a = \log_b [\frac{a^2}{x}(5a)] = \log_b \frac{5a^3}{x}.$$

61. Refer to the properties of logarithms (in the text) on page 242.

$5\log_b (2x) - \log_b x - \log_b (4x^2)$

$= \log_b (2x)^5 - \log_b x - \log_b (4x^2)$ Applying Property 3

$= \log_b \frac{(2x)^5}{x} - \log_b (4x^2) = \log_b (32x^4) - \log_b (4x^2)$ Applying Property 2

$= \log_b \frac{32x^4}{4x^2}$ Applying Property 2

$= \log_b (8x^2)$

65. Refer to the properties of logarithms (in the text) on page 242.

$\log_b \frac{x^2}{y} = \log_b x^2 - \log_b y$ Applying Property 2

$= 2\log_b x - \log_b y$ Applying Property 3

69. Refer to the properties of logarithms (in the text) on page 242.

$$\log_b \frac{5x^7 \sqrt[3]{y}}{\sqrt{z}} = \log_b [(5x^7)(\sqrt[3]{y})] - \log_b \sqrt{z} \qquad \text{Applying Property 2}$$

$$= \log_b 5 + \log_b x^7 + \log_b \sqrt[3]{y} - \log_b \sqrt{z} \qquad \text{Applying Property 1}$$
$$= \log_b 5 + \log_b x^7 + \log_b y^{1/3} - \log_b z^{1/2}$$
$$= \log_b 5 + 7\log_b x + \frac{1}{3}\log_b y - \frac{1}{2}\log_b z \qquad \text{Applying Property 3}$$

73. $\log_{10}\left(\frac{2}{3}\right) = \log_{10} 2 - \log_{10} 3 \approx 0.301 - 0.477 = -0.176$

77. $\log_{10} 30 = \log_{10}[(3)(10)] = \log_{10} 3 + \log_{10} 10 = \log_{10} 3 + 1 \approx 0.477 + 1 = 1.477$

81. Refer to Example 14 (page 248) in the text
for the general graph of $y = \ln |x| = f(x)$.
Recall from Section 2.3 that to graph
$y = f(x+1) = \ln |x + 1|$, we translate the
graph of $y = f(x)$ one unit to the left. There
is a vertical asymptote at $x = -1$.

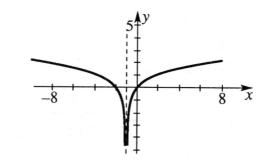

85. We are given that $N(r) = 1300$ and we need to determine r :
$$1300 = -5000 \ln r \implies \frac{1300}{-5000} = \ln r \implies r = e^{-13/50} \approx 0.77 \text{ or } 77\%$$

89. Let $r = \log_b P$; in exponential form this is equivalent to $P = b^r$. Consider $\log_b P^t$:
$$\log_b P^t = \log_b (b^r)^t = \log_b b^{rt} = tr = t\log_b P$$

93. $3^{\log_9 x} = (9^{1/2})^{\log_9 x} = 9^{(1/2)\log_9 x} = 9^{\log_9 \sqrt{x}} = \sqrt{x}$

Section 4.4

1. Keystrokes: $\qquad$ 43.15 $\boxed{\log}$ $\qquad\qquad\qquad\qquad$ Display: 1.6350

5. Using properties of logarithms, $\log(2.34 \times 10^{45}) = \log 2.34 + \log 10^{45} = \log 2.34 + 45$.
$\qquad$ Keystrokes: $\qquad$ 45 $\boxed{+}$ 2.34 $\boxed{\log}$ $\boxed{=}$ $\qquad\qquad$ Display: 45.3692

9. Using properties of logarithms, $\log(3.2 \times 10^{455}) = \log 3.2 + \log 10^{455} = \log 3.2 + 455$.
$\qquad$ Keystrokes: $\qquad$ 455 $\boxed{+}$ 3.2 $\boxed{\log}$ $\boxed{=}$ $\qquad\qquad$ Display: 455.5052

13. Using a property of logarithms we have

$$\ln 6 + \ln 2x = \ln 12x \Rightarrow \ln 12x = 3 \Rightarrow 12x = e^3 \Rightarrow x = \frac{e^3}{12}$$

Keystrokes: 3 $\boxed{e^x}$ $\boxed{\div}$ 12 $\boxed{=}$ Display: 1.6738

17. Using a property of logarithms we have

$$3\log^2 x - \log x^2 = 3\log^2 x - 2\log x \Rightarrow 3\log^2 x - 2\log x = 1.$$ This equation is quadratic

in $\log x \Rightarrow 3\log^2 x - 2\log x - 1 = 0 \Rightarrow (3\log x + 1)(\log x - 1) = 0 \Rightarrow \log x = 1$ or $-\frac{1}{3}$.

Therefore $x = 10$ or $10^{-1/3}$.

Keystrokes: 3 $\boxed{1/x}$ $\boxed{+/-}$ $\boxed{10^x}$ Display: 0.4642

21. $e^x = 19 \Rightarrow x = \ln 19 \approx 2.9444$

25. The logarithmic form of $2.5^{3t} = 24$ is $3t = \log_{2.5} 24 \Rightarrow t = \frac{1}{3}\log_{2.5} 24$. Applying the

change of base formula, we have $t = \frac{1}{3}\log_{2.5} 24 = \frac{1}{3}\frac{\log 24}{\log 2.5}$ or $\frac{\log 24}{3\log 2.5}$.

Keystrokes: 24 $\boxed{\log}$ $\boxed{\div}$ $\boxed{(}$ 3 $\boxed{\times}$ 2.5 $\boxed{\log}$ $\boxed{=}$ $\boxed{)}$ $\boxed{=}$

Display: 1.1561

29. $6^x = 8(5^x) \Rightarrow \frac{6^x}{5^x} = 8 \Rightarrow (\frac{6}{5})^x = 8.$ Taking the common logarithm of both sides and using a

property of logarithms, we have $\log(\frac{6}{5})^x = \log 8 \Rightarrow x \log(\frac{6}{5}) = \log 8 \Rightarrow x = \frac{\log 8}{\log(6/5)}$.

Keystrokes: 8 $\boxed{\log}$ $\boxed{\div}$ $\boxed{(}$ 6 $\boxed{\div}$ 5 $\boxed{=}$ $\boxed{\log}$ $\boxed{)}$ $\boxed{=}$

Display: 11.4054

33. If we multiply both sides by $2(7^x)$ we will obtain an equation that is quadratic in 7^x:

$$7^{2x} + 1 = 4(7^x) \Rightarrow (7^x)^2 - 4(7^x) + 1 = 0.$$ Using the quadratic formula, we have

$$7^x = \frac{4 \pm \sqrt{16 - 4}}{2} = 2 \pm \sqrt{3}.$$ Taking the common logarithm of both sides, we get

$\log 7^x = \log(2 \pm \sqrt{3}) \Rightarrow x \log 7 = \log(2 \pm \sqrt{3})$

$\Rightarrow x = \dfrac{\log(2 + \sqrt{3})}{\log 7}$ or $x = \dfrac{\log(2 - \sqrt{3})}{\log 7}$.

Keystrokes: 2 $\boxed{-}$ 3 $\boxed{\sqrt{}}$ $\boxed{=}$ $\boxed{\log}$ $\boxed{\div}$ 7 $\boxed{=}$ Display: -0.6768

                   2 $\boxed{+}$ 3 $\boxed{\sqrt{}}$ $\boxed{=}$ $\boxed{\log}$ $\boxed{\div}$ 7 $\boxed{=}$ Display: 0.6768

37. The equation is quadratic in 2^x :

$2^{2x} - 5(2^x) = 6 \Rightarrow (2^x)^2 - 5(2^x) - 6 = 0 \Rightarrow (2^x - 6)(2^x + 1) = 0 \Rightarrow 2^x = 6$ or $2^x = -1$.

This last equation has no real solutions since $2^x > 0$ for all real x. The equation $2^x = 6$ does have a solution, $x = \log_2 6$. Using the change of base formula, this is

$x = \log_2 6 = \dfrac{\log 6}{\log 2}$

Keystrokes: 6 [log] [÷] 2 [log] [=]   Display: 2.5850

41. The equation is equivalent to $4\log_3 x - 2\log_6 x = 1$. Using the change of base formula,

$4\dfrac{\log x}{\log 3} - 2\dfrac{\log x}{\log 6} = 1 \Rightarrow \log x \left(\dfrac{4}{\log 3} - \dfrac{2}{\log 6}\right) = 1$. Combining the fractions in the parentheses and simplifying with the properties of logarithms gives

$\log x\left(\dfrac{\log 6^4 - \log 3^2}{\log 3 \log 6}\right) = 1 \Rightarrow \log x\left(\dfrac{\log 144}{\log 3 \log 6}\right) = 1$

$\Rightarrow \log x = \dfrac{\log 3 \log 6}{\log 144} \Rightarrow x = 10^{(\log 3 \log 6)/\log 144}$

Keystrokes: 3 [log] [X] 6 [log] [=] [÷] 144 [log] [=] [10^x]

Display: 1.4860

45. Refer to Example 9 on page 256. We have $10 = 25\left(\frac{1}{2}\right)^{t/11} \Rightarrow 0.4 = \left(\frac{1}{2}\right)^{t/11}$. Taking the common logarithm of both sides gives $\log 0.4 = \dfrac{t}{11} \log\dfrac{1}{2}$

$\Rightarrow t = \dfrac{11\log(0.4)}{\log(1/2)} \approx 14.5$ years.

Keystrokes: 11 [X] 0.4 [log] [=] [÷] 2 [$1/x$] [log] [=]

49. Refer to Example 11 on page 257. We have $M = \log\dfrac{1500A_0}{A_0} = \log 1500 \approx 3.2$.

53. a) We have $5.8 = -\log[H^+] \Rightarrow [H^+] = 10^{-5.8} \approx 0.00000158$.

b) We have $pH = -\log(4.3 \times 10^{-4}) \Rightarrow pH = -\log 4.3 - \log 10^{-4} = -\log 4.3 - (-4) \approx 3.4$

c) Since $[H^+] = 10^{-pH} = \dfrac{1}{10^{pH}}$, increasing the pH will decrease $[H^+]$. In other words, a higher pH means a lower hydrogen ion concentration.

57. We are given $120 = 10\log\left(\dfrac{I}{I_0}\right) \Rightarrow 12 = \log\left(\dfrac{I}{I_0}\right) \Rightarrow \dfrac{I}{I_0} = 10^{12} \Rightarrow I = I_0 10^{12}$.

Therefore I must be 10^{12} times louder than I_0.

61. Substituting the given values into the equation, we get $1.1 = \frac{6}{4}(1 - e^{-4t/0.02})$

$\frac{(1.1)(4)}{6} = (1 - e^{-4t/0.02}) \Rightarrow \frac{2.2}{3} = (1 - e^{-200t}) \Rightarrow e^{-200t} = 1 - \frac{2.2}{3}$

$\Rightarrow -200t = \ln(1 - \frac{2.2}{3}) \Rightarrow t = \frac{\ln(4/15)}{-200} \approx 0.0066$ seconds.

Keystrokes: 4 $\boxed{\div}$ 15 $\boxed{=}$ $\boxed{\ln}$ $\boxed{\div}$ 200 $\boxed{+/-}$ $\boxed{=}$

65. The initial value of M is $M_0 = 160$. Substituting this into the given model gives $M = 160 \cdot 10^{-kt}$. We are given that $M = 110$ when $t = 6$. So $110 = 160(10)^{-6k}$ $\Rightarrow 10^{-k} = (\frac{11}{16})^{1/6}$. Therefore, $M = 160 \cdot 10^{-kt} \Rightarrow M = 160(\frac{11}{16})^{t/6}$. The question calls for t when $M = 0.4$, so we will substitute these values and solve for t.

$0.4 = 160(\frac{11}{16})^{t/6} \Rightarrow \frac{0.4}{160} = (\frac{11}{16})^{t/6} \Rightarrow (\frac{1}{400})^6 = (\frac{11}{16})^t$. Taking the log of both sides and applying a property of logarithms, we get $6\log(\frac{1}{400}) = t\log(\frac{11}{16})$

$\Rightarrow t = \frac{6\log(1/400)}{\log(11/16)} \approx 96$ days.

Keystrokes: 6 $\boxed{\times}$ 400 $\boxed{1/x}$ $\boxed{\log}$ $\boxed{\div}$ $\boxed{(}$ 11 $\boxed{\div}$ 16 $\boxed{)}$ $\boxed{\log}$ $\boxed{=}$

Miscellaneous Exercises for Chapter 4

1. Keystrokes: 5 $\boxed{y^x}$ $\boxed{(}$ 3 $\boxed{\sqrt{}}$ $\boxed{\div}$ 2 $\boxed{)}$ $\boxed{=}$ Display: 4.0302

5. Keystrokes: 87 $\boxed{\sqrt{}}$ $\boxed{\log}$ Display: 0.9698

9. First graph $y = f(x) = 2^x$. (Refer to page 220 of the text for the general shape of the graph of $f(x) = b^x$.) To graph $y = 2^{x+2} = f(x + 2)$, shift the graph of $y = f(x)$ two units to the left.

13. First notice that the function $f(x) = 2^{-x^2}$ is even, since $f(-x) = f(x)$. Thus, we can plot points for $x \geq 0$, and then complete the graph by noting that an even function is symmetric with respect to the y-axis. A few points are $(0, 1)$, $(1, \frac{1}{2})$, and $(2, \frac{1}{16})$. See page 222 Example 9.

17. First graph $y = f(x) = e^x$. (Refer to Figure 9 on page 229 of the text.) To graph $y = -e^{x-3}$ $= -f(x-3)$, reflect the graph of $y = f(x)$ through the x-axis, then shift the graph three units to the right.

21. First graph $y = \log_4 x = f(x)$. (See solution to Problem 41 in Section 4.3 of this manual.) The graph of $y = f(x + 5)$ $= \log_4 (x + 5)$ is a translation of the graph of $y = f(x)$ five units to the left. There is a vertical asymptote at $x = -5$.

25. Using a property of logarithms, we have $\ln x^2 = 2\ln x$ for $x > 0$, so we first graph $y = \ln x$ and then expand the graph away from the x-axis by a factor of 2. We complete the graph by noting that the graph is symmetric with respect to the y-axis, since for $f(x) = \ln x^2$, $f(-x) = f(x)$.

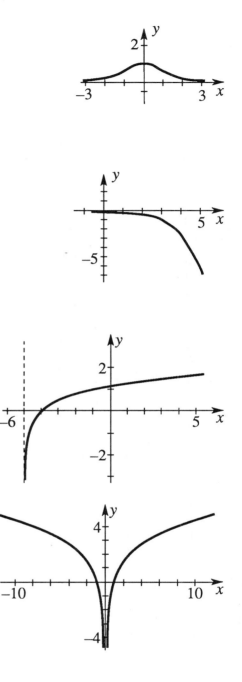

29. Since $\ln e^x = x$, we simply graph $y = x$.

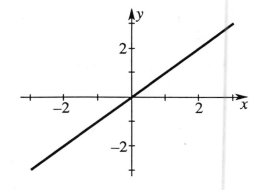

33. Answers vary. One possibility is to let $f(x) = \sqrt[4]{x}$ and $g(x) = \log x$. Then

$$f[g(x)] = f(\log x) = \sqrt[4]{\log x} \ .$$

37. We will use the fact that as $x \to \infty$, $(1 + \frac{1}{k})^k \to e$. This suggests letting $\frac{3}{x} = \frac{1}{k}$, which is

equivalent to $x = 3k$. Thus $(1 + \frac{3}{x})^x = (1 + \frac{1}{k})^{3k} = [(1 + \frac{1}{k})^k]^3 \to e^3$, as $x \to \infty$.

41. Refer to the properties of logarithms (in the text) on page 242. Applying Property 1,

$$\log_b (2t) + \log_b (t^2) = \log_b 2t^3.$$

45. Refer to the properties of logarithms (in the text) on page 242.

$$\ln \frac{3z^4 \sqrt{x}}{\sqrt{y^3}} = \ln [(3z^4)(\sqrt{x})] - \ln \sqrt{y^3} \qquad \text{Applying Property 2}$$

$$= \ln 3z^4 + \ln \sqrt{x} - \ln \sqrt{y^3} \qquad \text{Applying Property 1}$$
$$= \ln 3 + \ln z^4 + \ln x^{1/2} - \ln y^{3/2} \qquad \text{Applying Property 1}$$
$$= \ln 3 + 4 \ln z + \frac{1}{2} \ln x - \frac{3}{2} \ln y \qquad \text{Applying Property 3}$$

49. $\log_5 (x + 7) + \log_5 (13 - 2x) = 2 \Rightarrow \log_5 (x + 7)(13 - 2x) = 2 \Rightarrow (x + 7)(13 - 2x) = 5^2$

$\Rightarrow -2x^2 - x + 91 = 25 \Rightarrow -2x^2 - x + 66 = 0 \Rightarrow (x + 6)(-2x + 11) = 0 \Rightarrow x = -6$ or $\frac{11}{2}$.

53. $\ln |x - 5| = 2 \Rightarrow |x - 5| = e^2 \Rightarrow x - 5 = e^2$ or $x - 5 = -e^2 \Rightarrow x = 5 + e^2$ or $x = 5 - e^2$.
Approximations are $5 + e^2 \approx 12.3891$, $5 - e^2 \approx -2.3891$.
Keystrokes: $5 \boxed{+} 2 \boxed{e^x} \boxed{=}$ and $5 \boxed{-} 2 \boxed{e^x} \boxed{=}$

57. $4xe^x - 3e^x = 0 \Rightarrow e^x(4x - 3) = 0 \Rightarrow e^x = 0$ or $4x - 3 = 0$. Since $e^x > 0$, the only solution
is $x = \frac{3}{4}$.

61. Subtracting 72 from both sides gives $120 - 72 = 210e^{-0.2x} \Rightarrow \dfrac{48}{210} = e^{-0.2x} \Rightarrow \dfrac{8}{35} = e^{-0.2x}$

$\Rightarrow -0.2x = \ln(\dfrac{8}{35}) \Rightarrow x = -5\ln(\dfrac{8}{35}) \approx 7.3795$.

Keystrokes: 5 $\boxed{+/-}$ $\boxed{\times}$ $\boxed{(}$ 8 $\boxed{\div}$ 35 $\boxed{)}$ $\boxed{\ln}$ $\boxed{=}$

65. Multiplying both sides by e^x gives $e^{2x} - 1 = 3e^x \Rightarrow e^{2x} - 3e^x - 1 = 0 \Rightarrow (e^x)^2 - 3e^x - 1 = 0$.

This last equation is quadratic in e^x. By the quadratic formula, $e^x = \dfrac{3 \pm \sqrt{13}}{2}$. Taking

the natural logarithm of both sides gives $x = \ln(\dfrac{3 - \sqrt{13}}{2})$ which is not a real number,

or $x = \ln(\dfrac{3 + \sqrt{13}}{2}) \approx 1.1948$.

Keystrokes: 3 $\boxed{+}$ 13 $\boxed{\sqrt{}}$ $\boxed{=}$ $\boxed{\div}$ 2 $\boxed{=}$ $\boxed{\ln}$

69. Substituting $C_0 = 15$ into the given formula gives $C(t) = 15e^{-kt}$. We are given that

$C = 7.5$ when $t = 2.4$, so $7.5 = C(2.4) = 15e^{-k(2.4)} \Rightarrow e^{-k(12/5)} = \dfrac{7.5}{15} \Rightarrow e^{-k} = (\dfrac{1}{2})^{5/12}$

Thus, $C(t) = 15[(\dfrac{1}{2})^{5/12}]^t \Rightarrow C(t) = 15(\dfrac{1}{2})^{5t/12}$. After 4 hours the codeine remaining

will be $C(4) = 15(\dfrac{1}{2})^{5(4)/12} = 15(\dfrac{1}{2})^{5/3} \approx 4.7$ mg.

Keystrokes: 15 $\boxed{\times}$ 0.5 $\boxed{y^x}$ $\boxed{(}$ 5 $\boxed{\div}$ 3 $\boxed{)}$ $\boxed{=}$

73. Refer to the formula for continuously compounded interest on page 232 in the text. Substituting $P = 8500$, $r = 0.09$, and $t = 7$, we have $A = 8500e^{(0.09)7} \approx \$15959.69$

Keystrokes: 8500 $\boxed{\times}$ $\boxed{(}$ 0.09 $\boxed{\times}$ 7 $\boxed{)}$ $\boxed{e^x}$ $\boxed{=}$

77. Substituting $Q_0 = 62$ into the given formula gives $Q(t) = 62(\dfrac{1}{2})^{t/h}$. We are given that

$Q = 54$ when $t = 15$, so $54 = Q(15) = 62(\dfrac{1}{2})^{15/h} \Rightarrow \dfrac{54}{62} = (\dfrac{1}{2})^{15/h}$. Reducing the

fraction on the left side of the last equation and taking the log of both sides gives

$\log(\dfrac{27}{31}) = \log(\dfrac{1}{2})^{15/h} \Rightarrow \log(\dfrac{27}{31}) = (\dfrac{15}{h})\log(\dfrac{1}{2}) \Rightarrow h = \dfrac{15\log(1/2)}{\log(27/31)}$

≈ 75.26 minutes.(approximately 75 minutes and 16 seconds)

Keystrokes: 15 $\boxed{\times}$ 0.5 $\boxed{\log}$ $\boxed{\div}$ $\boxed{(}$ 27 $\boxed{\div}$ 31 $\boxed{)}$ $\boxed{\log}$ $\boxed{=}$

Chapter 5

Section 5.1

1. The hypotenuse is 50, the side adjacent to θ is 30, and the side opposite θ is 40. By the definitions on page 266, we get

$$\sin\theta = \frac{40}{50} = \frac{4}{5} \qquad \cos\theta = \frac{30}{50} = \frac{3}{5} \qquad \tan\theta = \frac{40}{30} = \frac{4}{3}$$

$$\csc\theta = \frac{50}{40} = \frac{5}{4} \qquad \sec\theta = \frac{50}{30} = \frac{5}{3} \qquad \cot\theta = \frac{30}{40} = \frac{3}{4}$$

5. The hypotenuse is 21 and the side opposite θ is 9. The side adjacent to θ is $\sqrt{21^2-9^2} = 6\sqrt{10}$. By the definitions on page 266, we get

$$\sin\theta = \frac{9}{21} = \frac{3}{7} \qquad \cos\theta = \frac{6\sqrt{10}}{21} = \frac{2\sqrt{10}}{7} \qquad \tan\theta = \frac{9}{6\sqrt{10}} = \frac{3}{2\sqrt{10}}$$

$$\csc\theta = \frac{21}{9} = \frac{7}{3} \qquad \sec\theta = \frac{21}{6\sqrt{10}} = \frac{7}{2\sqrt{10}} \qquad \cot\theta = \frac{6\sqrt{10}}{9} = \frac{2\sqrt{10}}{3}$$

9. a) Keystrokes: 32 $\boxed{\sin}$ $\boxed{1/x}$ Display: 1.8871

   b) Keystrokes: 58 $\boxed{\cos}$ $\boxed{1/x}$ Display: 1.8871

   c) Keystrokes: 49 $\boxed{\tan}$ $\boxed{1/x}$ Display: 0.8693

   d) Keystrokes: 41 $\boxed{\tan}$ Display: 0.8693

13. $\tan 35.4° = \frac{x}{12.0} \Rightarrow x = 12.0 \tan 35.4° = 8.5$. Keystrokes: 12.0 $\boxed{\times}$ 35.4 $\boxed{\tan}$ $\boxed{=}$

17. $\sin 40.1° = \frac{8.23}{x} \Rightarrow x = \frac{8.23}{\sin 40.1°} = 12.8$. Keystrokes: 8.23 $\boxed{\div}$ 40.1 $\boxed{\sin}$ $\boxed{=}$

21. The acute angle θ is shown in the triangle at right. By the Pythagorean theorem, the hypotenuse is $\sqrt{7^2+2^2} = \sqrt{53}$. By the definitions on page 266, we get

$$\sin\theta = \frac{7}{\sqrt{53}} \qquad \cos\theta = \frac{2}{\sqrt{53}} \qquad \csc\theta = \frac{\sqrt{53}}{7} \qquad \sec\theta = \frac{\sqrt{53}}{2} \qquad \cot\theta = \frac{2}{7}$$

25. $\sec\theta = \dfrac{\text{hyp}}{\text{adj}} = \dfrac{1}{\text{adj}/\text{hyp}} = \dfrac{1}{\cos\theta}$; $\cot\theta = \dfrac{\text{adj}}{\text{opp}} = \dfrac{1}{\text{opp}/\text{adj}} = \dfrac{1}{\tan\theta}$

29. The hypotenuse of the triangle is 1. The side opposite θ is x. By the Pythagorean theorem, the side adjacent to θ is $\sqrt{1-x^2}$ So,

$$\sin\theta = \frac{x}{1} = x \qquad \cos\theta = \frac{\sqrt{1-x^2}}{1} = \sqrt{1-x^2} \qquad \tan\theta = \frac{x}{\sqrt{1-x^2}}$$

$$\csc\theta = \frac{1}{x} \qquad \sec\theta = \frac{1}{\sqrt{1-x^2}} \qquad \cot\theta = \frac{\sqrt{1-x^2}}{x}$$

33. The side adjacent to θ is $2x$. The side opposite θ is 3. By the Pythagorean theorem, the hypotenuse of the triangle is $\sqrt{4x^2+9}$. So,

$$\sin\theta = \frac{3}{\sqrt{4x^2+9}} \qquad \cos\theta = \frac{2x}{\sqrt{4x^2+9}} \qquad \tan\theta = \frac{3}{2x}$$

$$\csc\theta = \frac{\sqrt{4x^2+9}}{3} \qquad \sec\theta = \frac{\sqrt{4x^2+9}}{2x} \qquad \cot\theta = \frac{2x}{3}$$

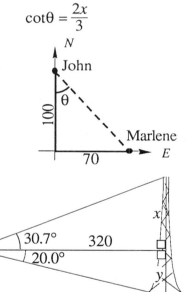

37. From the figure at right, $\tan\theta = \frac{70}{100}$, or $\theta = \tan^{-1}\frac{70}{100} = 35°$. The compass bearing therefore is S35°E.

Keystrokes: 70 ÷ 100 = inv tan

41. From the figure at right, the height of the tower is $x+y$. Using the definition of the tangent function and the two right triangles formed, we get

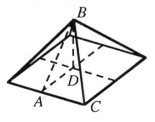

$$\tan 30.7° = \frac{x}{320} \Rightarrow x = 320 \tan 30.7°, \text{ and}$$

$$\tan 20.0° = \frac{y}{320} \Rightarrow y = 320 \tan 20.0°.$$

So, $x+y = 320 \tan 30.7° + 320 \tan 20.0° = 306$. The height of the tower is 306 m,

Keystrokes: 320 X 30.7° tan + 320 X 20.0° tan =

45. Triangle ABD is a right triangle with right angle at D and $BD = \sqrt{AB^2-DA^2}$. The length of DA is 5 ft, since it is half the length of the side the base. Because AB is a side of triangle ABC, $AB = \sqrt{BC^2 - AC^2} = \sqrt{10^2 - 5^2} = 5\sqrt{3}$. Thus, the pyramid's height is $BD = \sqrt{AB^2 - DA^2} = \sqrt{(5\sqrt{3})^2 - 5^2} = 5\sqrt{2}$.

49. From the figure, the tangent of the angle of elevation θ is $\dfrac{38t^2}{570} = \dfrac{t^2}{15}$. The function in question is $\theta(t) = \dfrac{t^2}{15}$.

Section 5.2

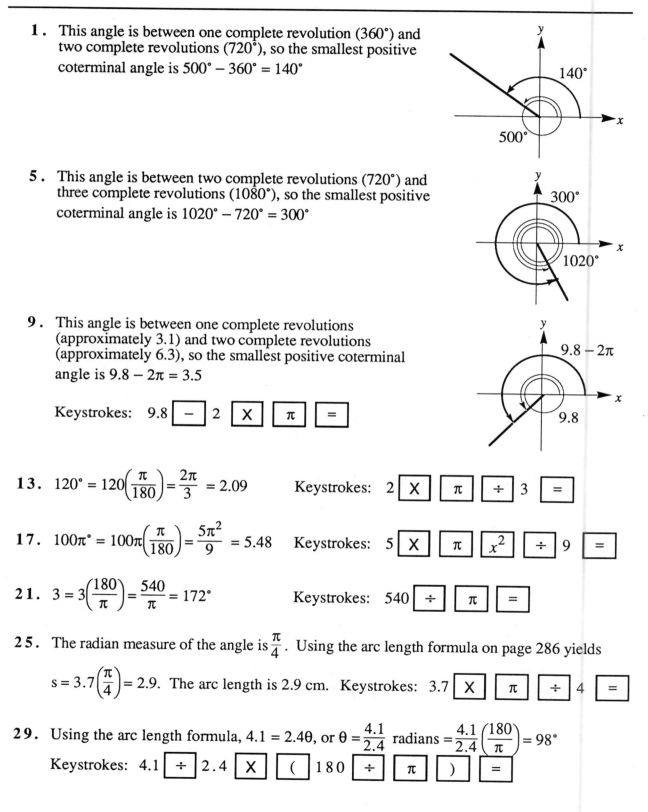

1. This angle is between one complete revolution (360°) and two complete revolutions (720°), so the smallest positive coterminal angle is 500° − 360° = 140°

5. This angle is between two complete revolutions (720°) and three complete revolutions (1080°), so the smallest positive coterminal angle is 1020° − 720° = 300°

9. This angle is between one complete revolutions (approximately 3.1) and two complete revolutions (approximately 6.3), so the smallest positive coterminal angle is $9.8 - 2\pi = 3.5$

Keystrokes: 9.8 $\boxed{-}$ 2 $\boxed{\times}$ $\boxed{\pi}$ $\boxed{=}$

13. $120° = 120\left(\dfrac{\pi}{180}\right) = \dfrac{2\pi}{3} = 2.09$ Keystrokes: 2 $\boxed{\times}$ $\boxed{\pi}$ $\boxed{\div}$ 3 $\boxed{=}$

17. $100\pi° = 100\pi\left(\dfrac{\pi}{180}\right) = \dfrac{5\pi^2}{9} = 5.48$ Keystrokes: 5 $\boxed{\times}$ $\boxed{\pi}$ $\boxed{x^2}$ $\boxed{\div}$ 9 $\boxed{=}$

21. $3 = 3\left(\dfrac{180}{\pi}\right) = \dfrac{540}{\pi} = 172°$ Keystrokes: 540 $\boxed{\div}$ $\boxed{\pi}$ $\boxed{=}$

25. The radian measure of the angle is $\dfrac{\pi}{4}$. Using the arc length formula on page 286 yields

$s = 3.7\left(\dfrac{\pi}{4}\right) = 2.9$. The arc length is 2.9 cm. Keystrokes: 3.7 $\boxed{\times}$ $\boxed{\pi}$ $\boxed{\div}$ 4 $\boxed{=}$

29. Using the arc length formula, $4.1 = 2.4\theta$, or $\theta = \dfrac{4.1}{2.4}$ radians $= \dfrac{4.1}{2.4}\left(\dfrac{180}{\pi}\right) = 98°$

Keystrokes: 4.1 $\boxed{\div}$ 2.4 $\boxed{\times}$ $\boxed{(}$ 180 $\boxed{\div}$ $\boxed{\pi}$ $\boxed{)}$ $\boxed{=}$

33. From our discussion back in Section 2.1, the points with coordinates $(-x, -y)$ and (x, y) are symmetric with respect to the origin. In the figure at right, we see that the angle in question is $\theta + \pi$.

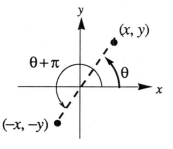

37. In 10 days, the earth travels $\dfrac{10}{365.25}$ of a revolution, or $\dfrac{10}{365.25} 360° = \dfrac{3600}{365.25}° = 10°$

Keystrokes: $3600 \boxed{\div} 365.25 \boxed{=}$.

In radian measure, the measure of this angle is $\dfrac{3600}{365.25}\left(\dfrac{\pi}{180}\right) = \dfrac{20\pi}{365.25}$. The arc length

intercepted by this angle therefore is $s = (93 \text{ million miles})\dfrac{20\pi}{365.25} = 16.0$ million miles

Keystrokes: $93 \boxed{\times} 20 \boxed{\times} \boxed{\pi} \boxed{\div} 365.25 \boxed{=}$.

41. In radian measure, the central angle that intercepts an arc length of one nautical mile is

$\dfrac{1}{60}\left(\dfrac{\pi}{180}\right) = \dfrac{\pi}{1080}$. The radius of the earth in feet is $(3960)(5280)$. Using the arc length

formula, we get $s = (3960)(5280)\left(\dfrac{\pi}{1080}\right) = 6082$ ft.

Keystrokes: $3960 \boxed{\times} 5280 \boxed{\times} \boxed{\pi} \boxed{\div} 1080 \boxed{=}$.

45. Solving the arc length formula for θ yields $\theta = \dfrac{s}{r}$. This can be used in the area formula:

$A = \frac{1}{2}r^2\theta = \frac{1}{2}r^2\left(\dfrac{s}{r}\right) = \frac{1}{2}rs$. The formula we seek is $A = \frac{1}{2}rs$.

49. Starting at $(0, -1)$ and walking 9 units is equivalent to walking $9 - \dfrac{\pi}{2}$ from $(0, 1)$. This is approximately 7.4 units. The smallest positive angle that is coterminal with this angle is $\left(9 - \frac{\pi}{2}\right) - 2\pi = 9 - \frac{5\pi}{2} \approx 1.14$.
This is in quadrant I.
Keystrokes: $9 \boxed{-} 5 \boxed{\times} \boxed{\pi} \boxed{\div} 2 \boxed{=}$

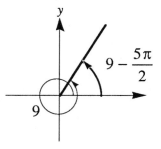

Section 5.3

1. The value of r is $\sqrt{3^2+4^2} = 5$. By the definitions on page 291, we get

$$\sin\theta = \frac{4}{5} \qquad\qquad \cos\theta = \frac{3}{5} \qquad\qquad \tan\theta = \frac{4}{3}$$

5. The value of r is $\sqrt{(-3)^2+(-4)^2} = 5$. By the definitions on page 291, we get

$$\sin\theta = \frac{-4}{5} = -\frac{4}{5} \qquad\qquad \cos\theta = \frac{-3}{5} = -\frac{3}{5} \qquad\qquad \tan\theta = \frac{-4}{-3} = \frac{4}{3}$$

9. The value of r is $\sqrt{\left(\frac{1}{2}\right)^2 + \left(-\frac{3}{4}\right)^2} = \sqrt{\frac{1}{4}+\frac{9}{16}} = \frac{\sqrt{13}}{4}$. By the definitions on page 291, we get

$$\sin\theta = \frac{\left(-\frac{3}{4}\right)}{\left(\frac{\sqrt{13}}{4}\right)} = -\frac{3}{\sqrt{13}} \qquad \cos\theta = \frac{\left(\frac{1}{2}\right)}{\left(\frac{\sqrt{13}}{4}\right)} = \frac{2}{\sqrt{13}} \qquad \tan\theta = \frac{\left(-\frac{3}{4}\right)}{\left(\frac{1}{2}\right)} = -\frac{3}{2}$$

13. From the figure at right the reference angle is 60°, and the terminal side is in the first quadrant. From the table on page 298, we get:

$$\sin(-300°) = \sin 60° = \frac{\sqrt{3}}{2} \qquad \cos(-300°) = \cos 60° = \frac{1}{2}$$

$$\tan(-300°) = \tan 60° = \sqrt{3}$$

17. The terminal side of 720° falls on the positive x–axis, so the values of the trigonometric functions are the same as they are for 0°. From the table on page 298, we get:

$$\sin 720° = \sin 0° = 0 \qquad \cos 720° = \cos 0° = 1$$

$$\tan 720° = \tan 0° = 1$$

21. From the figure at right, the reference angle is $\frac{\pi}{4}$ and the terminal side is in the second quadrant. From page 298, we get:

$$\sin\frac{11\pi}{4} = \sin\frac{\pi}{4} = \frac{1}{\sqrt{2}} \qquad \cos\frac{11\pi}{4} = -\cos\frac{\pi}{4} = -\frac{1}{\sqrt{2}}$$

$$\tan\frac{11\pi}{4} = -\tan\frac{\pi}{4} = -1$$

25. Your calculator should be in degree mode:

Keystrokes: 110 $\boxed{\sin}$ Display: 0.9397

29. Your calculator should be in radian mode:

Keystrokes: 2 $\boxed{\times}$ $\boxed{\pi}$ $\boxed{\div}$ 5 $\boxed{=}$ $\boxed{+\!/\!-}$ $\boxed{\sin}$ Display: −0.9511

33. Your calculator should be in radian mode:

Keystrokes: 13 $\boxed{\cos}$ Display: 0.9744

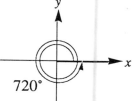

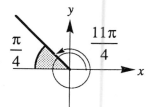

37. From $\cos\theta = \frac{3}{5}$, we get $x = 3$, $r = 5$, and $y = -\sqrt{5^2 - 3^2} = -4$

Using these values and the definitions on page 291 yields

$$\sin\theta = -\frac{4}{5} \qquad \tan\theta = -\frac{4}{3} \qquad \csc\theta = -\frac{5}{4}$$

$$\sec\theta = \frac{5}{3} \qquad \cot\theta = -\frac{3}{4}$$

41. Since $\sec\theta = \frac{7}{\sqrt{17}}$, we get $x = \sqrt{17}$, $r = 7$, and

$y = -\sqrt{7^2 - \sqrt{17}^2} = -4\sqrt{2}$. Using these values and the definitions on page 291 yields

$$\sin\theta = -\frac{4\sqrt{2}}{7} \qquad \cos\theta = \frac{\sqrt{17}}{7} \qquad \tan\theta = -\frac{4\sqrt{2}}{\sqrt{17}} = -\frac{4\sqrt{34}}{17}$$

$$\csc\theta = -\frac{7}{4\sqrt{2}} \qquad \cot\theta = -\frac{\sqrt{17}}{4\sqrt{2}} = -\frac{\sqrt{34}}{8}$$

45. If (x, y) is a point on the terminal side of θ, then $\sin\theta = \dfrac{y}{\sqrt{x^2+y^2}} = k$. In each case,

depending on the symmetry of the figure, the value of the sine is either k or $-k$.

a)

$$\sin(\pi - \alpha) = \frac{y}{\sqrt{(-x)^2 + y^2}}$$

$$= \frac{y}{\sqrt{x^2+y^2}} = k$$

b)

$$\sin(\pi + \alpha) = \frac{-y}{\sqrt{(-x)^2 + (-y)^2}}$$

$$= -\frac{y}{\sqrt{x^2+y^2}} = -k$$

c)

$$\sin(-\alpha) = \frac{-y}{\sqrt{x^2 + (-y)^2}}$$

$$= -\frac{y}{\sqrt{x^2+y^2}} = -k$$

d)

$$\sin(-\pi - \alpha) = \frac{y}{\sqrt{(-x)^2 + y^2}}$$

$$= \frac{y}{\sqrt{x^2+y^2}} = k$$

Section 5.4

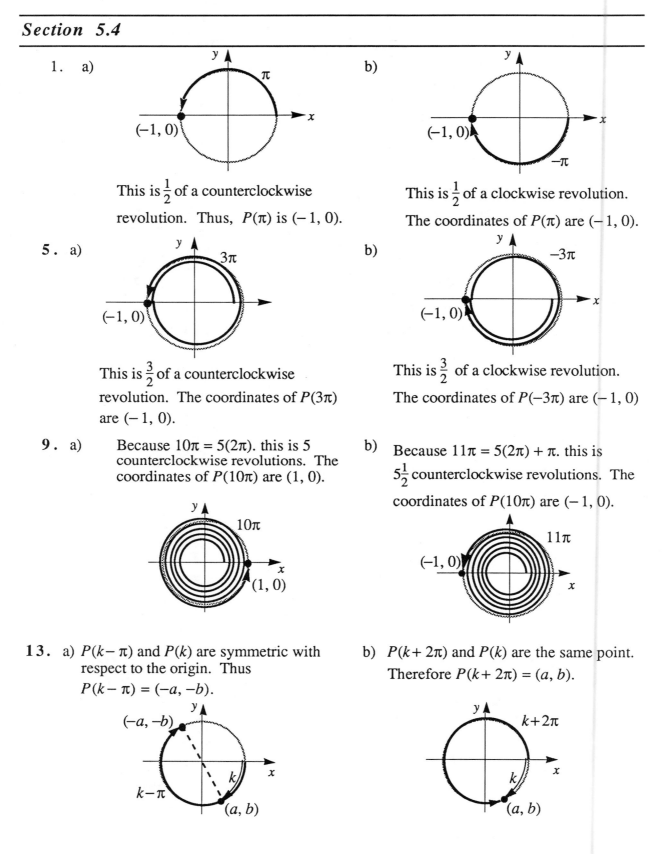

1. a) This is $\frac{1}{2}$ of a counterclockwise revolution. Thus, $P(\pi)$ is $(-1, 0)$.

    b) This is $\frac{1}{2}$ of a clockwise revolution. The coordinates of $P(\pi)$ are $(-1, 0)$.

5. a) This is $\frac{3}{2}$ of a counterclockwise revolution. The coordinates of $P(3\pi)$ are $(-1, 0)$.

    b) This is $\frac{3}{2}$ of a clockwise revolution. The coordinates of $P(-3\pi)$ are $(-1, 0)$

9. a) Because $10\pi = 5(2\pi)$. this is 5 counterclockwise revolutions. The coordinates of $P(10\pi)$ are $(1, 0)$.

    b) Because $11\pi = 5(2\pi) + \pi$. this is $5\frac{1}{2}$ counterclockwise revolutions. The coordinates of $P(10\pi)$ are $(-1, 0)$.

13. a) $P(k-\pi)$ and $P(k)$ are symmetric with respect to the origin. Thus $P(k-\pi) = (-a, -b)$.

    b) $P(k+2\pi)$ and $P(k)$ are the same point. Therefore $P(k+2\pi) = (a, b)$.

c) From the figure, $P(k+\pi)$ and $P(k)$ are symmetric with respect to the origin.
Therefore $P(k+\pi) = (-a, -b)$.

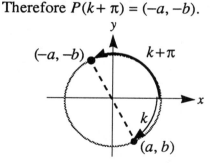

d) From the figure, $P(\pi - k)$ and $P(k)$ are symmetric with respect to the y-axis.
Therefore $P(\pi - k) = (-a, b)$.

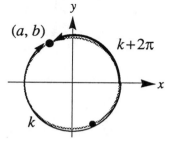

17. a) From the figure, $P(k-\pi)$ and $P(k)$ are symmetric with respect to the origin.
Therefore $P(k-\pi) = (-a, -b)$.

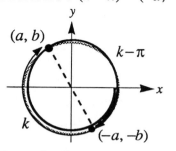

b) From the figure, $P(k+2\pi)$ and $P(k)$ are the same point. Therefore
$P(k+2\pi) = (a, b)$.

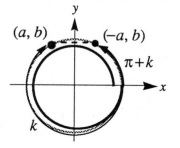

c) From the figure, $P(k+\pi)$ and $P(k)$ are symmetric with respect to the origin.
Therefore $P(k+\pi) = (-a, -b)$.

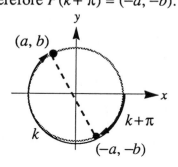

d) From the figure, $P(\pi - k)$ and $P(k)$ are symmetric with respect to the y–axis.
Therefore $P(\pi - k) = (-a, b)$.

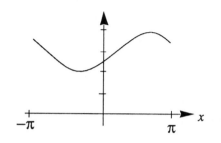

21. The graph of the function is the graph of $y = \cos t$ that is translated $\dfrac{2\pi}{3}$ units horizontally and 3 units vertically.

25. The graph of the function is the graph of $y = \sin t$ that is vertically expanded by a factor of two.

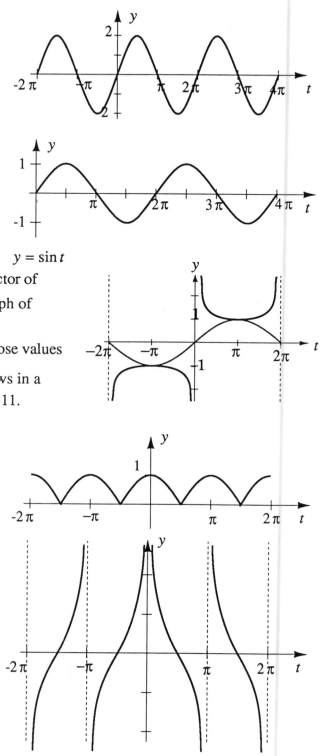

29. The graph of the function is the graph of $y = \sin t$ that is reflected through the x-axis and translated π units horizontally.

33. The graph of $y = \sin\frac{1}{2}t$ is the graph of $y = \sin t$ that is compressed horizontally by a factor of two. Because $\csc\frac{1}{2}t = \dfrac{1}{\sin\frac{1}{2}t}$, the graph of

$y = \csc\frac{1}{2}t$ has vertical asymptotes at those values of t where $\sin\frac{1}{2}t = 0$. The graph follows in a fashion similar to Example 6 on page 311.

37. The graph of $y = |\cos t|$ is the same as the graph of $y = \cos t$ except that those points below the x–axis are reflected through the x–axis. (See the Summary on page 131).

41. The graph of $y = \cot|t|$ is the same as the graph of $y = \cot t$ for $x > 0$. For $x < 0$, the graph is the reflection through the y–axis of the graph for $x > 0$. (See the Summary on page 131).

45. Because the sine function is odd, $y = \sin\left(-t - \dfrac{3\pi}{2}\right) = -\sin\left(t + \dfrac{3\pi}{2}\right)$. This is the graph of $y = -\sin t$ translated $-\dfrac{3\pi}{2}$ units horizontally. This graph is the graph of $y = -\cos t$.

49. Because the period of the cosine function is 2π, $y = \cos(4\pi + t) = \cos(2\pi + t) = \cos t$. This is the graph of $y = \cos t$.

53. Because the period of the cotangent function is π, $y = \cot(t - 2\pi) = \cot t$. Thus the graph of $y = \cot(t - 2\pi)$ is the graph of $y = \cot t$.

57. The period of the cotangent function is π, so $y = \cot(3\pi - t) = \cot(-t)$. Because the cotangent function is odd, $y = \cot(-t) = -\cot t$. Thus the graph of $y = \cot(3\pi - t)$ is the graph of $y = -\cot t$.

61. Starting at $(0, -1)$ on the unit circle, the bug walks $\frac{\pi}{2}$ units to $(1, 0)$, then continues on for $9 - \frac{\pi}{2}$ units to $P\left(9 - \frac{\pi}{2}\right)$. The coordinates of this point are $\left(\cos\left(9 - \frac{\pi}{2}\right), \sin\left(9 - \frac{\pi}{2}\right)\right)$. Using a calculator, we approximate these as $(0.4121, 0.9111)$.

Keystrokes: 9 $\boxed{-}$ $\boxed{\pi}$ $\boxed{\div}$ 2 $\boxed{=}$ $\boxed{\cos}$ and 9 $\boxed{-}$ $\boxed{\pi}$ $\boxed{\div}$ 2 $\boxed{=}$ $\boxed{\sin}$

65. $\sin^2 t + \cos^2 t = 1 \quad\Rightarrow\quad \left(\dfrac{1}{\cos^2 t}\right)\left(\sin^2 t + \cos^2 t\right) = (1)\left(\dfrac{1}{\cos^2 t}\right)$

$\Rightarrow \quad \dfrac{\sin^2 t}{\cos^2 t} + 1 = \dfrac{1}{\cos^2 t} \quad\Rightarrow\quad \left(\dfrac{\sin t}{\cos t}\right)^2 + 1 = \left(\dfrac{1}{\cos t}\right)^2 \quad\Rightarrow\quad \tan^2 t + 1 = \sec^2 t$

Section 5.5

1. The amplitude is $A = 2$, the period is $\dfrac{2\pi}{B} = \dfrac{2\pi}{3}$. The box is formed by horizontal lines at $y = \pm 2$, and the vertical lines $x = 0$ and $x = \dfrac{2\pi}{3}$. A sine curve (as shown on page 317) is sketched in the box.

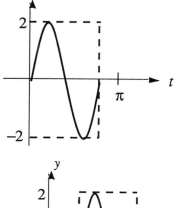

5. This is a translation of the graph of $y = 2\sin 3t$ which has amplitude $A = 2$, and period $\dfrac{2\pi}{B} = \dfrac{2\pi}{3}$. The box is formed by horizontal lines at $y = \pm 2$, and the vertical lines $x = \dfrac{\pi}{3}$ and $x = \pi$. A sine curve (as shown on page 317) is sketched in the box.

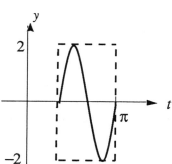

9. First, note that $y = \sin\left(2t - \dfrac{\pi}{2}\right) = \sin 2\left(t - \dfrac{\pi}{4}\right)$

This is a horizontal translation (of $\dfrac{\pi}{4}$ units) of the graph of $y = 2\sin 2t$ which has amplitude $A = 2$, and period $\dfrac{2\pi}{B} = \dfrac{2\pi}{2} = \pi$. The box is formed by horizontal lines at $y = \pm 2$, and the vertical lines $x = \dfrac{\pi}{4}$ and $x = \dfrac{5\pi}{4}$. A sine curve is sketched in the box.

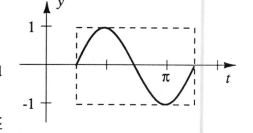

13. The amplitude is $\dfrac{3}{2}$ and the period is π. There is no translation. We get four complete periods over the interval $-2\pi \le t \le 2\pi$.

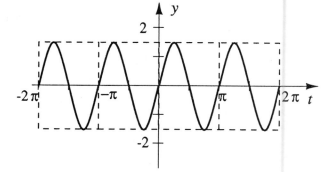

17. The amplitude is 2 and the period is $\dfrac{2\pi}{3}$. There is a horizontal translation of $\dfrac{\pi}{3}$ units. We get four complete periods over the interval $-2\pi \le t \le 2\pi$

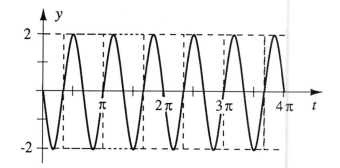

21. First, note that $y = 2\sin\left(2t - \dfrac{\pi}{2}\right) = 2\sin 2\left(t - \dfrac{\pi}{4}\right)$
The amplitude is 2 and the period is π. There is a horizontal translation of $\dfrac{\pi}{4}$.

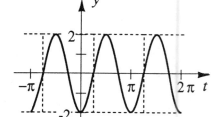

25. The graph in the figure has period π and amplitude 2. Of those functions in the list, VI and XI each has this amplitude and period.

29. The graph in the figure has period 2 and amplitude 3. Of those functions in the list, I, IV, VII, and VIII each has this amplitude and period. However, if we substitute $x = 0$ in each of these four functions, we see that only the graphs of the functions in I and VIII pass through the origin. These two correctly describe the graph.

33. First we sketch $y = t$ and $y = -\sin t$. Using the technique of adding y-coordinates, we get the graph shown.

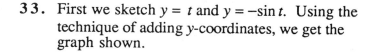

37. First we sketch $y = \cos\frac{1}{2}t$ and $y = \sin t$.

Using the technique of adding y-coordinates, we get the graph shown.

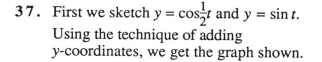

41. First we sketch $y = |t|$ and $y = \cos t - 1$. Using the technique of adding y-coordinates, we get the graph shown.

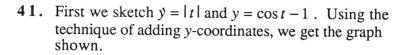

45. First we sketch $y = \left|\frac{1}{2}x\right|$ and $y = -\left|\frac{1}{2}x\right|$. Let $f(x) = \frac{1}{2}x \sin x$. Then

$$f(-x) = \frac{1}{2}(-x)\sin(-x)$$

$$= \frac{1}{2}(-x)(-\sin x)$$

$$= \frac{1}{2}x \sin x = f(x)$$

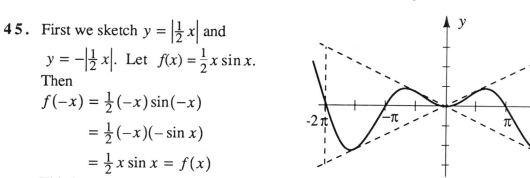

This implies that the function is even and that the graph is symmetric with respect to the y-axis. The quasiperiod is 2π.

49. Because the the radicand of a square root must be nonnegative, the domain of the function is $x \geq 0$. This implies that the function is neither even nor odd.

First we sketch $y = \sqrt{x}$ and $y = -\sqrt{x}$. The quasiperiod is 2π.

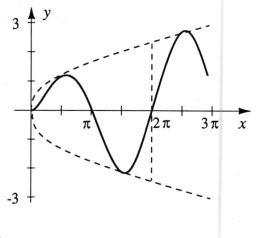

53. The function is neither even nor odd. First we sketch $y = 2^{-x} + 1$ and $y = -(2^{-x} + 1)$. The quasiperiod is π.

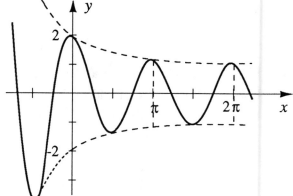

57. The period of the function is $\dfrac{2\pi}{B} = 4$. Solving this equation for B yields $\dfrac{\pi}{2}$.

61. The amplitude is $A = 36$. The period of the function is $\dfrac{2\pi}{B} = \dfrac{2\pi}{(2\pi / 365)} = 365$. The translation is 101 days horizontally and 14° vertically.

The maximum temperature occurs when $\dfrac{2\pi}{365}(t - 101) = \dfrac{\pi}{2}$:

$$\frac{2\pi}{365}(t - 101) = \frac{\pi}{2} \Rightarrow t - 101 = 91.25 \Rightarrow t = 192.25$$

$\Rightarrow$ Warmest day is day 192 (July 12)

The minimum temperature occurs when $\dfrac{2\pi}{365}(t - 101) = \dfrac{3\pi}{2}$:

$$\frac{2\pi}{365}(t - 101) = \frac{3\pi}{2} \Rightarrow t - 101 = 273.75 \Rightarrow t = 374.75$$

However, since the year is 365 days, we make an adjustment: $374 - 365 = 9$

$\Rightarrow$ Coldest day is day 9. (January 9)

Miscellaneous Exercises for Chapter 5

1. From the figure, the hypotenuse is 9 and the side opposite θ is 3. Using the Pythagorean theorem, we find that the side adjacent to θ is $\sqrt{9^2 - 3^2} = 6\sqrt{2}$. By the definitions on page 266, we get

$$\sin\theta = \frac{3}{9} = \frac{1}{3} \qquad \cos\theta = \frac{6\sqrt{2}}{9} = \frac{2\sqrt{2}}{3} \qquad \tan\theta = \frac{3}{6\sqrt{2}} = \frac{1}{2\sqrt{2}}$$

$$\csc\theta = \frac{9}{3} = 3 \qquad \sec\theta = \frac{9}{6\sqrt{2}} = \frac{3}{2\sqrt{2}} \qquad \cot\theta = \frac{6\sqrt{2}}{3} = 2\sqrt{2}$$

5. Refer to the figure at right. Because triangle *CDB* is a 45°-45° right triangle, $CD = DB = 6$. Thus, in triangle *ADB*, the side opposite θ is 6 and the side adjacent to θ is 4. Using the Pythagorean theorem, we find that the hypotenuse is $\sqrt{4^2 + 6^2} = 2\sqrt{13}$. By the definitions on page 266, we get

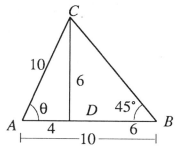

$$\sin\theta = \frac{6}{2\sqrt{13}} = \frac{3}{\sqrt{13}} \qquad \csc\theta = \frac{2\sqrt{13}}{6} = \frac{\sqrt{13}}{3}$$

$$\cos\theta = \frac{4}{2\sqrt{13}} = \frac{2}{\sqrt{13}} \qquad \sec\theta = \frac{2\sqrt{13}}{4} = \frac{\sqrt{13}}{2}$$

$$\tan\theta = \frac{6}{4} = \frac{3}{2} \qquad \cot\theta = \frac{4}{6} = \frac{2}{3}$$

9. From the figure, the side opposite θ is $2n$ and the side adjacent to θ is n. This implies that $\tan\theta = \frac{2n}{n} = 2$, or $\theta = \tan^{-1}2 = 63°$. Keystrokes (in degree mode): 2 $\boxed{\text{inv}}$ $\boxed{\text{tan}}$

13. a) The reference angle for $\frac{7\pi}{4}$ is $\frac{\pi}{4}$. Since this angle is in Quadrant IV and the sine is negative there, $\sin\frac{7\pi}{4} = -\sin\frac{\pi}{4} = -\frac{1}{\sqrt{2}}$.

b) The reference angle for $-\frac{5\pi}{3}$ is $\frac{\pi}{3}$. Since this angle is in Quadrant I and the cosine is positive there, $\cos\left(-\frac{5\pi}{3}\right) = \cos\frac{\pi}{3} = \frac{1}{2}$.

c) The reference angle for $-\frac{5\pi}{6}$ is $\frac{\pi}{6}$. Since this angle is in Quadrant III and the sine is negaitive there, $\sin\left(-\frac{5\pi}{6}\right) = -\sin\frac{\pi}{6} = -\frac{1}{2}$.

d) Because $\frac{13\pi}{2} - 6\pi = \frac{\pi}{2}$, $\frac{13\pi}{2}$ is coterminal with $\frac{\pi}{2}$. Thus $\cos\frac{13\pi}{2} = \cos\frac{\pi}{2} = 0$

17. a) The coordinates of $P\left(-\frac{\pi}{4}\right)$ are $\left(\cos\left(-\frac{\pi}{4}\right), \sin\left(-\frac{\pi}{4}\right)\right) = \left(\frac{1}{\sqrt{2}}, -\frac{1}{\sqrt{2}}\right)$

   b) The coordinates of $P\left(-\frac{13\pi}{3}\right)$ are $\left(\cos\left(-\frac{13\pi}{3}\right), \sin\left(-\frac{13\pi}{3}\right)\right) = \left(\frac{1}{2}, -\frac{\sqrt{3}}{2}\right)$

   c) The coordinates of $P\left(\frac{11\pi}{6}\right)$ are $\left(\cos\frac{11\pi}{6}, \sin\frac{11\pi}{6}\right) = \left(\frac{\sqrt{3}}{2}, -\frac{1}{2}\right)$

   d) The coordinates of $P\left(-\frac{17\pi}{3}\right)$ are $\left(\cos\left(-\frac{17\pi}{3}\right), \sin\left(-\frac{17\pi}{3}\right)\right) = \left(\frac{1}{2}, \frac{\sqrt{3}}{2}\right)$

21. The amplitude is 2 and the period is π. There is no translation. We get two complete periods of a cosine graph over the interval $0 \le t \le 4\pi$.

25. The amplitude is 1 and the period is 4π. There is a vertical translation of -4 units. We get $1\frac{1}{2}$ complete periods of a cosine graph over the interval $-2\pi \le t \le 4\pi$.

29. The amplitude is 1 and the period is 2π. There is a horizontal translation of $\frac{\pi}{3}$ units and a vertical translation of 3 units.

33. Because $\cos(3t+\pi) = \cos 3\left(t+\frac{\pi}{3}\right)$, the amplitude is 1 and the period is 2π. There is a horizontal translation of $-\frac{\pi}{3}$ units.

37. Because $2\sin(\pi t + 3\pi) = 2\sin\pi(t+3)$, the amplitude is 2 and the period is 2. There is a horizontal translation of -3 units.

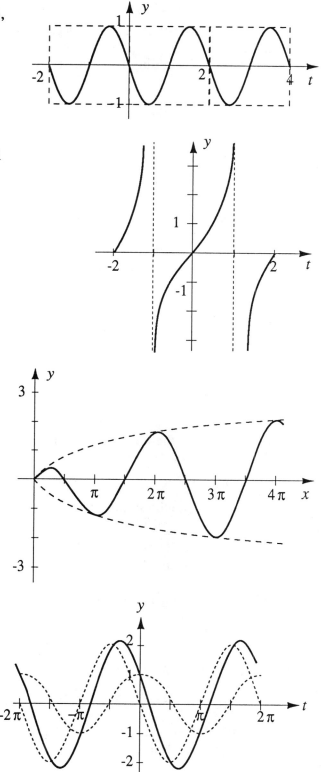

41. Because $\tan\left(\dfrac{\pi t}{2}\right) = \tan\dfrac{\pi}{2}t$, the period

is $\dfrac{\pi}{(\pi/2)} = 2$. .

45. First we sketch $y = \pm\ln(x+1)$. (These are horizontal translations of $y = \pm\ln x$) The quasiperiod is 2π.

49. First we sketch $y = \cos t$ (amplitude 1 and period 2π) and $y = -2\sin t$ (amplitude 2 and period 2π) over the interval. The graph in question is determined by adding y-coordinates. The period of this graph is 2π.

53. First we sketch $y = \log_2 t$ and

$y = -\sin \pi t$ (amplitude 1 and period 2) over the interval. The graph in question is determined by adding y-coordinates.

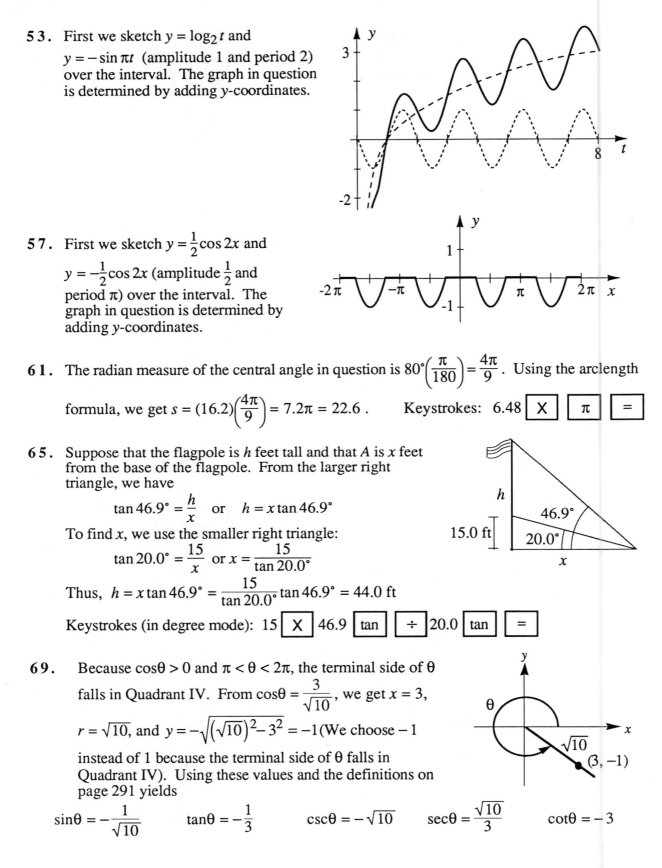

57. First we sketch $y = \frac{1}{2}\cos 2x$ and

$y = -\frac{1}{2}\cos 2x$ (amplitude $\frac{1}{2}$ and period π) over the interval. The graph in question is determined by adding y-coordinates.

61. The radian measure of the central angle in question is $80° \left(\dfrac{\pi}{180}\right) = \dfrac{4\pi}{9}$. Using the arclength

formula, we get $s = (16.2)\left(\dfrac{4\pi}{9}\right) = 7.2\pi = 22.6$.   Keystrokes: 6.48 $\boxed{\times}$ $\boxed{\pi}$ $\boxed{=}$

65. Suppose that the flagpole is h feet tall and that A is x feet from the base of the flagpole. From the larger right triangle, we have

$$\tan 46.9° = \frac{h}{x} \quad \text{or} \quad h = x \tan 46.9°$$

To find x, we use the smaller right triangle:

$$\tan 20.0° = \frac{15}{x} \quad \text{or} \quad x = \frac{15}{\tan 20.0°}$$

Thus, $h = x \tan 46.9° = \dfrac{15}{\tan 20.0°} \tan 46.9° = 44.0$ ft

Keystrokes (in degree mode): 15 $\boxed{\times}$ 46.9 $\boxed{\tan}$ $\boxed{\div}$ 20.0 $\boxed{\tan}$ $\boxed{=}$

69. Because $\cos\theta > 0$ and $\pi < \theta < 2\pi$, the terminal side of θ

falls in Quadrant IV. From $\cos\theta = \dfrac{3}{\sqrt{10}}$, we get $x = 3$,

$r = \sqrt{10}$, and $y = -\sqrt{(\sqrt{10})^2 - 3^2} = -1$ (We choose -1

instead of 1 because the terminal side of θ falls in Quadrant IV). Using these values and the definitions on page 291 yields

$$\sin\theta = -\frac{1}{\sqrt{10}} \qquad \tan\theta = -\frac{1}{3} \qquad \csc\theta = -\sqrt{10} \qquad \sec\theta = \frac{\sqrt{10}}{3} \qquad \cot\theta = -3$$

73. a) Because C is a right triangle, the length of line segment $BD = 12\sqrt{2}$. Since triangle CBD is equilateral, $AB = AD = BD = 12\sqrt{2}$. The radius of the larger circle is $12\sqrt{2}$.

b) From the figure in the text, the shaded area is

$$\begin{pmatrix} \text{shaded} \\ \text{area} \end{pmatrix} = \begin{pmatrix} \text{area of sectors} \\ CBD \text{ and } ABD \end{pmatrix} - \begin{pmatrix} \text{area of triangles} \\ CBD \text{ and } ABD \end{pmatrix}$$

The angle at C is a right triangle, so the area of sector ABD is $\dfrac{1}{2}(12)^2\left(\dfrac{\pi}{2}\right) = 36\pi$ and

the area of triangle ABD is $\dfrac{1}{2}(12)^2 = 72$. The area of sector CBD is

$\dfrac{1}{2}\left(12\sqrt{2}\right)^2\left(\dfrac{\pi}{3}\right) = 48\pi$ and the area of triangle CBD is $\dfrac{\sqrt{3}}{4}\left(12\sqrt{2}\right)^2 = 72\sqrt{3}$ (see

Exercise 41 of Section 2.2). So,

$$\begin{pmatrix} \text{shaded} \\ \text{area} \end{pmatrix} = (36\pi + 48\pi) - \left(72 + 72\sqrt{3}\right)$$

$$= 84\pi - 72 - 72\sqrt{3}$$

(This is approximately 67.2)

77. The solid generated can be considered a cylinder of radius 9 and height 3 with a frustum with radii 9 and 5 and height 3 removed. Using the formulas available on the endpapers inside the front cover, we get

$$\begin{pmatrix} \text{Volume} \\ \text{of solid} \end{pmatrix} = \begin{pmatrix} \text{Volume} \\ \text{of cylinder} \end{pmatrix} - \begin{pmatrix} \text{Volume} \\ \text{of frustum} \end{pmatrix}$$

$$= \pi(9)^2(3) - \frac{\pi}{3}(3)\left((5)^2 + (5)(9) + (9)^2\right)$$

$$= 243\pi - 151\pi$$

$$= 92\pi$$

Chapter 6

Section 6.1

1. $\dfrac{\tan\theta}{\sec\theta} = \dfrac{\sin\theta/\cos\theta}{1/\cos\theta} = \sin\theta$

5. $(\sec^2 t - \tan^2 t)^3 = (1)^3 = 1$

9. $\dfrac{\cos^4 x - \cos^2 x}{\sin x} = \dfrac{\cos^2 x(\cos^2 x - 1)}{\sin x} = \dfrac{\cos^2 x(-\sin^2 x)}{\sin x} = -\cos^2 x \sin x$

13. $\dfrac{\tan^3\alpha + 1}{\tan\alpha + 1} = \dfrac{(\tan\alpha + 1)(\tan^2\alpha - \tan\alpha + 1)}{\tan\alpha + 1} = \tan^2\alpha + 1 - \tan\alpha = \sec^2\alpha - \tan\alpha$

17. $\sqrt{16\tan^2 x + 16} = 4\sqrt{\tan^2 x + 1} = 4\sqrt{\sec^2 x} = 4|\sec x|$

21. $\dfrac{\tan^2\gamma}{\sec\gamma - 1} = \dfrac{\sec^2\gamma - 1}{\sec\gamma - 1} = \dfrac{(\sec\gamma - 1)(\sec\gamma + 1)}{\sec\gamma - 1} = \sec\gamma + 1$

25. $\cos\theta(\sec\theta - \cos\theta) = \cos\theta\sec\theta - \cos\theta\cos\theta = 1 - \cos^2\theta = \sin^2\theta$

29. $\cos^2\beta - \sin^2\beta = \cos^2\beta - (1 - \cos^2\beta) = 2\cos^2\beta - 1$

33. $\dfrac{\csc\psi - \sec\psi}{\csc\psi + \sec\psi} = \dfrac{\dfrac{1}{\sin\psi} - \dfrac{1}{\cos\psi}}{\dfrac{1}{\sin\psi} + \dfrac{1}{\cos\psi}} = \dfrac{\dfrac{\cos\psi - \sin\psi}{\cos\psi\sin\psi}}{\dfrac{\cos\psi + \sin\psi}{\cos\psi\sin\psi}} = \dfrac{\cos\psi - \sin\psi}{\cos\psi + \sin\psi}$

37. $\tan^4\lambda - 2\sec^2\lambda\tan^2\lambda + \sec^4\lambda = (\tan^2\lambda - \sec^2\lambda)^2 = (-1)^2 = 1$

41. $\dfrac{\tan^3\theta - \cot^3\theta}{\tan\theta - \cot\theta} = \dfrac{(\tan\theta - \cot\theta)(\tan^2\theta + \tan\theta\cot\theta + \cot^2\theta)}{\tan\theta - \cot\theta}$

$= \tan^2\theta + 1 + \cot^2\theta = (\tan^2\theta + 1) + \cot^2\theta = \sec^2\theta + \cot^2\theta$

45. $\dfrac{\sin\theta + \cos\theta}{\cos\theta - \sin\theta} - \dfrac{\tan\theta - 1}{\tan\theta + 1} = \dfrac{\sin\theta + \cos\theta}{\cos\theta - \sin\theta} - \dfrac{(\sin\theta/\cos\theta) - 1}{(\sin\theta/\cos\theta) + 1}$

$= \dfrac{\sin\theta + \cos\theta}{\cos\theta - \sin\theta} - \dfrac{\sin\theta - \cos\theta}{\cos\theta + \sin\theta}$

$= \dfrac{(\sin\theta + \cos\theta)^2 - (\sin\theta - \cos\theta)(\cos\theta - \sin\theta)}{(\cos\theta - \sin\theta)(\cos\theta + \sin\theta)}$

$= \dfrac{\sin^2\theta + 2\sin\theta\cos\theta + \cos^2\theta + \sin^2\theta - 2\sin\theta\cos\theta + \cos^2\theta}{\cos^2\theta - \sin^2\theta} = \dfrac{2}{\cos^2\theta - \sin^2\theta}$

49. Let, for example, $\theta = \dfrac{\pi}{6}$. The left side of the equation is $\sin(\dfrac{\pi}{6} + \dfrac{\pi}{2}) = \sin(\dfrac{2\pi}{3}) = \dfrac{\sqrt{3}}{2}$.

The right side of the equation is $\sin\dfrac{\pi}{6} + \sin\dfrac{\pi}{2} = \dfrac{1}{2} + 1 = \dfrac{3}{2}$. Therefore,

$\sin(\dfrac{\pi}{6} + \dfrac{\pi}{2}) \neq \sin\dfrac{\pi}{6} + \sin\dfrac{\pi}{2}$.

53. Let, for example, $\theta = \dfrac{\pi}{4}$. The left side of the equation is $1 + \tan\dfrac{\pi}{4} = 2$. The right side of

the equation is $\sec\dfrac{\pi}{4} = \sqrt{2}$. Therefore, $1 + \tan\dfrac{\pi}{4} \neq \sec\dfrac{\pi}{4}$.

57. Let, for example, $\theta = 0$. The left side of the equation is $\dfrac{\sin 0}{\sin 0 + \cos 0} = \dfrac{0}{1} = 0$. The

right side of the equation is $\tan 0 + 1 = 1$. Therefore, $\dfrac{\sin 0}{\sin 0 + \cos 0} \neq \tan 0 + 1$.

61. $\ln(1 - \cos\theta) + \ln(1 + \cos\theta) = \ln[(1 - \cos\theta)(1 + \cos\theta)] = \ln[1 - \cos^2\theta] = \ln[\sin^2\theta]$

65. $\cos t \, \log(100^{\sec t}) = \cos t \, \sec t \, \log 100 = 1 \, \log 100 = 2$

69. $(\cos\alpha - \cos\beta)^2 + (\sin\alpha - \sin\beta)^2$

$= (\cos^2\alpha - 2\cos\alpha\cos\beta + \cos^2\beta) + (\sin^2\alpha - 2\sin\alpha\sin\beta + \sin^2\beta)$

$= (\cos^2\alpha + \sin^2\alpha) + (\cos^2\beta + \sin^2\beta) - 2\cos\alpha\cos\beta - 2\sin\alpha\sin\beta$

$= 2 - 2(\cos\alpha\cos\beta + \sin\alpha\sin\beta)$

73. First note that $\sin\theta$ is positive, since $0 < \theta < \pi$. Thus, $\sin\theta = \sqrt{1 - \cos^2\theta} = \sqrt{1 - x^2}$,

$\tan\theta = \dfrac{\sin\theta}{\cos\theta} = \dfrac{\sqrt{1 - x^2}}{x}$, $\sec\theta = \dfrac{1}{\cos\theta} = \dfrac{1}{x}$,

$\csc\theta = \dfrac{1}{\sin\theta} = \dfrac{1}{\sqrt{1 - x^2}}$, $\cot\theta = \dfrac{1}{\tan\theta} = \dfrac{x}{\sqrt{1 - x^2}}$

77. First note that $\sin\theta$ is negative, since $-\frac{\pi}{2} < \theta < 0$. Thus, $\sin\theta = -\sqrt{1 - \cos^2\theta}$

$$= -\sqrt{1 - \frac{1}{x^2}} = -\frac{\sqrt{x^2 - 1}}{x}, \ \tan\theta = \frac{\sin\theta}{\cos\theta} = \frac{-\sqrt{x^2 - 1}/x}{1/x} = -\sqrt{x^2 - 1},$$

$$\sec\theta = \frac{1}{\cos\theta} = x, \ \csc\theta = \frac{1}{\sin\theta} = -\frac{x}{\sqrt{x^2 - 1}}, \ \cot\theta = \frac{1}{\tan\theta} = -\frac{1}{\sqrt{x^2 - 1}}$$

81. $\dfrac{(a^2 - u^2)^2}{u^4} = \dfrac{(a^2 - a^2\sin^2\theta)^2}{(a\sin\theta)^4} = \dfrac{a^4(1 - \sin^2\theta)^2}{a^4\sin^4\theta} = \dfrac{(\cos^2\theta)^2}{\sin^4\theta} = \dfrac{\cos^4\theta}{\sin^4\theta} = \cot^4\theta.$

85. $\sqrt{u^2 + a^2} = \sqrt{(a\tan\theta)^2 + a^2} = a\sqrt{(\tan\theta)^2 + 1} = a\sqrt{(\sec\theta)^2} = a\sec\theta$

89. $(a^2 + u^2)^{5/2} = (a^2 + a^2\tan^2\theta)^{5/2} = a^5(1 + \tan^2\theta)^{5/2} = a^5(\sec^2\theta)^{5/2} = a^5\sec^5\theta$

93. $\dfrac{1}{u\sqrt{u^2 - a^2}} = \dfrac{1}{a\sec\theta\sqrt{a^2\sec^2\theta - a^2}} = \dfrac{1}{a^2\sec\theta\sqrt{\sec^2\theta - 1}} = \dfrac{1}{a^2\sec\theta\sqrt{\tan^2\theta}}$

$$= \dfrac{1}{a^2\sec\theta\tan\theta} = \dfrac{\cos\theta\cot\theta}{a^2}$$

Section 6.2

1. a) $\sin 15° = \sin(60° - 45°) = \sin 60° \cos 45° - \cos 60° \sin 45° = (\frac{\sqrt{3}}{2})(\frac{\sqrt{2}}{2}) - (\frac{1}{2})(\frac{\sqrt{2}}{2})$

$$= \frac{\sqrt{6} - \sqrt{2}}{4}$$

b) $\csc 15° = \dfrac{1}{\sin\theta} = \dfrac{4}{\sqrt{6} - \sqrt{2}} = \dfrac{4}{\sqrt{6} - \sqrt{2}} \cdot \dfrac{\sqrt{6} + \sqrt{2}}{\sqrt{6} + \sqrt{2}} = \dfrac{4(\sqrt{6} + \sqrt{2})}{6 - 2} = \sqrt{6} + \sqrt{2}$

5. a) $\tan(-\frac{5\pi}{12}) = -\tan(\frac{5\pi}{12}) = -\tan(\frac{\pi}{6} + \frac{\pi}{4}) = -\dfrac{\tan\frac{\pi}{6} + \tan\frac{\pi}{4}}{1 - \tan\frac{\pi}{6}\tan\frac{\pi}{4}} = -\dfrac{\frac{\sqrt{3}}{3} + 1}{1 - \frac{\sqrt{3}}{3}}$

$$= -2 - \sqrt{3}$$

b) $\cot(-\frac{5\pi}{12}) = \dfrac{1}{\tan(-\frac{5\pi}{12})} = \dfrac{1}{-2 - \sqrt{3}} = -2 + \sqrt{3}$

9. a) $\quad \cos 17° \cos 43° - \sin 17° \sin 43° = \cos(17° + 43°) = \cos 60° = \dfrac{1}{2}$

b) $\quad \sin 17° \cos 43° - \sin 17° \cos 43° = \sin(17° + 43°) = \sin 60° = \dfrac{\sqrt{3}}{2}$

13. $\cos(\theta + \dfrac{\pi}{2}) = \cos\theta \cos\dfrac{\pi}{2} - \sin\theta \sin\dfrac{\pi}{2} = \cos\theta \, (0) - \sin\theta(1) = -\sin\theta$

17. $2\cos(\theta + \dfrac{\pi}{4}) = 2[\cos\theta \cos\dfrac{\pi}{4} - \sin\theta \sin\dfrac{\pi}{4}] = 2[\cos\theta \, (\dfrac{\sqrt{2}}{2}) - \sin\theta(\dfrac{\sqrt{2}}{2})]$

$$= \sqrt{2}\cos\theta - \sqrt{2}\sin\theta$$

21. $\sin(\theta + \dfrac{\pi}{2}) - \cos(\theta + \dfrac{\pi}{3}) = [\sin\theta \cos\dfrac{\pi}{2} + \cos\theta \sin\dfrac{\pi}{2}] - [\cos\theta \cos\dfrac{\pi}{3} - \sin\theta \sin\dfrac{\pi}{3}]$

$$= [\sin\theta \cdot 0 + \cos\theta \cdot 1] - [\cos\theta \cdot \dfrac{1}{2} - \sin\theta \cdot \dfrac{\sqrt{3}}{2}] = \cos\theta - [\cos\theta\dfrac{1}{2} - \sin\theta\dfrac{\sqrt{3}}{2}]$$

$$= \dfrac{1}{2}\cos\theta + \dfrac{\sqrt{3}}{2}\sin\theta$$

25. $\tan(\theta + \dfrac{\pi}{4}) = \dfrac{\tan\theta + \tan\dfrac{\pi}{4}}{1 - \tan\theta \tan\dfrac{\pi}{4}} = \dfrac{1 + \tan\theta}{1 - \tan\theta}$

29. $\cot(\theta + \dfrac{\pi}{4}) = \dfrac{1}{\tan(\theta + \dfrac{\pi}{4})} = \dfrac{1 - \tan\theta \tan\dfrac{\pi}{4}}{\tan\theta + \tan\dfrac{\pi}{4}} = \dfrac{1 - \tan\theta}{\tan\theta + 1} \cdot \dfrac{\cot\theta}{\cot\theta} = \dfrac{\cot\theta - 1}{1 + \cot\theta}$

33. $\sin(x + y) + \sin(x - y) = (\sin x \cos y + \cos x \sin y) + (\sin x \cos y - \cos x \sin y) = 2\sin x \cos y$

37. $\cos\alpha = \sqrt{1 - \sin^2\alpha} = \sqrt{1 - (\dfrac{2}{3})^2} = \dfrac{\sqrt{5}}{3}$. Similarly, $\sin\beta = \sqrt{1 - \cos^2\beta} = \dfrac{2\sqrt{6}}{7}$.

a) $\quad \sin(\alpha + \beta) = \sin\alpha \cos\beta + \cos\alpha \sin\beta = (\dfrac{2}{3})(\dfrac{5}{7}) + (\dfrac{\sqrt{5}}{3})(\dfrac{2\sqrt{6}}{7}) = \dfrac{10 + 2\sqrt{30}}{21}$

b) $\quad \sin(\alpha - \beta) = \sin\alpha \cos\beta - \cos\alpha \sin\beta = (\dfrac{2}{3})(\dfrac{5}{7}) - (\dfrac{\sqrt{5}}{3})(\dfrac{2\sqrt{6}}{7}) = \dfrac{10 - 2\sqrt{30}}{21}$

c) $\quad \cos(\alpha + \beta) = \cos\alpha \cos\beta - \sin\alpha \sin\beta = (\dfrac{\sqrt{5}}{3})(\dfrac{5}{7}) - (\dfrac{2}{3})(\dfrac{2\sqrt{6}}{7}) = \dfrac{5\sqrt{5} - 4\sqrt{6}}{21}$

d) $\quad \cos(\alpha - \beta) = \cos\alpha \cos\beta + \sin\alpha \sin\beta = (\dfrac{\sqrt{5}}{3})(\dfrac{5}{7}) + (\dfrac{2}{3})(\dfrac{2\sqrt{6}}{7}) = \dfrac{5\sqrt{5} + 4\sqrt{6}}{21}$

41. First note that θ is in the first quadrant (since $\tan\theta > 0$ and $0 < \theta < \pi$). Since

$\sec\theta = \sqrt{1 + \tan^2\theta}$, $\sec\theta = \sqrt{1 + 2^2} = \sqrt{5}$. So $\cos\theta = \dfrac{1}{\sqrt{5}}$. Similarly,

$\csc\theta = \sqrt{1 + \cot^2\theta} = \sqrt{1 + (\tfrac{1}{2})^2} = \dfrac{\sqrt{5}}{2}$. So $\sin\theta = \dfrac{2}{\sqrt{5}}$.

a) $\tan2\theta = \dfrac{2\tan\theta}{1 - \tan^2\theta} = \dfrac{2(2)}{1 - 2^2} = -\dfrac{4}{3}$

b) $\cos(\dfrac{\pi}{3} + \theta) = \cos\dfrac{\pi}{3}\cos\theta - \sin\dfrac{\pi}{3}\sin\theta = (\dfrac{1}{2})(\dfrac{1}{\sqrt{5}}) - (\dfrac{\sqrt{3}}{2})(\dfrac{2}{\sqrt{5}}) = \dfrac{\sqrt{5} - 2\sqrt{15}}{10}$

c) $\sin(\theta + \dfrac{\pi}{4}) = \sin\theta\cos\dfrac{\pi}{4} + \cos\theta\sin\dfrac{\pi}{4} = (\dfrac{2}{\sqrt{5}})(\dfrac{\sqrt{2}}{2}) + (\dfrac{1}{\sqrt{5}})(\dfrac{\sqrt{2}}{2}) = \dfrac{3\sqrt{10}}{10}$

d) $\tan(\theta - \dfrac{\pi}{4}) = \dfrac{\tan\theta - \tan\dfrac{\pi}{4}}{1 + \tan\theta\,\tan\dfrac{\pi}{4}} = \dfrac{2 - 1}{1 + 2(1)} = \dfrac{1}{3}$

45. a) $\cos2\theta = \cos^2\theta - \sin^2\theta = \cos^2\theta - (1 - \cos^2\theta) = 2\cos^2\theta - 1$

  b) $\cos2\theta = \cos^2\theta - \sin^2\theta = (1 - \sin^2\theta) - \sin^2\theta = 1 - 2\sin^2\theta$

49. Refer to the box entitled "Linear Combinations of $\cos Bt$

and $\sin Bt$" on page 347. In this case $A = \sqrt{a^2 + b^2}$

$= \sqrt{(\sqrt{3})^2 + 1^2} = 2$. The point $(a, b) = (\sqrt{3}, 1)$ is on the

terminal side of angle $\phi = \tan^{-1}\dfrac{1}{\sqrt{3}} = \dfrac{\pi}{6}$. Thus,

$$\sqrt{3}\cos t + \sin t = 2\cos(t - \dfrac{\pi}{6}).$$

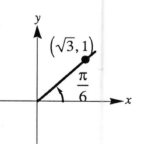

53. Start with a sketch of the angle λ with point $P(b,a)$ on the

terminal side of λ. We have

$\cos\lambda = \dfrac{b}{\sqrt{a^2 + b^2}}$ and $\sin\lambda = \dfrac{a}{\sqrt{a^2 + b^2}}$. Therefore

$A\sin(Bt + \lambda) = \sqrt{a^2 + b^2}\,[\sin Bt\cos\lambda + \cos Bt\sin\lambda]$

$= \sqrt{a^2 + b^2}\,[\sin Bt\,\dfrac{b}{\sqrt{a^2 + b^2}} + \cos Bt\,\dfrac{a}{\sqrt{a^2 + b^2}}]$

$= b\sin Bt + a\cos Bt$

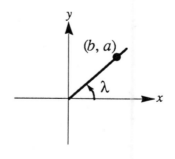

57. $\cos 3x = \cos(2x + x) = \cos 2x \cos x - \sin 2x \sin x = (2\cos^2 x - 1)\cos x - (2\sin x \cos x)\sin x$

$\quad = 2\cos^3 x - \cos x - 2\sin^2 x \cos x = 2\cos^3 x - \cos x - 2(1 - \cos^2 x)\cos x = 4\cos^3 x - 3\cos x$

61. $\dfrac{\cos(x + \Delta x) - \cos x}{\Delta x} = \dfrac{\cos x \cos \Delta x - \sin x \sin \Delta x - \cos x}{\Delta x}$

$\quad = \dfrac{\cos x (\cos \Delta x - 1) - \sin x \sin \Delta x}{\Delta x} = \cos x \cdot \dfrac{\cos \Delta x - 1}{\Delta x} - \sin x \cdot \dfrac{\sin \Delta x}{\Delta x}$

Section 6.3

1. a) $\quad \sin 15° = \sqrt{\dfrac{1 - \cos 30°}{2}} = \sqrt{\dfrac{1 - \sqrt{3}/2}{2}} = \dfrac{1}{2}\sqrt{2 - \sqrt{3}}$

$\quad$ b) $\quad \csc 15° = \dfrac{1}{\sin 15°} = \dfrac{2}{\sqrt{2 - \sqrt{3}}}$

5. a) $\quad \tan\left(-\dfrac{5\pi}{8}\right) = \dfrac{1 - \cos\left(-\dfrac{5\pi}{4}\right)}{\sin\left(-\dfrac{5\pi}{4}\right)} = \dfrac{1 + \sqrt{2}/2}{\sqrt{2}/2} = \sqrt{2} + 1$

$\quad$ b) $\quad \cot\left(-\dfrac{5\pi}{8}\right) = \dfrac{1}{\tan\left(-\dfrac{5\pi}{8}\right)} = \dfrac{1}{\sqrt{2} + 1} = \sqrt{2} - 1$

9. a) $\quad \tan\left[\dfrac{1}{2}\left(-\dfrac{3\pi}{4}\right)\right] = \dfrac{1 - \cos\left(-\dfrac{3\pi}{4}\right)}{\sin\left(-\dfrac{3\pi}{4}\right)} = \dfrac{1 + \sqrt{2}/2}{-\sqrt{2}/2} = -\sqrt{2} - 1$

$\quad$ b) $\quad \dfrac{1}{2}\tan\left(-\dfrac{3\pi}{4}\right) = \dfrac{1}{2}(1) = \dfrac{1}{2}$

13. $\cos\dfrac{2}{3}\alpha \sin\dfrac{4}{3}\alpha = \sin\dfrac{4}{3}\alpha \cos\dfrac{2}{3}\alpha = \dfrac{1}{2}\left[\sin\left(\dfrac{4}{3}\alpha + \dfrac{2}{3}\alpha\right) + \sin\left(\dfrac{4}{3}\alpha - \dfrac{2}{3}\alpha\right)\right]$

$\quad\quad\quad\quad\quad\quad\quad\quad\quad\quad\quad\quad\quad\quad\quad\quad\quad = \dfrac{1}{2}\left[\sin 2\alpha + \sin\dfrac{2}{3}\alpha\right]$

17. $\cos 2x + \cos 3x = 2\cos\left(\dfrac{2x + 3x}{2}\right)\cos\left(\dfrac{2x - 3x}{2}\right) = 2\cos\left(\dfrac{5x}{2}\right)\cos\left(\dfrac{x}{2}\right)$

21. $\cos\left(x - \dfrac{3\pi}{2}\right) + \cos\left(\dfrac{\pi}{2} - x\right)$

$\quad = 2\cos\left(\dfrac{\left(x - \dfrac{3\pi}{2}\right) + \left(\dfrac{\pi}{2} - x\right)}{2}\right)\cos\left(\dfrac{\left(x - \dfrac{3\pi}{2}\right) - \left(\dfrac{\pi}{2} - x\right)}{2}\right) = 2\cos\left(-\dfrac{\pi}{2}\right)\cos(x - \pi) = 0$

25. $\sin^2\dfrac{\alpha}{2} - \cos^2\dfrac{\alpha}{2} = -\left(\cos^2\dfrac{\alpha}{2} - \sin^2\dfrac{\alpha}{2}\right) = -\cos\alpha$

29. $\dfrac{\sin2\theta - \sin6\theta}{\cos6\theta - \cos2\theta} = \dfrac{2\cos4\theta\sin(-2\theta)}{-2\sin4\theta\sin2\theta} = \cot4\theta$

33. $\dfrac{\sin\alpha + \sin5\alpha + \sin3\alpha}{\cos\alpha + \cos5\alpha + \cos3\alpha} = \dfrac{2\sin3\alpha\cos2\alpha + \sin3\alpha}{2\cos3\alpha\cos2\alpha + \cos3\alpha} = \dfrac{\sin3\alpha(2\cos2\alpha + 1)}{\cos3\alpha(2\cos2\alpha + 1)} = \tan3\alpha$

37. First note that ϕ is in the third quadrant (since $\cos\phi < 0$ and $\pi < \phi < 2\pi$). Furthermore,

$\pi < \phi < \dfrac{3\pi}{2} \Rightarrow \dfrac{\pi}{2} < \dfrac{\phi}{2} < \dfrac{3\pi}{4}$.

a) $\cos\dfrac{1}{2}\phi = -\sqrt{\dfrac{1 + \cos\phi}{2}} = -\sqrt{\dfrac{1 + (-2/\sqrt{5})}{2}} = -\dfrac{1}{10}\sqrt{50 - 20\sqrt{5}}$

b) $\cos2\phi = 2\cos^2\phi - 1 = 2(-\dfrac{2}{\sqrt{5}})^2 - 1 = \dfrac{3}{5}$

c) $\sec\dfrac{1}{2}\phi = \dfrac{1}{\cos\dfrac{1}{2}\phi} = \dfrac{-10}{\sqrt{50 - 20\sqrt{5}}}$

41. $f(t) = \cos 2t \cos t = \dfrac{1}{2}[\cos(2t + t) + \cos(2t - t)] = \dfrac{1}{2}\cos 3t + \dfrac{1}{2}\cos t$.

To graph $f(t) = \dfrac{1}{2}\cos 3t + \dfrac{1}{2}\cos t$, graph $y = \dfrac{1}{2}\cos 3t$ and $y = \dfrac{1}{2}\cos t$ on the same axes. Then use the technique of adding y-coordinates to graph $y = f(t)$.

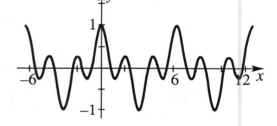

45. $f(t) = \sin\dfrac{1}{4}t \cos\dfrac{3}{4}t = \dfrac{1}{2}[\sin(\dfrac{1}{4}t + \dfrac{3}{4}t) + \sin(\dfrac{1}{4}t - \dfrac{3}{4}t)] = \dfrac{1}{2}[\sin t + \sin(-\dfrac{1}{2}t)]$

$= \dfrac{1}{2}\sin t - \dfrac{1}{2}\sin(\dfrac{1}{2}t)$

To graph $f(t) = \dfrac{1}{2}\sin t - \dfrac{1}{2}\sin\dfrac{1}{2}t$, graph $y = \dfrac{1}{2}\sin t$ and $y = -\dfrac{1}{2}\sin\dfrac{1}{2}t$ on the same axes. Then use the technique of adding y-coordinates to graph $y = f(t)$.

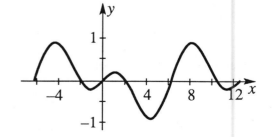

49. $\cos^3\theta = \cos^2\theta\,\cos\theta = \dfrac{1 + \cos2\theta}{2}\cos\theta = \dfrac{1}{2}[\cos\theta + \cos2\theta\cos\theta]$

$= \dfrac{1}{2}[\cos\theta + \dfrac{1}{2}(\cos3\theta + \cos\theta)] = \dfrac{1}{2}[\dfrac{3}{2}\cos\theta + \dfrac{1}{2}\cos3\theta] = \dfrac{3}{4}\cos\theta + \dfrac{1}{4}\cos3\theta$

53. $\frac{1}{2}[\cos(\alpha + \beta) + \cos(\alpha - \beta)] = \frac{1}{2}[\cos\alpha\cos\beta - \sin\alpha\sin\beta + \cos\alpha\cos\beta + \sin\alpha\sin\beta]$

$$= \frac{1}{2}[2\cos\alpha\cos\beta] = \cos\alpha\cos\beta$$

$\frac{1}{2}[\cos(\alpha - \beta) - \cos(\alpha + \beta)] = \frac{1}{2}[\cos\alpha\cos\beta + \sin\alpha\sin\beta - (\cos\alpha\cos\beta - \sin\alpha\sin\beta)]$

$$= \frac{1}{2}[2\sin\alpha\sin\beta] = \sin\alpha\sin\beta$$

57. $\dfrac{1}{1 - \sin\theta} = \dfrac{1}{1 - \dfrac{2u}{1 + u^2}} = \dfrac{1}{1 - \dfrac{2u}{1 + u^2}} \cdot \dfrac{1 + u^2}{1 + u^2} = \dfrac{1 + u^2}{(1 + u^2) - 2u} = \dfrac{1 + u^2}{(u - 1)^2}$

61. $\dfrac{\sec\theta}{\tan\theta - \csc\theta} = \dfrac{\dfrac{1}{\cos\theta}}{\dfrac{\sin\theta}{\cos\theta} - \dfrac{1}{\sin\theta}} = \dfrac{\dfrac{1 + u^2}{1 - u^2}}{\dfrac{2u}{1 - u^2} - \dfrac{1 + u^2}{2u}}$

$$= \dfrac{\dfrac{1 + u^2}{1 - u^2}}{\dfrac{2u}{1 - u^2} - \dfrac{1 + u^2}{2u}} \cdot \dfrac{2u(1 - u^2)}{2u(1 - u^2)} = \dfrac{2u(1 + u^2)}{u^4 + 4u^2 - 1}$$

Section 6.4

---

1. a) $\sin\theta = \dfrac{1}{2}$ for all angles coterminal with $\dfrac{\pi}{6}$ or $\dfrac{5\pi}{6}$. Thus $\theta = \dfrac{\pi}{6} \pm 2n\pi$ or $\dfrac{5\pi}{6} \pm 2n\pi$

  b) $-\dfrac{\pi}{2} < \sin^{-1}x < \dfrac{\pi}{2} \Rightarrow \sin^{-1}\dfrac{1}{2}$ is the unique angle θ such that $\sin\theta = \dfrac{1}{2}$, and

between $-\dfrac{\pi}{2}$ and $\dfrac{\pi}{2}$. Therefore $\sin^{-1}\dfrac{1}{2} = \dfrac{\pi}{6}$.

5. a) $\cos\theta = -\dfrac{1}{\sqrt{2}}$ for all angles coterminal with $\dfrac{3\pi}{4}$ or $\dfrac{5\pi}{4}$. Thus $\theta = \dfrac{3\pi}{4} \pm 2n\pi$

  or $\dfrac{5\pi}{4} \pm 2n\pi$

  b) $0 < \cos^{-1}x < \pi \Rightarrow \cos^{-1}\left(-\dfrac{1}{\sqrt{2}}\right)$ is the unique angle θ such that $\cos\theta = -\dfrac{1}{\sqrt{2}}$,

and between 0 and π. Therefore $\cos^{-1}\left(-\dfrac{1}{\sqrt{2}}\right) = \dfrac{3\pi}{4}$.

9. arccos $\left(-\frac{\sqrt{3}}{2}\right)$ is the unique angle θ such that $\cos\theta = -\frac{\sqrt{3}}{2}$, and between 0 and π.

Therefore $\cos^{-1}\left(-\frac{\sqrt{3}}{2}\right) = \frac{5\pi}{6}$.

13. arccot 1 is the unique angle θ such that $\cot\theta = 1$, and between 0 and π. Therefore

arccot $1 = \frac{\pi}{4}$.

17. $\csc^{-1}(-1)$ is the unique angle θ such that $\csc\theta = -1$, and between $-\frac{\pi}{2}$ and $\frac{\pi}{2}$,

excluding 0. Therefore $\csc^{-1}(-1) = -\frac{\pi}{2}$.

21. Keystrokes: .56 $\boxed{+/-}$ $\boxed{\text{inv}}$ $\boxed{\cos}$ $\boxed{=}$ Display: 2.17

25. Keystrokes: 500 $\boxed{\text{inv}}$ $\boxed{\tan}$ $\boxed{=}$ Display: 1.57

29. Keystrokes: 1.3 $\boxed{1/x}$ $\boxed{\text{inv}}$ $\boxed{\sin}$ $\boxed{=}$ Display: 0.88

33. a) We need to determine the number in the interval $[-\frac{\pi}{2}, \frac{\pi}{2}]$ whose tangent is the same as $\tan\frac{2\pi}{9}$.

b) We need to determine the number in the interval $[0, \pi]$ whose cotangent is the same as $\cot(-2)$. The answer is not -2, since -2 is not in $[0, \pi]$.

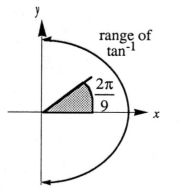

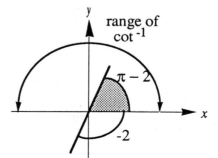

The answer is clearly $\frac{2\pi}{9}$.

The value that does satisfy the requirement is $-2 + \pi$. Thus, $\cot^{-1}(\cot(-2)) = \pi - 2$.

37. a)   Let $\theta = \tan^{-1}(-\frac{2}{7})$. It follows that $\tan\theta = -\frac{2}{7}$. So $\cot(\tan^{-1}(-\frac{2}{7})) = \cot\theta = -\frac{7}{2}$.

   b)   Let $\theta = \cot^{-1}(\frac{9}{14})$. It follows that $\cot\theta = \frac{9}{14}$. So $\tan(\cot^{-1}(\frac{9}{14})) = \tan\theta = \frac{14}{9}$.

41. Let $\theta = \cos^{-1}(\frac{2}{7})$. Then $0 \le \theta \le \pi$ and $\cos\theta = \frac{2}{7}$. So $\sin[\frac{\pi}{4} + \cos^{-1}(\frac{2}{7})] = \sin[\frac{\pi}{4} + \theta]$

$= \sin\frac{\pi}{4}\cos\theta + \cos\frac{\pi}{4}\sin\theta$. We know $\cos\theta$ and can find $\sin\theta$:

$\sin\theta = \sqrt{1 - \cos^2\theta} = \sqrt{1 - (\frac{2}{7})^2} = \frac{3\sqrt{5}}{7}$. So,

$\sin\frac{\pi}{4}\cos\theta + \cos\frac{\pi}{4}\sin\theta = (\frac{\sqrt{2}}{2})(\frac{2}{7}) + (\frac{\sqrt{2}}{2})(\frac{3\sqrt{5}}{7}) = \frac{2\sqrt{2} + 3\sqrt{10}}{14}$.

Thus, $\sin[\frac{\pi}{4} + \cos^{-1}(\frac{2}{7})] = \frac{2\sqrt{2} + 3\sqrt{10}}{14}$.

45. Let $\theta = \cos^{-1}(\frac{2}{7})$. Then

$\sin[\frac{1}{2}\cos^{-1}(\frac{2}{7})] = \sin[\frac{1}{2}\theta] = \sqrt{\frac{1 - \cos\theta}{2}} = \sqrt{\frac{1 - (2/7)}{2}} = \frac{\sqrt{70}}{14}$

49. First graph $f(x) = \cos^{-1}x$ (see page 363 in the text). Recall from Section 2.5 that the graph of $y = f(\frac{x}{2}) = \cos^{-1}\frac{x}{2}$ is an expansion of $y = f(x)$ away from the y-axis by a factor of 2.

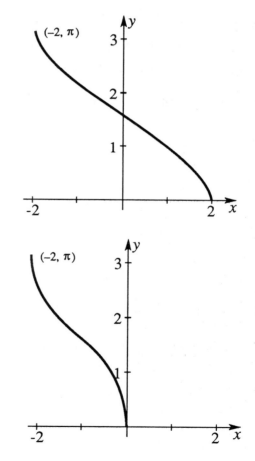

53. First graph $f(x) = \cos^{-1}x$ (see page 363 in the text). Recall from Section 2.3 that the graph of $y = f(x + 1)$ $= \cos^{-1}(x + 1)$ is a translation of $y = f(x)$ one unit to the left.

57. First graph $f(x) = \tan^{-1}x$ (see page 365 in the text). Recall from Section 2.5 that the graph of $y = -f(x)$ $= -\tan^{-1}x$ is a reflection of $y = f(x)$ through the x-axis. Finally, to graph

$$y = \frac{\pi}{2} - \tan^{-1}x \Leftrightarrow y - \frac{\pi}{2} = -\tan^{-1}x,$$

translate $\frac{\pi}{2}$ units up.

61. $\cos(\sec^{-1}(\frac{2n}{3})) = \cos(\cos^{-1}(\frac{3}{2n})) = \frac{3}{2n}$

65. By sketching $y = \sec^{-1}x$ and $y = \csc^{-1}x$ on the same axes, we can add y-coordinates to sketch $y = \sec^{-1}x + \csc^{-1}x$. From the graph, it appears that $\sec^{-1}x + \csc^{-1}x = \frac{\pi}{2}$. To prove this, let $\theta = \sec^{-1}x$. Then $x = \sec\theta$, and since $\sec\theta = \csc(\frac{\pi}{2} - \theta)$,

$$\Rightarrow x = \csc(\frac{\pi}{2} - \theta) \Rightarrow \csc^{-1}x = \frac{\pi}{2} - \theta \Rightarrow \csc^{-1}x = \frac{\pi}{2} - \sec^{-1}x$$

$$\Rightarrow \sec^{-1}x + \csc^{-1}x = \frac{\pi}{2}.$$

69. a)   Let $\beta = \tan^{-1}(\frac{1}{x}) \Rightarrow \frac{1}{x} = \tan\beta \Rightarrow x = \tan(\frac{\pi}{2} - \beta) \Rightarrow \tan^{-1}x = \frac{\pi}{2} - \beta$

$\Rightarrow \tan^{-1}x = \frac{\pi}{2} - \tan^{-1}(\frac{1}{x})$. It follows that $\tan^{-1}x + \tan^{-1}(\frac{1}{x}) = \frac{\pi}{2}$, for $x > 0$.

b)   We use the fact that $y = \tan^{-1}x$ is an odd function. (This is suggested by the fact that the graph of $y = \tan^{-1}x$ on page 365: the proof is left to you.) Let $x = -y$. Because x is a negative number, y must be positive. Thus,

$$\tan^{-1}x + \tan^{-1}(\frac{1}{x}) = \tan^{-1}(-y) + \tan^{-1}(\frac{1}{-y}) = -\tan^{-1}(y) - \tan^{-1}(\frac{1}{y})$$

$$= -(\tan^{-1}y + \tan^{-1}(\frac{1}{y})). \text{ This last expression, by part a) is } -\frac{\pi}{2}. \text{ Thus,}$$

$$\tan^{-1}x + \tan^{-1}(\frac{1}{x}) = -\frac{\pi}{2}.$$

73. The result of Problem 72 that we need is $\tan[4\tan^{-1}\frac{1}{5}] = \dfrac{4(1/5) - 4(1/5)^3}{1 - 6(1/5)^2 + (1/5)^4} = \dfrac{120}{119}$.

Now, let $\alpha = \tan^{-1}(\frac{1}{5})$ and $\beta = \tan^{-1}(\frac{1}{239})$. We have established that $\tan 4\alpha = \dfrac{120}{119}$.

Thus, $\tan(4\alpha - \beta) = \dfrac{\tan 4\alpha - \tan\beta}{1 + \tan 4\alpha \, \tan\beta} = \dfrac{(120/119) - (1/239)}{1 + (120/119)(1/239)} = 1$. Since

$0 < \tan^{-1}\frac{1}{5} < \dfrac{\pi}{4}$, and $0 < \tan^{-1}\dfrac{1}{239} < \dfrac{\pi}{4}$, and $\alpha > \beta$, $0 < 4\alpha - \beta < \pi$. This last inequality,

together with $\tan(4\alpha - \beta) = 1 \Rightarrow 4\alpha - \beta = \dfrac{\pi}{4}$.

Section 6.5

1. a) $2\sin x = \sqrt{3} \Rightarrow \sin x = \dfrac{\sqrt{3}}{2} \Rightarrow x = \dfrac{\pi}{3} \pm 2n\pi$ or $\dfrac{2\pi}{3} \pm 2n\pi$

    b) Those solutions in part a) that are in the interval $[0, 2\pi)$ are $x = \dfrac{\pi}{3}$ or $\dfrac{2\pi}{3}$.

5. a) $\sqrt{2}\sec x = -2 \Rightarrow \sec x = \dfrac{-2}{\sqrt{2}} \Rightarrow \cos x = -\dfrac{\sqrt{2}}{2} \Rightarrow x = \dfrac{3\pi}{4}$ or $\dfrac{5\pi}{4}$

    b) $\sqrt{2}\sec x = 2 \Rightarrow \sec x = \dfrac{2}{\sqrt{2}} \Rightarrow \cos x = \dfrac{\sqrt{2}}{2} \Rightarrow x = \dfrac{\pi}{4}$ or $\dfrac{7\pi}{4}$.

9. $2\sin^2 x - 1 = 0 \Rightarrow \sin^2 x = \dfrac{1}{2} \Rightarrow \sin x = \pm\dfrac{1}{\sqrt{2}} \Rightarrow x = \dfrac{\pi}{4}, \dfrac{3\pi}{4}, \dfrac{5\pi}{4}$, or $\dfrac{7\pi}{4}$

13. $2\sin^2 x - \sin x = 0 \Rightarrow \sin x(2\sin x - 1) = 0 \Rightarrow \sin x = 0$ or $2\sin x - 1 = 0 \Rightarrow \sin x = 0$ or $\sin x = \dfrac{1}{2}$. The solutions to $\sin x = 0$ are $x = 0$ or π. The solutions to $\sin x = \dfrac{1}{2}$ are $x = \dfrac{\pi}{6}$ or $\dfrac{5\pi}{6}$.

17. $3\sin x - 5 = 1 \Rightarrow \sin x = 2$. There is no solution to this equation, since for all values of x, $-1 \le \sin x \le 1$.

21. $\cos(x + \dfrac{\pi}{3}) = -1 \Rightarrow (x + \dfrac{\pi}{3}) = \pi \Rightarrow x = \dfrac{2\pi}{3}$

25. First, note that $0 \le x < 2\pi \Rightarrow 0 \le 2x < 4\pi$. Now, $2\sin 2x = 1 \Rightarrow \sin 2x = \dfrac{1}{2}$

$\Rightarrow 2x = \dfrac{\pi}{6}, \dfrac{5\pi}{6}, \dfrac{13\pi}{6}$, or $\dfrac{17\pi}{6}$. Dividing both sides of each equation by 2 gives

$x = \dfrac{\pi}{12}, \dfrac{5\pi}{12}, \dfrac{13\pi}{12}$, or $\dfrac{17\pi}{12}$.

29. First, note that $-\pi \le x \le \pi \Rightarrow -2\pi \le 2x \le 2\pi$. Now, $\sqrt{2}\sec 2x = 2 \Rightarrow \sec 2x = \dfrac{2}{\sqrt{2}}$

$\Rightarrow \cos 2x = \dfrac{\sqrt{2}}{2} \Rightarrow 2x = -\dfrac{7\pi}{4}, -\dfrac{\pi}{4}, \dfrac{\pi}{4}$, or $\dfrac{7\pi}{4} \Rightarrow x = -\dfrac{7\pi}{8}, -\dfrac{\pi}{8}, \dfrac{\pi}{8}$, or $\dfrac{7\pi}{8}$.

33. $\tan^2 x = \dfrac{\sin x}{1 - \sin x} \Rightarrow \tan^2 x = \dfrac{\sin x}{1 - \sin x} \cdot \dfrac{1 + \sin x}{1 + \sin x}$

$\Rightarrow \tan^2 x = \dfrac{\sin x(1 + \sin x)}{1 - \sin^2 x} \Rightarrow \dfrac{\sin^2 x}{\cos^2 x} = \dfrac{\sin x(1 + \sin x)}{\cos^2 x} \Rightarrow \sin^2 x = \sin x(1 + \sin x)$

$\Rightarrow \sin^2 x = \sin x + \sin^2 x \Rightarrow 0 = \sin x \Rightarrow x = 0$ or π

37. Using a double-angle formula gives $\cos 2t = 2\cos t \Rightarrow 2\cos^2 t - 1 = 2\cos t$

$\Rightarrow 2\cos^2 t - 2\cos t - 1 = 0$. This last equation is quadratic in $\cos t$, but does not factor. According to the quadratic formula,

$\cos t = \dfrac{-b \pm \sqrt{b^2 - 4ac}}{2a} = \dfrac{-(-2) \pm \sqrt{(-2)^2 - 4(2)(-1)}}{2(2)} = \dfrac{1 \pm \sqrt{3}}{2}$. Since

$\cos t$ cannot exceed 1, we discard the case of $\cos t = \dfrac{1 + \sqrt{3}}{2}$. On the interval $[-\pi, \pi]$ the

solutions to $\cos t = \dfrac{1 - \sqrt{3}}{2}$ are

$t = \cos^{-1}\left(\dfrac{1 - \sqrt{3}}{2}\right) \approx 1.9$ or $t = -\cos^{-1}\left(\dfrac{1 - \sqrt{3}}{2}\right) \approx -1.9$

Keystrokes: 1 $\boxed{-}$ 3 $\boxed{\sqrt{}}$ $\boxed{=}$ $\boxed{\div}$ 2 $\boxed{=}$ $\boxed{\text{inv}}$ $\boxed{\cos}$ (The other value is the opposite of this.)

41. Since $\sin 2t = 2\sin t\cos t$, $4\cos t\sin t = 1 \Rightarrow 2\sin 2t = 1 \Rightarrow \sin 2t = \dfrac{1}{2} \Rightarrow 2t = \dfrac{\pi}{6}$ or $\dfrac{5\pi}{6}$

$\Rightarrow t = \dfrac{\pi}{12}$ or $\dfrac{5\pi}{12}$.

45. First consider the left hand side of the equation, $-\sqrt{3}\cos x + \sin x$. Refer to the box entitled "Linear Combinations of $\cos Bt$ and $\sin Bt$" on page 347. In this case, $A = \sqrt{a^2 + b^2}$ $= \sqrt{(-\sqrt{3})^2 + 1^2} = 2$. The point $(a, b) = (-\sqrt{3}, 1)$ is on the terminal side of angle $\phi = \dfrac{5\pi}{6}$. Thus, $-\sqrt{3}\cos t + \sin t = 2\cos(t - \dfrac{5\pi}{6})$. Therefore, solving

$-\sqrt{3}\cos t + \sin t = 1$ is equivalent to solving $2\cos(t - \dfrac{5\pi}{6}) = 1 \Rightarrow \cos(t - \dfrac{5\pi}{6}) = \dfrac{1}{2}$

$\Rightarrow t - \dfrac{5\pi}{6} = -\dfrac{\pi}{3}$ or $\dfrac{\pi}{3} \Rightarrow t = \dfrac{\pi}{2}$ or $\dfrac{7\pi}{6}$.

49. $\ln(\sin t) - \ln(\cos t) = 0 \Rightarrow \ln(\dfrac{\sin t}{\cos t}) = 0 \Rightarrow \ln(\tan t) = 0 \Rightarrow \tan t = 1 \Rightarrow t = \dfrac{\pi}{4}$

53. $2\cos 2t - \sin t = 0 \Rightarrow 2(1 - 2\sin^2 t) - \sin t = 0 \Rightarrow 4\sin^2 t + \sin t - 2 = 0$. This last equation is quadratic in $\sin t$, but does not factor. According to the quadratic formula,

$$\sin t = \frac{-b \pm \sqrt{b^2 - 4ac}}{2a} = \frac{-(1) \pm \sqrt{(1)^2 - 4(4)(-2)}}{2(4)} = \frac{-1 \pm \sqrt{33}}{8}.$$

On the interval $[0, 2\pi)$ the solutions to $\sin t$ $= \dfrac{-1 + \sqrt{33}}{8}$ are

$t = \sin^{-1}(\dfrac{-1 + \sqrt{33}}{8}) \approx 0.6349$ or

$t = \pi - \sin^{-1}(\dfrac{-1 + \sqrt{33}}{8}) \approx 2.5067$.

The solutions to $\sin t = \dfrac{-1 - \sqrt{33}}{8}$ are

$t = 2\pi + \sin^{-1}(\dfrac{-1 - \sqrt{33}}{8}) \approx 5.2802$

or $t = \pi - \sin^{-1}(\dfrac{-1 - \sqrt{33}}{8}) \approx 4.1446$. The graph of $y = \sin 2t + \cos t$ is shown.

Miscellaneous Exercises for Chapter 6

1. $\sin\theta\tan\theta + \cos\theta = \sin\theta \cdot \dfrac{\sin\theta}{\cos\theta} + \cos\theta = \dfrac{\sin^2\theta}{\cos\theta} + \dfrac{\cos^2\theta}{\cos\theta}$

$= \dfrac{\sin^2\theta + \cos^2\theta}{\cos\theta} = \dfrac{1}{\cos\theta} = \sec\theta$

5. $\cos(\frac{\pi}{2} - \theta)\cot\theta = \sin\theta\cot\theta = \sin\theta \cdot \dfrac{\cos\theta}{\sin\theta} = \cos\theta$

9. $\dfrac{\csc\theta(1 + \cos\theta)(1 - \cos\theta)}{\cos\theta} = \dfrac{\csc\theta(1 - \cos^2\theta)}{\cos\theta} = \dfrac{\csc\theta\sin^2\theta}{\cos\theta} = \dfrac{\sin\theta}{\cos\theta} = \tan\theta$

13. $\cos\theta(\sec\theta - \cos\theta) = \cos\theta\sec\theta - \cos^2\theta = 1 - \cos^2\theta = \sin^2\theta$

17. Consider the left side: $\dfrac{\tan\theta}{\tan\theta + 1} = \dfrac{\dfrac{\sin\theta}{\cos\theta}}{\dfrac{\sin\theta}{\cos\theta} + 1} = \dfrac{\dfrac{\sin\theta}{\cos\theta}}{\dfrac{\sin\theta}{\cos\theta} + 1} \cdot \dfrac{\cos\theta}{\cos\theta} = \dfrac{\sin\theta}{\sin\theta + \cos\theta}$

21. $\cos^4\theta - \sin^4\theta = (\cos^2\theta + \sin^2\theta)(\cos^2\theta - \sin^2\theta) = (1)(\cos^2\theta - \sin^2\theta) = \cos2\theta$

25. $\dfrac{(\sin^3\theta + \cos^3\theta)}{1 - \sin\theta\cos\theta} = \dfrac{(\sin\theta + \cos\theta)(\sin^2\theta - \sin\theta\cos\theta + \cos^2\theta)}{1 - \sin\theta\cos\theta}$

$= \dfrac{(\sin\theta + \cos\theta)(\sin^2\theta + \cos^2\theta - \sin\theta\cos\theta)}{1 - \sin\theta\cos\theta} = \dfrac{(\sin\theta + \cos\theta)(1 - \sin\theta\cos\theta)}{1 - \sin\theta\cos\theta}$

$= \sin\theta + \cos\theta$

29. $\dfrac{\cos\theta + \sin\theta}{\cos\theta - \sin\theta} = \dfrac{(\cos\theta + \sin\theta)(\cos\theta - \sin\theta)}{(\cos\theta - \sin\theta)(\cos\theta - \sin\theta)}$

$= \dfrac{\cos^2\theta - \sin^2\theta}{\cos^2\theta - 2\sin\theta\cos\theta + \sin^2\theta} = \dfrac{\cos2\theta}{1 - 2\sin\theta\cos\theta} = \dfrac{\cos2\theta}{1 - \sin2\theta}$

33. We first use the sum to product identities:

$\dfrac{\sin(2\alpha - \beta) + \sin\beta}{\cos(2\alpha - \beta) + \cos\beta} = \dfrac{2\sin\alpha\cos(\alpha - \beta)}{2\cos\alpha\cos(\alpha - \beta)} = \dfrac{\sin\alpha}{\cos\alpha} = \tan\alpha$

37. $\dfrac{\cos\frac{1}{2}\theta + \sin\frac{1}{2}\theta}{\cos\frac{1}{2}\theta - \sin\frac{1}{2}\theta} = \dfrac{(\cos\frac{1}{2}\theta + \sin\frac{1}{2}\theta)(\cos\frac{1}{2}\theta + \sin\frac{1}{2}\theta)}{(\cos\frac{1}{2}\theta - \sin\frac{1}{2}\theta)(\cos\frac{1}{2}\theta + \sin\frac{1}{2}\theta)}$

$= \dfrac{\cos^2\frac{1}{2}\theta + 2\sin\frac{1}{2}\theta\cos\frac{1}{2}\theta + \sin^2\frac{1}{2}\theta}{\cos^2\frac{1}{2}\theta - \sin^2\frac{1}{2}\theta} = \dfrac{1 + 2\sin\frac{1}{2}\theta\cos\frac{1}{2}\theta}{\cos\theta} = \dfrac{1 + \sin\theta}{\cos\theta}$

41. $\cot\dfrac{\theta}{2} - \tan\dfrac{\theta}{2} = \dfrac{1}{\tan\dfrac{\theta}{2}} - \tan\dfrac{\theta}{2} = \dfrac{1}{\left(\dfrac{\sin\theta}{1 + \cos\theta}\right)} - \dfrac{1 - \cos\theta}{\sin\theta}$

$$= \dfrac{1 + \cos\theta}{\sin\theta} - \dfrac{1 - \cos\theta}{\sin\theta} = \dfrac{2\cos\theta}{\sin\theta} = 2\cot\theta$$

45. $-\cos\left(\dfrac{\alpha}{2} + \dfrac{3\beta}{2} + \dfrac{\gamma}{2}\right) = -\cos\left(\dfrac{\alpha}{2} + \dfrac{\beta}{2} + \beta + \dfrac{\gamma}{2}\right)$

$$= -\cos\left(\dfrac{\pi}{2} + \beta\right) = -\cos\left(\dfrac{\pi}{2} - (-\beta)\right) = -\sin(-\beta) = \sin\beta$$

49. Since $\cos\alpha$ is positive and $0 < \alpha < \pi$, α is in the first quadrant. Since $\cos\beta$ is negative and $0 < \beta < \pi$, β is in the second quadrant. Now,

$\sin\alpha = \sqrt{1 - \cos^2\alpha} = \sqrt{1 - \left(\dfrac{2}{5}\right)^2} = \dfrac{\sqrt{21}}{5}$. Similarly, $\sin\beta = \dfrac{\sqrt{39}}{8}$.

a) $\sin(\alpha + \beta) = \sin\alpha\,\cos\beta + \cos\alpha\,\sin\beta = \left(\dfrac{\sqrt{21}}{5}\right)\left(\dfrac{-5}{8}\right) + \left(\dfrac{2}{5}\right)\left(\dfrac{\sqrt{39}}{8}\right) = \dfrac{2\sqrt{39} - 5\sqrt{21}}{40}$

b) $\sin(\alpha - \beta) = \sin\alpha\,\cos\beta - \cos\alpha\,\sin\beta = \left(\dfrac{\sqrt{21}}{5}\right)\left(\dfrac{-5}{8}\right) - \left(\dfrac{2}{5}\right)\left(\dfrac{\sqrt{39}}{8}\right) = \dfrac{-2\sqrt{39} - 5\sqrt{21}}{40}$

c) $\cos(\alpha + \beta) = \cos\alpha\,\cos\beta - \sin\alpha\,\sin\beta = \left(\dfrac{2}{5}\right)\left(\dfrac{-5}{8}\right) - \left(\dfrac{\sqrt{21}}{5}\right)\left(\dfrac{\sqrt{39}}{8}\right) = \dfrac{-10 - 3\sqrt{91}}{40}$

d) $\cos(\alpha - \beta) = \cos\alpha\,\cos\beta + \sin\alpha\,\sin\beta = \left(\dfrac{2}{5}\right)\left(\dfrac{-5}{8}\right) + \left(\dfrac{\sqrt{21}}{5}\right)\left(\dfrac{\sqrt{39}}{8}\right) = \dfrac{-10 + 3\sqrt{91}}{40}$

53. a) $\sin\theta = \sqrt{1 - \cos^2\theta} = \sqrt{1 - x^2}$

b) $\cos\left(\dfrac{\pi}{2} - \theta\right) = \sin\theta = \sqrt{1 - \cos^2\theta} = \sqrt{1 - x^2}$

c) $\cot\theta = \dfrac{\cos\theta}{\sin\theta} = \dfrac{x}{\sqrt{1 - x^2}}$

d) $\cos(\pi - \theta) = \cos\pi\,\cos\theta + \sin\pi\,\sin\theta = (-1)\cos\theta + (0)\sin\theta = -\cos\theta = -x$

57. First note that $\frac{\pi}{2} < \theta < \pi \Rightarrow \frac{\pi}{4} < \frac{1}{2}\theta < \frac{\pi}{2}$.

a) $\sin\frac{1}{2}\theta = \sqrt{\dfrac{1 - \cos\theta}{2}} = \sqrt{\dfrac{1 - (-7/25)}{2}} = \dfrac{4}{5}$

b) $\cos\frac{1}{2}\theta = \sqrt{\dfrac{1 + \cos\theta}{2}} = \sqrt{\dfrac{1 + (-7/25)}{2}} = \dfrac{3}{5}$

c) $2\cos\frac{1}{2}(\theta - \frac{\pi}{3}) = 2\cos(\frac{1}{2}\theta - \frac{\pi}{6}) = 2[\cos\frac{1}{2}\theta \cos\frac{\pi}{6} + \sin\frac{1}{2}\theta \sin\frac{\pi}{6}]$

$$= 2[(\frac{3}{5})(\frac{\sqrt{3}}{2}) + (\frac{4}{5})(\frac{1}{2})] = \frac{3\sqrt{3} + 4}{5}$$

d) $2\sin\frac{1}{2}(\theta - \frac{\pi}{2}) = 2\sin(\frac{1}{2}\theta - \frac{\pi}{4}) = 2[\sin\frac{1}{2}\theta \cos\frac{\pi}{4} - \cos\frac{1}{2}\theta \sin\frac{\pi}{4}]$

$$= 2[(\frac{4}{5})(\frac{\sqrt{2}}{2}) - (\frac{3}{5})(\frac{\sqrt{2}}{2})] = \frac{\sqrt{2}}{5}$$

61. Since $\cos(\frac{\pi}{2} - \theta) = \sin\theta \Rightarrow \cos[\frac{\pi}{2} - \sin^{-1}(\frac{3}{5})] = \sin[\sin^{-1}(\frac{3}{5})] = \frac{3}{5}$

65. Since $\sin\frac{1}{2}\theta = \pm\sqrt{\dfrac{1 - \cos\theta}{2}} \Rightarrow \sin[\frac{1}{2}\cos^{-1}(\frac{3}{5})] = \sqrt{\dfrac{1 - \cos(\cos^{-1}(\frac{3}{5}))}{2}}$

$= \sqrt{\dfrac{1 - \frac{3}{5}}{2}} = \dfrac{\sqrt{5}}{5}$ (We know that $\theta = \frac{1}{2}\cos^{-1}(\frac{3}{5})$ is in the first quadrant

$\Rightarrow \sin[\frac{1}{2}\cos^{-1}(\frac{3}{5})]$ is positive.)

69. $4\sin^2\theta - 3 = 0 \Rightarrow \sin\theta = \pm\frac{\sqrt{3}}{2} \Rightarrow \theta = \frac{\pi}{3}$ or $\frac{2\pi}{3}$

73. $\cos^2\theta - \sin^2\theta = -1 \Rightarrow \cos 2\theta = -1 \Rightarrow 2\theta = -\pi$ or $\pi \Rightarrow \theta = -\frac{\pi}{2}$ or $\frac{\pi}{2}$

77. $\sin\theta - \sin\frac{1}{2}\theta = 0 \Rightarrow 2\sin\frac{1}{2}\theta\cos\frac{1}{2}\theta - \sin\frac{1}{2}\theta = 0 \Rightarrow \sin\frac{1}{2}\theta(2\cos\frac{1}{2}\theta - 1) = 0$

$\Rightarrow \sin\frac{1}{2}\theta = 0$ or $2\cos\frac{1}{2}\theta - 1 = 0 \Rightarrow \sin\frac{1}{2}\theta = 0$ or $\cos\frac{1}{2}\theta = \frac{1}{2}$. The solution to $\sin\frac{1}{2}\theta = 0$

on the interval $[0,\pi]$ is $\frac{1}{2}\theta = 0 \Rightarrow \theta = 0$. The solution to $\cos\frac{1}{2}\theta = \frac{1}{2}$ on the interval $[0,\pi]$ is

$\frac{1}{2}\theta = \frac{\pi}{3} \Rightarrow \theta = \frac{2\pi}{3}$.

81. Refer to the box entitled "Linear Combinations of cos*Bt* and sin*Bt*" on page 347. Consider the left side of the equation: $3\cos\theta - 4\sin\theta$. In this case $A = \sqrt{a^2 + b^2}$

$= \sqrt{3^2 + (-4)^2} = 5$. The point $(a, b) = (3, -4)$ is on the terminal side of angle $\phi = \tan^{-1}(\frac{-4}{3})$. Thus, $3\cos\theta -$

$4\sin\theta = 5\cos(\theta - \tan^{-1}(\frac{-4}{3}))$. Therefore, solving $3\cos\theta -$

$4\sin\theta = 5$ is equivalent to solving $5\cos(\theta - \tan^{-1}(\frac{-4}{3})) = 5$

$\Rightarrow \cos(\theta - \tan^{-1}(\frac{-4}{3})) = 1 \Rightarrow \theta - \tan^{-1}(\frac{-4}{3}) = 2n\pi \Rightarrow \theta = 2n\pi + \tan^{-1}(\frac{-4}{3})$. The value

for θ that falls in the interval $[0, 2\pi]$ is $\theta = 2\pi + \tan^{-1}(\frac{-4}{3}) \approx 5.4$.

Keystrokes: 2 $\boxed{\times}$ $\boxed{\pi}$ $\boxed{+}$ $\boxed{(}$ $4\boxed{+/-}$ $\boxed{\div}$ 3 $\boxed{)}$ $\boxed{\text{inv}}$ $\boxed{\tan}$

85. Set $y = 0$: $(\tan\theta)x - (\frac{16}{v_0^2\cos^2\theta})x^2 = 0 \Rightarrow x[\tan\theta - \frac{16x}{v_0^2\cos^2\theta}] = 0$

$\Rightarrow x = 0$ or $\tan\theta - \frac{16x}{v_0^2\cos^2\theta} = 0$. The distance corresponds to the latter case, so we solve

$\tan\theta - \frac{16x}{v_0^2\cos^2\theta} = 0$ for x: $\frac{16x}{v_0^2\cos^2\theta} = \tan\theta$

$\Rightarrow x = \frac{v_0^2\cos^2\theta\tan\theta}{16} = \frac{v_0^2\sin\theta\cos\theta}{16} = \frac{v_0^2 2\sin\theta\cos\theta}{32} = \frac{v_0^2\sin2\theta}{32}$

89. Solve $72 = \frac{1}{2}(12)^2(\sin\theta - \cos\theta + 1) \Rightarrow 1 = \sin\theta - \cos\theta + 1$

$\Rightarrow \sin\theta = \cos\theta \Rightarrow \tan\theta = 1 \Rightarrow \theta = \frac{\pi}{4}$

93. Since $\alpha + \beta = 180° - \gamma$, $\cos(\alpha + \beta) = -\cos\gamma \Rightarrow \cos^2(\alpha + \beta) = \cos^2\gamma$. Thus,

$\cos^2\alpha + \cos^2\beta + \cos^2\gamma = \cos^2\alpha + \cos^2\beta + \cos^2(\alpha + \beta)$

$= \cos^2\alpha + \cos^2\beta + (\cos\alpha\cos\beta - \sin\alpha\sin\beta)^2$

$= \cos^2\alpha + \cos^2\beta + (\cos^2\alpha\cos^2\beta - 2\cos\alpha\cos\beta\sin\alpha\sin\beta + \sin^2\alpha\sin^2\beta)$

$= \cos^2\alpha + \cos^2\beta + \cos^2\alpha\cos^2\beta - 2\cos\alpha\cos\beta\sin\alpha\sin\beta + (1 - \cos^2\alpha)(1 - \cos^2\beta)$

$= \cos^2\alpha + \cos^2\beta + \cos^2\alpha\cos^2\beta - 2\cos\alpha\cos\beta\sin\alpha\sin\beta + 1 - \cos^2\alpha - \cos^2\beta + \cos^2\alpha\cos^2\beta$

$= 1 + 2\cos^2\alpha\cos^2\beta - 2\cos\alpha\cos\beta\sin\alpha\sin\beta = 1 + 2\cos\alpha\cos\beta[\cos\alpha\cos\beta - \sin\alpha\sin\beta]$

$= 1 + 2\cos\alpha\cos\beta[\cos(\alpha + \beta)] = 1 + 2\cos\alpha\cos\beta[\cos(\alpha + \beta)] = 1 + 2\cos\alpha\cos\beta[-\cos\gamma]$

$= 1 - 2\cos\alpha\cos\beta\cos\gamma$.

Chapter 7

Section 7.1

1. Draw a picture of the data. Since $\alpha + \beta + \gamma = 180°$, $\gamma = 180° - (\alpha + \beta) = 180° - (40° + 53°) = 87°$. By the law of sines $\dfrac{b}{\sin 53°} = \dfrac{16.2}{\sin 40°} \Rightarrow b = \dfrac{16.2 \sin 53°}{\sin 40°} \approx 20.1$.

   Keystrokes: 16.2 $\boxed{X}$ 53 $\boxed{\sin}$ $\boxed{\div}$ 40 $\boxed{\sin}$ $\boxed{=}$

   Also, $\dfrac{c}{\sin 87°} = \dfrac{16.2}{\sin 40°} \Rightarrow c = \dfrac{16.2 \sin 87°}{\sin 40°} \approx 25.2$.

   Keystrokes: 16.2 $\boxed{X}$ 87 $\boxed{\sin}$ $\boxed{\div}$ 40 $\boxed{\sin}$ $\boxed{=}$

5. Draw a picture of the data. We are given two sides and a non-included angle, so we need to be alert to a possible ambiguous case. By the law of sines $\dfrac{\sin \beta}{3.0} = \dfrac{\sin 70°}{5.0}$

   $\Rightarrow \sin \beta = \dfrac{3.0 \sin 70°}{5.0} = 0.56381557....$

   $\Rightarrow \beta = \sin^{-1}\left(\dfrac{3.0 \sin 70°}{5.0}\right) \Rightarrow \beta \approx 34°$ or $146°$.

   Keystrokes: 3 $\boxed{X}$ 70 $\boxed{\sin}$ $\boxed{\div}$ 5 $\boxed{=}$ $\boxed{inv}$ $\boxed{\sin}$ (The other angle is $180° - 34°$.) But β cannot be $146°$, since $\alpha + \beta = 70° + 146° \geq 180°$. Thus $\beta \approx 34°$, and there is only one triangle to solve. The remaining angle is $\gamma = 180° - (\alpha + \beta)$ $= 180° - (70° + 34°) = 76°$. Also, $\dfrac{c}{\sin 76°} = \dfrac{5.0}{\sin 70°} \Rightarrow c = \dfrac{5.0 \sin 76°}{\sin 70°} \approx 5.2$.

   Keystrokes: 5 $\boxed{X}$ 76 $\boxed{\sin}$ $\boxed{\div}$ 70 $\boxed{\sin}$ $\boxed{=}$

9. Draw a picture of the data. We are given two sides and a non-included angle, so we need to be alert to a possible ambiguous case. By the law of sines

   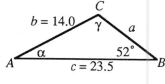

   $\dfrac{\sin \gamma}{23.5} = \dfrac{\sin 52°}{14.0} \Rightarrow \sin \gamma = \dfrac{23.5 \sin 52°}{14.0} = 1.3227....$

   Keystrokes: 23.5 $\boxed{X}$ 52 $\boxed{\sin}$ $\boxed{\div}$ 14 $\boxed{=}$

   This is not possible, since $-1 \leq \sin \gamma \leq 1$. No triangle can be formed for which $b = 14.0$, $c = 23.5$, $\beta = 52°$.

13. $\dfrac{x}{\sin 45°} = \dfrac{24}{\sin 30°} \Rightarrow x = \dfrac{24 \sin 45°}{\sin 30°} = \dfrac{24(\sqrt{2}/2)}{1/2} = 24\sqrt{2}$

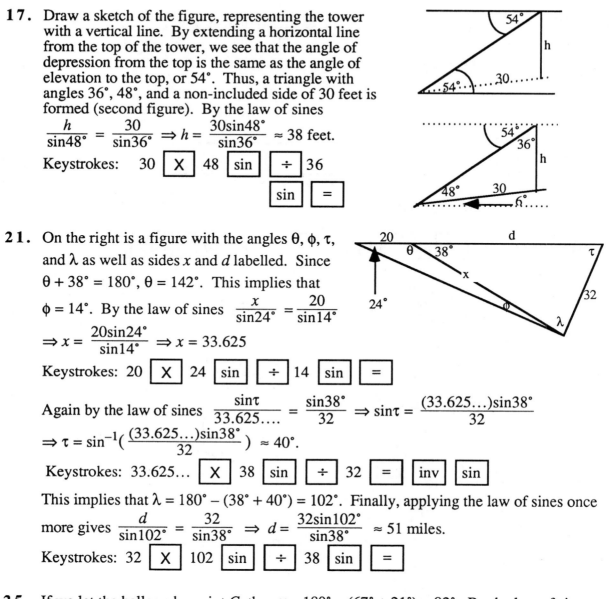

17. Draw a sketch of the figure, representing the tower with a vertical line. By extending a horizontal line from the top of the tower, we see that the angle of depression from the top is the same as the angle of elevation to the top, or 54°. Thus, a triangle with angles 36°, 48°, and a non-included side of 30 feet is formed (second figure). By the law of sines

$$\frac{h}{\sin 48°} = \frac{30}{\sin 36°} \Rightarrow h = \frac{30\sin 48°}{\sin 36°} \approx 38 \text{ feet.}$$

Keystrokes: 30 [X] 48 [sin] [÷] 36 [sin] [=]

21. On the right is a figure with the angles θ, φ, τ, and λ as well as sides x and d labelled. Since $θ + 38° = 180°$, $θ = 142°$. This implies that $φ = 14°$. By the law of sines $\frac{x}{\sin 24°} = \frac{20}{\sin 14°}$

$$\Rightarrow x = \frac{20\sin 24°}{\sin 14°} \Rightarrow x = 33.625$$

Keystrokes: 20 [X] 24 [sin] [÷] 14 [sin] [=]

Again by the law of sines $\frac{\sin τ}{33.625....} = \frac{\sin 38°}{32} \Rightarrow \sin τ = \frac{(33.625...)\sin 38°}{32}$

$$\Rightarrow τ = \sin^{-1}\left(\frac{(33.625...)\sin 38°}{32}\right) \approx 40°.$$

Keystrokes: 33.625... [X] 38 [sin] [÷] 32 [=] [inv] [sin]

This implies that $λ = 180° - (38° + 40°) = 102°$. Finally, applying the law of sines once more gives $\frac{d}{\sin 102°} = \frac{32}{\sin 38°} \Rightarrow d = \frac{32\sin 102°}{\sin 38°} \approx 51$ miles.

Keystrokes: 32 [X] 102 [sin] [÷] 38 [sin] [=]

25. If we let the balloon be point C, then $γ = 180° - (67° + 21°) = 92°$. By the law of sines,

$$\frac{b}{\sin 21°} = \frac{11}{\sin 92°} \Rightarrow b = \frac{11\sin 21°}{\sin 92°} = 3.944445....$$

Keystrokes: 11 [X] 21 [sin] [÷] 92 [sin] [=]

Draw a perpendicular from C to side AB, and consider the right triangles formed. For the "left" right triangle, the height h is opposite the 67° angle and the hypotenuse is $b = 3.94445....$ Since $\sin α = \frac{\text{opposite side}}{\text{hypotenuse}}$, $h = b \sin α = (3.94445...)\sin 67° \approx 3.6$ miles.

29. $\dfrac{a}{\sin\alpha} = \dfrac{b}{\sin\beta} \Rightarrow \dfrac{a}{b} = \dfrac{\sin\alpha}{\sin\beta} \Rightarrow \dfrac{a}{b} + 1 = \dfrac{\sin\alpha}{\sin\beta} + 1 \Rightarrow \dfrac{a}{b} + \dfrac{b}{b} = \dfrac{\sin\alpha}{\sin\beta} + \dfrac{\sin\beta}{\sin\beta}$

$\Rightarrow \dfrac{a+b}{b} = \dfrac{\sin\alpha + \sin\beta}{\sin\beta}$

Section 7.2

---

1. $c^2 = 10.0^2 + 15.0^2 - 2(10.0)(15.0)\cos 24° \Rightarrow c = \sqrt{10.0^2 + 15.0^2 - 2(10.0)(15.0)\cos 24°}$
≈ 7.1.

Keystrokes: 10 $\boxed{x^2}$ $\boxed{+}$ 15 $\boxed{x^2}$ $\boxed{-}$ 2 $\boxed{\times}$ 10 $\boxed{\times}$ 15 $\boxed{\times}$ 24

$\boxed{\cos}$ $\boxed{=}$ $\boxed{\sqrt{}}$

By the law of sines $\dfrac{\sin\alpha}{10.0} = \dfrac{\sin 24°}{7.1} \Rightarrow \sin\alpha = \dfrac{10\sin 24°}{7.1}$

$\Rightarrow \alpha = \sin^{-1}\left(\dfrac{10\sin 24°}{7.1}\right) \approx 35°$.

Keystrokes: 10 $\boxed{\times}$ 24 $\boxed{\sin}$ $\boxed{\div}$ 7.1 $\boxed{=}$ $\boxed{\text{inv}}$ $\boxed{\sin}$

Since $\alpha + \beta + \gamma = 180°$, $\beta \approx 121°$.

5. Draw a picture of the data. We are given two sides and a non-included angle, so we need to be alert to a possible ambiguous case. By the law of sines

$\dfrac{\sin\alpha}{5.0} = \dfrac{\sin 70°}{3.0} \Rightarrow \sin\alpha = \dfrac{5.0\sin 70°}{3.0} = 1.566...$

Keystrokes: 5 $\boxed{\times}$ 70 $\boxed{\sin}$ $\boxed{\div}$ 3 $\boxed{=}$

This is not possible, since $-1 \le \sin\alpha \le 1$. No triangle can be formed for which $a = 5.0$, $c = 3.0$, $\gamma = 70°$.

9. $\cos\gamma = \dfrac{11.0^2 + 10.0^2 - 13.7^2}{2(11.0)(10.0)} \Rightarrow \gamma = \cos^{-1}(0.1514...) \approx 81.3°$. By the law of sines,

$\dfrac{\sin\alpha}{28.2} = \dfrac{\sin 81.3°}{13.7} \Rightarrow \sin\alpha = \dfrac{28.2\sin 81.3°}{13.7} = 2.03...$

Keystrokes: 28.2 $\boxed{\times}$ 81.3 $\boxed{\sin}$ $\boxed{\div}$ 13.7 $\boxed{=}$

This is not possible, since $-1 \le \sin\alpha \le 1$. No triangle can be formed for which $a = 28.2$, $b = 11.0$, $c = 13.7$. This may also have been determined by observing that the longest side is greater than the sum of the other two sides (see Example 3).

13. The largest angle θ is opposite the longest side. Thus, $\cos\theta = \dfrac{12.3^2 + 14.0^2 - 15.7^2}{2(12.3)(14.0)}$

$\Rightarrow \theta = \cos^{-1}\left(\dfrac{12.3^2 + 14.0^2 - 15.7^2}{2(12.3)(14.0)}\right) \approx 73°.$

Keystrokes: 12.3 $\boxed{x^2}$ $\boxed{+}$ 14 $\boxed{x^2}$ $\boxed{-}$ 15.7 $\boxed{x^2}$ $\boxed{=}$ $\boxed{\div}$ $\boxed{(}$ 2 $\boxed{X}$

12.3 $\boxed{X}$ 14 $\boxed{)}$ $\boxed{=}$ $\boxed{\text{inv}}$ $\boxed{\cos}$

17. The area is $A = \frac{1}{2}ab\sin\gamma = \frac{1}{2}(20.0)(14.0)\sin31° \approx 72.1$ square inches.

Keystrokes: 0.5 $\boxed{X}$ 20 $\boxed{X}$ 14 $\boxed{X}$ 31 $\boxed{\sin}$ $\boxed{=}$

21. The semiperimeter is $s = \frac{1}{2}(a + b + c) = \frac{1}{2}(36.0 + 11.0 + 26.2) = 36.6.$ The area is

$A = \sqrt{s(s-a)(s-b)(s-c)} = \sqrt{36.6(36.6 - 36.0)(36.6 - 11.0)(36.6 - 26.2)}$
$= \sqrt{36.6(0.6)(25.6)(10.4)} = \sqrt{5846.6304} \approx 76.5$ square units.

25. The sides of the triangle are 8, 9, and 11 units. Let α be the angle opposite the side of

length 11. By the law of cosines, $\cos\alpha = \dfrac{8^2 + 9^2 - 11^2}{2(8)(9)} = \dfrac{1}{6} \Rightarrow \alpha = \cos^{-1}\left(\dfrac{1}{6}\right) \approx 80°.$

Let β be the angle opposite the side of length 9. By the law of cosines,

$\cos\beta = \dfrac{8^2 + 11^2 - 9^2}{2(8)(11)} = \dfrac{13}{22} \Rightarrow \beta = \cos^{-1}\left(\dfrac{13}{22}\right) \approx 54°.$ Since the sum of the angles of a

triangle is 180°, the remaining angle is 46°.

29. a) $\angle AOB = \frac{1}{8}(360°) = 45°,$ and $OA = AB = 1.$ Thus the area of triangle AOB is

$\frac{1}{2}(1)(1)\sin45° = \dfrac{\sqrt{2}}{4}$

b) The area of the octagon is eight times the area of triangle AOB: $8\left(\dfrac{\sqrt{2}}{4}\right) = 2\sqrt{2}$ square

units. The area of the unit circle is $\pi r^2 = \pi(1)^2 = \pi$ square units.

33. Knowing the measures of the three angles of a triangle does not determine the size of the triangle. For example, a triangle can have angles of 30°, 60°, and 90° but the sides may be 1, $\sqrt{3}$, and 2 or the sides may be 5, $5\sqrt{3}$, and 10.

37. Since the helicopters have been in flight for 1 hour 6 minutes = 1.1 hours, the distances Sky King and Seahunt have travelled are $(100)(1.1) = 110$ miles and $(110)(1.1) = 121$ miles respectively. Draw a sketch of the situation. Since the Seahunt helicopter heads N20°W, the angle between the lines of flight is 70°. By the law of cosines, the distance d between the helicopters is determined by

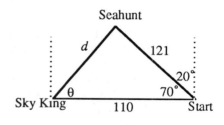

$d^2 = 110^2 + 121^2 - 2(110)(121)\cos 70° = 17636.42$

$\Rightarrow d = \sqrt{110^2 + 121^2 - 2(110)(121)\cos 70° = 17636.42} \approx 133$ miles.

Keystrokes: 110 $\boxed{x^2}$ $\boxed{+}$ 121 $\boxed{x^2}$ $\boxed{-}$ 2 $\boxed{\times}$ 110 $\boxed{\times}$ 121 $\boxed{\times}$ 70

$\boxed{\cos}$ $\boxed{=}$ $\boxed{\sqrt{}}$

Referring to the figure, let θ be the angle opposite the side with length 121 (miles). By the

law of sines $\dfrac{\sin\theta}{121} = \dfrac{\sin 70°}{133} \Rightarrow \sin\theta = \dfrac{121\sin 70°}{133} \Rightarrow \theta = \sin^{-1}(\dfrac{121\sin 70°}{133}) \approx 59°.$

Keystrokes: 121 $\boxed{\times}$ 70 $\boxed{\sin}$ $\boxed{\div}$ 133 $\boxed{=}$ $\boxed{\text{inv}}$ $\boxed{\sin}$

Thus, the direction that Sky King needs to take is N31°E.

41. We need the answer to Problem 40(b), which is the distance between the joggers. This

distance is $d(t) = \sqrt{49t^2 - 60t + 63}$. According to the hint (Example 11, Section

2.4), the minimum of $d(t)$ will occur at the same value of t as for $[d(t)]^2 = 49t^2 - 60t +$

63. Therefore, the minimum occurs when $t = -\dfrac{b}{2a} = \dfrac{60}{98} = \dfrac{30}{49}$ hours (approximately 36

minutes, 44seconds).

45. Let the equilateral triangle be labelled as shown in the first figure. Copy triangle *AOB* and rotate so that AB coincides with AC (second figure). $\angle OAO' = 60°$, so triangle *AOO'* is equilateral. Triangle COO' has sides 10, 14, and 16. By the law of cosines, $\angle COO' = \cos^{-1}(\dfrac{1}{7}) \approx 81.8°$. Triangle AO'C has sides 10 and 14, with included angle $\angle AO'C = 60° + \cos^{-1}(\dfrac{1}{7})$. By the law of cosines

$(AC)^2 = 14^2 + 10^2 - 2(14)(10)\cos(60° + \cos^{-1}(\dfrac{1}{7})).$

It follows that $AC = \sqrt{516} = 2\sqrt{129} \approx 22.7$.

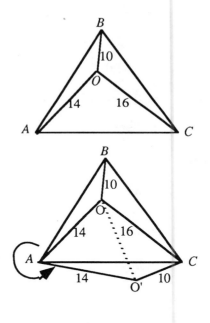

Section 7.3

1. Since standard position means that the initial point of the vector is placed at the origin of the coordinate system, the vector **e** is the only vector in standard position.

5. The vector **f** represents a horizontal displacement of 4 units (to the right) and a vertical displacement of 2 units (upwards). The vector **d** represents a displacement of 4 units (to the right), and the vector **b** represents a vertical displacement of 2 units (upwards). Therefore vector **f** can be represented as the sum of **d** and **b**.

9. The vector **e** – **b** can be considered to be the vector that must be added to vector **b** to get the vector **e**. Thus, we first graph vectors **e** and **b** so that their initial points coincide. Then graph the vector from the terminal point of **b** to the terminal point of **e**. This is **e** – **b**.

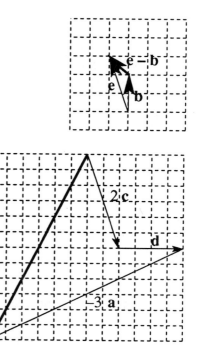

13. The vector 2**c** has the same direction as **c**, but is twice as long. The vector –3**a** is three times as long as vector **a**, and has a direction that is precisely opposite that of **a**. Now sketch 2**c** and then sketch vector **d** so that the initial point of **d** coincides with the terminal point of 2**c**. Finally, place the initial point of vector –3**a** at the terminal point of **d**. The vector from the initial point of 2**c** to the terminal point of –3**a** is the vector

$$2\mathbf{c} + \mathbf{d} + (-3\mathbf{a}) = 2\mathbf{c} + \mathbf{d} - 3\mathbf{a}.$$

17. $\mathbf{v} = \langle 0 - 7, 2 - (-8) \rangle = \langle -7, 10 \rangle$ **21.** $\|\mathbf{v}\| = \sqrt{5^2 + (-12)^2} = 13$

25. $\|\mathbf{v}\| = \sqrt{2^2 + (2\sqrt{3})^2} = 4$ **29.** $\|3\mathbf{u}\| = \|3\langle 4, 3 \rangle\| = \|\langle 12, 9 \rangle\| = \sqrt{12^2 + 9^2} = 15$

33. $\|\mathbf{v}\| + \|\mathbf{w}\| = \sqrt{5^2 + (-5)^2} + \sqrt{(-12)^2 + 10^2} = 5\sqrt{2} + 2\sqrt{61}$

37. $2\mathbf{x} + \mathbf{v} = 5\mathbf{u} \Rightarrow 2\mathbf{x} = 5\mathbf{u} - \mathbf{v} \Rightarrow \mathbf{x} = \dfrac{1}{2}(5\mathbf{u} - \mathbf{v}) = \dfrac{1}{2}(5\langle 4, 3 \rangle - \langle 5, -5 \rangle) = \dfrac{1}{2}\langle 15, 20 \rangle$

$$= \langle \tfrac{15}{2}, 10 \rangle$$

41. The magnitude of **v** is $\|\langle 2, 2 \rangle\| = \sqrt{2^2 + 2^2} = 2\sqrt{2}$. The vector having the same direction as **v** but with magnitude (length) 20 is $\dfrac{20}{\|\mathbf{v}\|}\mathbf{v} = \dfrac{20}{2\sqrt{2}}\langle 2, 2 \rangle = 5\sqrt{2}\langle 2, 2 \rangle$

$$= \langle 10\sqrt{2}, 10\sqrt{2} \rangle.$$

45. Refer to page 418 of the text between Example 6 and Example 7. In this case,
$\mathbf{v} = \|\mathbf{v}\| \langle \cos\theta, \sin\theta \rangle = 10 \langle \cos 135°, \sin 135° \rangle = \langle -5\sqrt{2}, 5\sqrt{2} \rangle.$

49. Refer to page 418 of the text between Example 6 and Example 7. In this case,
$\mathbf{v} = \|\mathbf{v}\| \langle \cos\theta, \sin\theta \rangle = \sqrt{3} \langle \cos 210°, \sin 210° \rangle = \langle -\dfrac{3}{2}, \dfrac{\sqrt{3}}{2} \rangle.$

53. If we sketch **v** in standard position, the terminal point is $P(1, -\sqrt{3})$. The distance from the origin to P is $r = \sqrt{(-\sqrt{3})^2 + 1^2} = 2$, so $\|\mathbf{v}\| = 2$. The reference angle is $\tan^{-1}(\frac{\sqrt{3}}{1}) = 60°$; so the smallest positive angle that **v** makes with the x-axis is $\theta = 300°$.

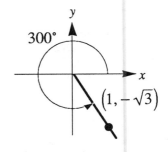

57. The midpoints of the quadrilateral $ABCD$ have the following coordinates:

Segment	AB	BC	CD	AD
Midpoint	$M_1(\frac{b}{2}, 0)$	$M_2(\frac{b+c}{2}, \frac{d}{2})$	$M_3(\frac{c+e}{2}, \frac{d+f}{2})$	$M_4(\frac{e}{2}, \frac{f}{2})$

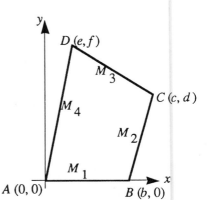

The vector from M_1 to M_2 is $\langle \frac{b+c}{2} - \frac{b}{2}, \frac{d}{2} - 0 \rangle$

$= \langle \frac{c}{2}, \frac{d}{2} \rangle$, and the vector from M_4 to M_3 is

$\langle \frac{c+e}{2} - \frac{e}{2}, \frac{d+f}{2} - \frac{f}{2} \rangle = \langle \frac{c}{2}, \frac{d}{2} \rangle$. Therefore, segments M_1M_2 and M_3M_4 are parallel and have the same length. Similarly, the vector from M_1 to M_4 is

$\langle \frac{e-b}{2}, \frac{f}{2} \rangle$, and the vector from M_2 to M_3 is $\langle \frac{e-b}{2}, \frac{f}{2} \rangle$. Therefore, segments M_1M_4 and M_2M_3 are parallel and have the same length. This proves that the quadrilateral $M_1M_2M_3M_4$ is a parallelogram.

61. Since the forces $\mathbf{F}_1$ and $\mathbf{F}_2$ are perpendicular, the angle between $\mathbf{F}_2$ and **w** is 85°, and the angle between $\mathbf{F}_1$ and **w** is 5°. Since the length of **w** is $\|\mathbf{w}\| = 400$, the length of $\mathbf{F}_2$ is $\|\mathbf{F}_2\| = 400\sin 5° \approx 35$. This magnitude represents 35 pounds of force in the direction of the incline.

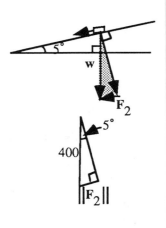

65. $\mathbf{v} \cdot \mathbf{w} = \langle 3, \sqrt{3} \rangle \cdot \langle 4\sqrt{3}, -4 \rangle = 3(4\sqrt{3}) + (\sqrt{3})(-4) = 8\sqrt{3}$

69. The angle between **v** and **w** is determined by $\cos\theta = \frac{\mathbf{v} \cdot \mathbf{w}}{\|\mathbf{v}\| \|\mathbf{w}\|}$

$$\Rightarrow \cos\theta = \frac{(7)(3) + (-1)(-4)}{\sqrt{7^2 + (-1)^2} \sqrt{3^2 + (-4)^2}} = \frac{25}{\sqrt{50} \sqrt{25}} = \frac{25}{5\sqrt{2} \cdot 5} = \frac{1}{\sqrt{2}} \Rightarrow \theta = 45°.$$

73. If $\mathbf{v \cdot w} = 0$, then $\dfrac{\mathbf{v \cdot w}}{\|\mathbf{v}\| \, \|\mathbf{w}\|} = 0$. Since $\dfrac{\mathbf{v \cdot w}}{\|\mathbf{v}\| \, \|\mathbf{w}\|} = \cos\theta$, we have $\cos\theta = 0 \Rightarrow \mathbf{v}$ and $\mathbf{w}$ are perpendicular. On the other hand, suppose vectors $\mathbf{v}$ and $\mathbf{w}$ are perpendicular. Then $\cos\theta = 0$. Since $\dfrac{\mathbf{v \cdot w}}{\|\mathbf{v}\| \, \|\mathbf{w}\|} = \cos\theta$, we have $\dfrac{\mathbf{v \cdot w}}{\|\mathbf{v}\| \, \|\mathbf{w}\|} = 0 \Rightarrow \mathbf{v \cdot w} = 0$. Thus, if $\mathbf{v \cdot w} = 0$, then $\mathbf{v}$ and $\mathbf{w}$ are perpendicular; and if $\mathbf{v}$ and $\mathbf{w}$ are perpendicular, then $\mathbf{v \cdot w} = 0$. This last statement says that $\mathbf{v}$ and $\mathbf{w}$ are perpendicular if and only if $\mathbf{v \cdot w} = 0$.

Section 7.4

1. The complex number $5 + 3i$ corresponds to the point $(5, 3)$.

5. The complex number $-10 = -10 + 0i$ corresponds to the point $(-10, 0)$.

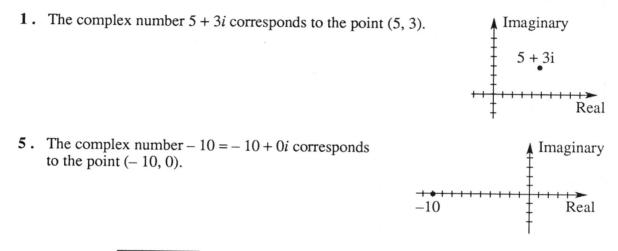

9. $|8 - 4i| = \sqrt{8^2 + (-4)^2} = 4\sqrt{5}$

13. a) $|x| \le 4$ represents all real numbers x that are no more than 4 units from the origin (0 on the real number line):

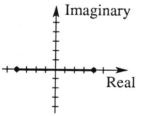

b) $|z| \le 4$ represents all complex numbers z that are no more than 4 units from the origin (in the complex plane):

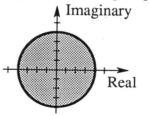

17. a) $|x + 5| \le 3 \Leftrightarrow |x - (-5)| \le 3$
represents all real numbers x that are no more than 3 units from -5 (on the real number line):

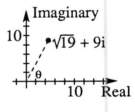

b) $|z + 5| \le 3 \Leftrightarrow |z - (-5 + 0i)| \le 3$
represents all real numbers z that are no more than 3 units from $-5 + 0i$ (in the complex plane):

21. Refer to the box at the top of page 425 in the text. First we consider the graph of $z = 1 + i$. Identifying $a = 1$ and $b = 1$, we have $r = \sqrt{1^2 + 1^2} = \sqrt{2}$ and $\tan\theta = 1 \Rightarrow \theta = 45°$. Thus $1 + i = \sqrt{2}(\cos45° + i\sin45°)$. This is abbreviated as $\sqrt{2}\text{cis}45°$.

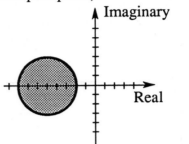

25. Refer to the box at the top of page 425 in the text. First we consider the graph of $z = \sqrt{19} + 9i$. Identifying $a = \sqrt{19}$ and $b = 9$, we have $r = \sqrt{\sqrt{19}^2 + 9^2} = 10$ and $\tan\theta = \dfrac{9}{\sqrt{19}}$
$\Rightarrow \theta \approx 64.16°$. So $\sqrt{19} + 9i = 10(\cos64.16° + i\sin64.16°)$. This is abbreviated as $10\text{cis}64.16°$.

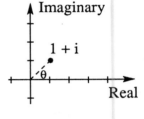

29. Consider the graph of $z = 20i$; it lies on the imaginary axis 20 units above the origin. Clearly $r = 20$ and $\theta = 90°$. Thus $20i = 20(\cos90° + i\sin90°)$ or $20\text{cis}90°$.

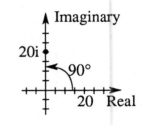

33. $8\text{cis}135° = 8(\cos135° + i\sin135°)$
$$= 8(-\frac{\sqrt{2}}{2} + i(\frac{\sqrt{2}}{2})) = -4\sqrt{2} + 4\sqrt{2}i.$$

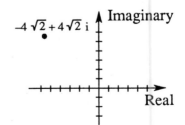

37. $10\text{cis}\dfrac{5\pi}{3} = 10(\cos\dfrac{5\pi}{3} + i\sin\dfrac{5\pi}{3}) = 10(\dfrac{1}{2} + i(-\dfrac{\sqrt{3}}{2}))$

$$= 5 - 5\sqrt{3}i.$$

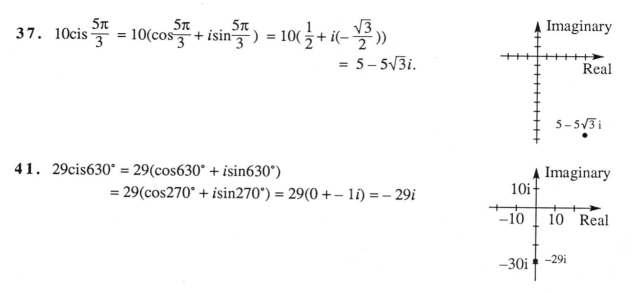

41. $29\text{cis}630° = 29(\cos630° + i\sin630°)$

$\qquad = 29(\cos270° + i\sin270°) = 29(0 + -1i) = -29i$

45. Refer to the box "Multiplication and Division in Trigonometric Form" on page 427.
$(3\text{cis}120°)(5\text{cis}195°) = (3)(5)\text{cis}(120° + 195°) = 15\text{cis}315°$ or $15(\cos315° + i\sin315°)$

49. $6(\cos24° + i\sin24°)\cdot\sqrt{5}(\cos219° + i\sin219°) = 6\sqrt{5}(\cos(24° + 219°) + i\sin(24° + 219°)$

$\qquad\qquad\qquad\qquad\qquad\qquad\qquad = 6\sqrt{5}(\cos243° + i\sin243°)$ or $6\sqrt{5}\text{cis}243°$

53. Refer to the box "Multiplication and Division in Trigonometric Form" on page 427.
$\dfrac{15\text{cis}80°}{24\text{cis}62°} = \dfrac{15}{24}\text{cis}(80° - 62°) = \dfrac{5}{8}\text{cis}18°$ or $\dfrac{5}{8}(\cos18° + i\sin18°)$.

57. $(4\text{cis}\dfrac{\pi}{3})^3 = (4\text{cis}\dfrac{\pi}{3})(4\text{cis}\dfrac{\pi}{3})(4\text{cis}\dfrac{\pi}{3}) = (4)(4)(4)\text{cis}(\dfrac{\pi}{3} + \dfrac{\pi}{3} + \dfrac{\pi}{3})$

$\qquad = 64\text{cis}\,\pi$ or $64(\cos\pi + i\sin\pi)$

61. If we first graph $2 - 2\sqrt{3}\,i$ in the complex plane, we have $r = \sqrt{2^2 + (-2\sqrt{3})^2} = 4$ and
$\tan\theta = \dfrac{-2\sqrt{3}}{2} \Rightarrow \theta = 300° \Rightarrow 2 - 2\sqrt{3}\,i = 4\text{cis}300°$. Similarly, (see solution to Problem
21 in this section) $1 + i = \sqrt{2}\text{cis}45°$. Thus, $(2 - 2\sqrt{3}\,i)(1 + i) = (4\text{cis}300°)(\sqrt{2}\text{cis}45°)$
$= 4\sqrt{2}\text{cis}345° = 4\sqrt{2}(\cos345° + i\sin345°)$. Since $\cos345° = \cos(300° + 45°)$
$= \cos300°\cos45° - \sin300°\sin45° = \dfrac{1}{2}(\dfrac{\sqrt{2}}{2}) - (-\dfrac{\sqrt{3}}{2})(\dfrac{\sqrt{2}}{2}) = \dfrac{\sqrt{2} + \sqrt{6}}{4}$. Similarly,
$\sin345° = \sin(300° + 45°) = \sin300°\cos45° + \cos300°\sin45° = \dfrac{-\sqrt{6} + \sqrt{2}}{4}$. Therefore
the rectangular form of $4\sqrt{2}(\cos345° + i\sin345°)$ is $4\sqrt{2}(\dfrac{\sqrt{2} + \sqrt{6}}{4} + i\,\dfrac{-\sqrt{6} + \sqrt{2}}{4})$
$= (2 + 2\sqrt{3}) + i(-2\sqrt{3} + 2)$.

65. By picturing $1 - \sqrt{3}\, i$ in the complex plane, we see that the trigonometric form is
$1 - \sqrt{3}\, i = 2\text{cis}300°$. Thus,
$(1 - \sqrt{3}\, i)^6 = (2\text{cis}300°)^6 = (2\text{cis}300°)(2\text{cis}300°)(2\text{cis}300°)(2\text{cis}300°)(2\text{cis}300°)(2\text{cis}300°)$
$= 64\text{cis}1800° = 64\text{cis}0° = 64.$

69. The impedance is $Z = R + (R_L - R_C)i = 16 + (8 - 12)i = 16 - 4i$. The voltage is
$V = IZ = 4(16 - 4i) = 64 - 16i$. The magnitude of the impedance is $\|Z\| = \sqrt{16^2 + 4^2}$
$= 4\sqrt{17}$; the magnitude of the voltage is $\|V\| = \sqrt{64^2 + 16^2} = 16\sqrt{17}$. The phase angle
for both is $\tan^{-1}(\frac{-4}{16}) = \tan^{-1}(\frac{-1}{4}) \approx -14.04°.$

Section 7.5

1. $(4\text{cis}20°)^6 = 4^6\text{cis}(6(20°)) = 4096\text{cis}120°$. In rectangular form this is
$$4096(\cos120° + i\sin120°) = 4096(-\frac{1}{2} + i(\frac{\sqrt{3}}{2})) = -2048 + 2048\sqrt{3}\, i$$

5. The trigonometric form of $-1 + i$ is $\sqrt{2}\,\text{cis}135°$. (See the
graph of $-1 + i$ in the complex plane on the right.) Thus,
$(-1 + i)^{14} = (\sqrt{2}\text{cis}135°)^{14} = (\sqrt{2})^{14}\text{cis}(14(135°))$
$= 128\text{cis}1890° = 128\text{cis}90°$. In rectangular form this is
$128(\cos90° + i\sin90°) = 128(0 + 1i) = 128i$.

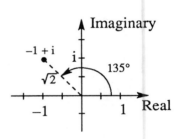

9. The trigonometric form of $-3\sqrt{3} + 2i$ is $r\text{cis}\theta$, where
$r = \sqrt{(-3\sqrt{3})^2 + 2^2} = \sqrt{31}$ (see figure at right), and
$\theta = 180° + \tan^{-1}(\frac{2}{-3\sqrt{3}}) \approx 158.9483°$. Thus,
$(-3\sqrt{3} + 2i)^5 \approx (\sqrt{31}\text{cis}158.9483°)^5$
$= (\sqrt{31})^5\text{cis}[5(158.9483°)] = 961\sqrt{31}\text{cis}794.74°$
$= 961\sqrt{31}\text{cis}74.74°$. In rectangular form this is
$961\sqrt{31}(\cos74.74° + i\sin74.74°) \approx 1408.16 + 5162i$.

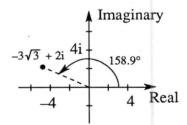

13. Refer to the box on page 434. The four roots are of the form $\sqrt[4]{81}\,\text{cis}(\frac{132° + 360°k}{4})$
$= 3\text{cis}(33° + 90°k)$. Letting $k = 0, 1, 2,$ and 3 gives all four roots:
$3\text{cis}(33° + 90°(0)) = 3\text{cis}33°$; $3\text{cis}(33° + 90°(1)) = 3\text{cis}123°$; $3\text{cis}(33° + 90°(2))$
$= 3\text{cis}213°$; and $3\text{cis}(33° + 90°(3)) = 3\text{cis}303°$.

17. Refer to the box on page 434. The trigonometric form of i is 1cis90°. The five roots are of the form $\sqrt[5]{1}\ cis(\dfrac{90° + 360°k}{5}) = 1cis(18° + 72°k)$. Letting $k = 0, 1, 2, 3$, and 4 gives all five roots: $1cis(18° + 72°(0)) = cis18°$; $1cis(18° + 72°(1)) = cis90°$; $1cis(18° + 72°(2)) = cis162°$; $1cis(18° + 72°(3)) = cis234°$; and $1cis(18° + 72°(4)) = cis306°$.

21. The trigonometric form of $\sqrt{2} - \sqrt{2}\ i$ is 2cis315°. The nine roots are of the form $\sqrt[9]{2}\ cis(\dfrac{315° + 360°k}{9}) = \sqrt[9]{2}\ cis(35° + 40°k)$. Letting $k = 0, 1, 2, 3, 4, 5, 6, 7$, and 8 gives all nine roots: $\sqrt[9]{2}cis(35° + 40°(0)) = \sqrt[9]{2}cis35°$; $\sqrt[9]{2}cis(35° + 40°(1)) = \sqrt[9]{2}cis75°$; $\sqrt[9]{2}cis(35° + 40°(2)) = \sqrt[9]{2}cis115°$; $\sqrt[9]{2}cis(35° + 40°(3)) = \sqrt[9]{2}cis155°$; $\sqrt[9]{2}cis(35° + 40°(4)) = \sqrt[9]{2}cis195°$; $\sqrt[9]{2}cis(35° + 40°(5)) = \sqrt[9]{2}cis235°$; $\sqrt[9]{2}cis(35° + 40°(6)) = \sqrt[9]{2}cis275°$; $\sqrt[9]{2}cis(35° + 40°(7)) = \sqrt[9]{2}cis315°$; and $\sqrt[9]{2}cis(35° + 40°(8)) = \sqrt[9]{2}cis355°$.

25. The sixth roots of unity are the sixth roots of 1. The trigonometric form of 1 is 1cis0°. The six roots are of the form $\sqrt[6]{1}\ cis(\dfrac{0° + 360°k}{6}) = cis\ 60°k$. Letting $k = 0, 1, 2, 3, 4$, and 5 gives all six roots: $cis(60°(0)) = cis0°$; $cis(60°(1)) = cis60°$; $cis(60°(2)) = cis120°$; $cis(60°(3)) = cis180°$; $cis(60°(4)) = cis240°$, and $cis(60°(5)) = cis300°$.

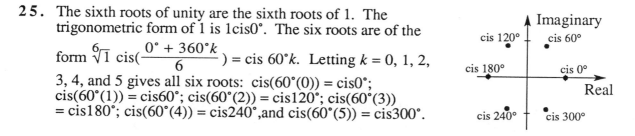

29. $(8cis51°)^{2/3} = 8^{2/3}cis[\dfrac{2}{3}(51° + 360°k)] = 4cis(34° + 240°k)$. Letting $k = 0, 1$, and 2 will generate all three values: $4cis(34° + 240°(0)) = 4cis34°$; $4cis(34° + 240°(1)) = 4cis274°$; and $4cis(34° + 240°(2)) = 4cis514° = 4cis154°$.

33. The trigonometric form for $(-\sqrt{3} + i)$ is 2cis150°. Thus $(-\sqrt{3} + i)^{5/4} = (2cis150°)^{5/4}$ $= 2^{5/4}cis[\dfrac{5}{4}(150° + 360°k)] = 2\sqrt[4]{2}\ cis(187.5° + 450°k)$. Letting $k = 0, 1, 2$, and 3 will generate all four values: $2\sqrt[4]{2}\ cis(187.5° + 450°(0)) = 2\sqrt[4]{2}\ cis187.5°$; $2\sqrt[4]{2}\ cis(187.5° + 450°(1)) = 2\sqrt[4]{2}\ cis637.5° = 2\sqrt[4]{2}\ cis277.5°$; $2\sqrt[4]{2}\ cis(187.5° + 450°(2)) = 2\sqrt[4]{2}\ cis1087.5° = 2\sqrt[4]{2}\ cis7.5°$; and $2\sqrt[4]{2}\ cis(187.5° + 450°(3)) = 2\sqrt[4]{2}\ cis1537.5° = 2\sqrt[4]{2}\ cis97.5°$.

37. $x^6 - 64i = 0 \Rightarrow x^6 = 64i$. Solving this equation is equivalent to finding all the sixth roots of $64i$. The trigonometric form of $64i$ is $64\text{cis}90°$; so the six roots have the form $\sqrt[6]{64} \text{ cis}(\dfrac{90° + 360°k}{6}) = 2\text{cis}(15° + 60°k)$. Letting $k = 0, 1, 2, 3, 4,$ and 5 gives all six roots: $2\text{cis}(15° + 60°(0)) = 2\text{cis}15°$; $2\text{cis}(15° + 60°(1)) = 2\text{cis}75°$; $2\text{cis}(15° + 60°(2)) = 2\text{cis}135°$; $2\text{cis}(15° + 60°(3)) = 2\text{cis}195°$; $2\text{cis}(15° + 60°(4)) = 2\text{cis}255°$, and $2\text{cis}(15° + 60°(5)) = 2\text{cis}315°$.

41. If we multiply both sides of the equation by $x - 1$, we get $(x - 1)(x^3 + x^2 + x + 1) = 0 \Rightarrow x^4 - 1 = 0 \Rightarrow x^4 = 1$. The four solutions to this equation are the fourth roots of 1, which have the form $\text{cis}(\dfrac{0° + 360°k}{4}) = \text{cis}(90°k)$. Letting $k = 0, 1, 2,$ and 3 we get $\text{cis}90°(0) = \text{cis } 0°$; $\text{cis}90°(1) = \text{cis } 90°$; $\text{cis}90°(2) = \text{cis } 180°$; and $\text{cis}90°(3) = \text{cis } 270°$. The root $\text{cis}0° = 1$ was introduced when we multiplied both sides by $x - 1$, so the roots of the original equation, $x^3 + x^2 + x + 1$, are the other three: $\text{cis}90° = i$, $\text{cis}180° = -1$, and $\text{cis}270° = -i$.

45. The trigonometric form of $1 + i$ is $\sqrt{2}$ cis 45°. Referring to the box "Euler's Identity" on page 438 of the text, $\text{cis}\theta = e^{i\theta}$, where θ must be in radians. Thus

$$\sqrt{2} \text{ cis } 45° = \sqrt{2} \text{ cis } \frac{\pi}{4} = \sqrt{2} \, e^{i\pi/4}.$$

49. The trigonometric form of $\sqrt{19} + 9i \approx 10\text{cis}64.16°$ (See solution to Problem 25 in Section 7.4). In radians, the trigonometric form is $10\text{cis}(1.12)$. Thus, an approximation is $10\text{cis } 64.16° = 10\text{cis}(1.12) = 10e^{i(1.12)}$. The exact value is $10e^{i\theta}$, where $\theta = \tan^{-1}(\dfrac{9}{\sqrt{19}})$.

53. The trigonometric form of $20i$ is $20\text{cis } 90°$. In radians, the trigonometric form is $20\text{cis}\dfrac{\pi}{2}$.

Thus $20i = 20\text{cis}\dfrac{\pi}{2} = 20 \, e^{i\pi/2}$.

Miscellaneous Exercises for Chapter 7

1. A sketch might be helpful. We are given two sides and the included angle, so by the law of cosines,

$$a^2 = 12^2 + 15^2 - 2(12)(15)\cos 105°$$

$$\Rightarrow a = \sqrt{12^2 + 15^2 - 2(12)(15)\cos 105°} = 21.498....$$

Keystrokes: 12 $\boxed{x^2}$ $\boxed{+}$ 15 $\boxed{x^2}$ $\boxed{-}$ 2 $\boxed{\times}$

12 $\boxed{\times}$ 15 $\boxed{\times}$ 105 $\boxed{\cos}$ $\boxed{=}$ $\boxed{\sqrt{}}$

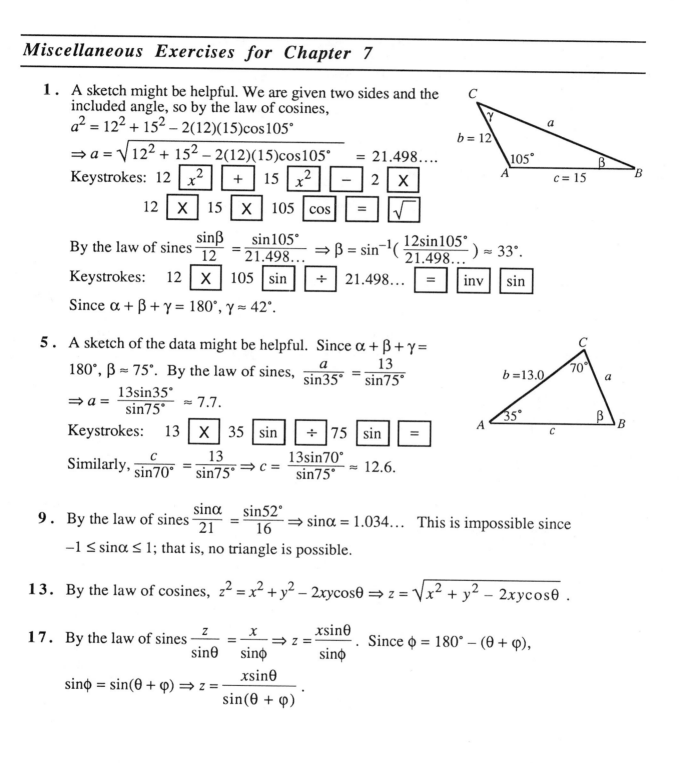

By the law of sines $\dfrac{\sin\beta}{12} = \dfrac{\sin 105°}{21.498...} \Rightarrow \beta = \sin^{-1}\left(\dfrac{12\sin 105°}{21.498...}\right) \approx 33°.$

Keystrokes: 12 $\boxed{\times}$ 105 $\boxed{\sin}$ $\boxed{\div}$ 21.498... $\boxed{=}$ $\boxed{\text{inv}}$ $\boxed{\sin}$

Since $\alpha + \beta + \gamma = 180°$, $\gamma \approx 42°$.

5. A sketch of the data might be helpful. Since $\alpha + \beta + \gamma = 180°$, $\beta \approx 75°$. By the law of sines, $\dfrac{a}{\sin 35°} = \dfrac{13}{\sin 75°}$

$$\Rightarrow a = \dfrac{13\sin 35°}{\sin 75°} \approx 7.7.$$

Keystrokes: 13 $\boxed{\times}$ 35 $\boxed{\sin}$ $\boxed{\div}$ 75 $\boxed{\sin}$ $\boxed{=}$

Similarly, $\dfrac{c}{\sin 70°} = \dfrac{13}{\sin 75°} \Rightarrow c = \dfrac{13\sin 70°}{\sin 75°} \approx 12.6.$

9. By the law of sines $\dfrac{\sin\alpha}{21} = \dfrac{\sin 52°}{16} \Rightarrow \sin\alpha = 1.034...$ This is impossible since

$-1 \leq \sin\alpha \leq 1$; that is, no triangle is possible.

13. By the law of cosines, $z^2 = x^2 + y^2 - 2xy\cos\theta \Rightarrow z = \sqrt{x^2 + y^2 - 2xy\cos\theta}$.

17. By the law of sines $\dfrac{z}{\sin\theta} = \dfrac{x}{\sin\phi} \Rightarrow z = \dfrac{x\sin\theta}{\sin\phi}$. Since $\phi = 180° - (\theta + \varphi)$,

$\sin\phi = \sin(\theta + \varphi) \Rightarrow z = \dfrac{x\sin\theta}{\sin(\theta + \varphi)}$.

21. First, we need a sketch of the situation . Since the segment from B to C makes a 20° angle with the vertical, the segment from B to C makes a 70° angle with the horizontal. The segment from A to B makes a 56° angle with the vertical, so the segment rom A to B makes a 34° angle with the horizontal. Therefore, $\angle ABC = 70° + 34° = 104°$. By the law of cosines,

$AC^2 = 320^2 + 550^2 - 2(320)(550)\cos 104°$

$\Rightarrow AC = \sqrt{320^2 + 550^2 - 2(320)(550)\cos 104°} \approx 700$ miles.

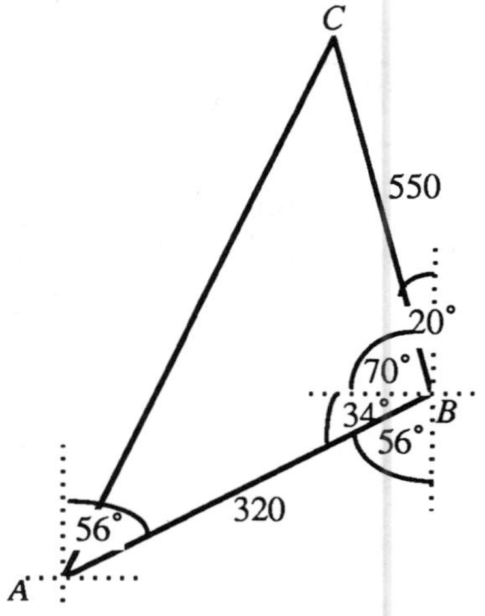

Keystrokes: 690 [x^2] [+] 780 [x^2] [−] 2 [X] 690 [X] 780 [X] 67 [cos] [=] [√]

25. We need a sketch of the situation . Since the jet planes have been in flight for 1.5 hours, the distances they have travelled are (460)(1.5) = 690 miles and (520)(1.5) = 780 miles. The angle between the lines of flight is 15° + 52° = 67°. By the law of cosines, the distance d between the jets is determined by $d^2 = 690^2 + 780^2 - 2(690)(780)\cos 67°$

$\Rightarrow d = \sqrt{690^2 + 780^2 - 2(690)(780)\cos 67°} \approx 815$ miles.

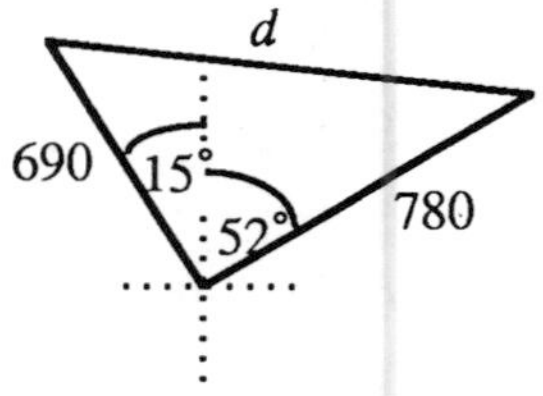

Keystrokes: 690 [x^2] [+] 780 [x^2] [−] 2 [X] 690 [X] 780 [X] 67 [cos] [=] [√]

29. The triangle formed by home plate, the mound, and first base has sides of 60.5 feet and 90 feet, and an included angle of 45°. By the law of cosines, the distance, d, from the mound to first base is determined by $d^2 = 60.5^2 + 90^2 - 2(60.5)(90)\cos 45° \Rightarrow d = \sqrt{4059.857\ldots} \approx 63.7$ feet.

33. Since $\alpha + \beta + \gamma = 180°$, $\beta \approx 75°$. Refer to the box between Example 7 and Example 8 on page 406. The area is $A = \dfrac{b^2 \sin\alpha \sin\beta}{2\sin\beta} = \dfrac{13.0^2 \sin 35° \sin 70°}{2\sin 75°} \approx 47.2$ square units.

Keystrokes: 13 [x^2] [X] 35 [sin] [X] 70 [sin] [÷] [(] 2 [X] 75 [sin] [)] [=]

37. By the law of sines $\dfrac{\sin\theta}{t} = \dfrac{\sin 30°}{x} \Rightarrow \dfrac{\sin\theta}{t} = \dfrac{1/2}{x} \Rightarrow \sin\theta = \dfrac{t}{2x}$. By the law of cosines

$t^2 = x^2 + (2t)^2 - 2x(2t)\cos\theta \Rightarrow \cos\theta = \dfrac{x^2 + 3t^2}{4xt}$. Also by the law of cosines

$x^2 = t^2 + (2t)^2 - 2t(2t)\cos 30° \Rightarrow x^2 = (5 - 2\sqrt{3})t^2$. Therefore, $\tan\theta = \sin\theta \cdot \dfrac{1}{\cos\theta}$

$= \dfrac{t}{2x} \cdot \dfrac{4xt}{x^2 + 3t^2} = \dfrac{2t^2}{x^2 + 3t^2} = \dfrac{2t^2}{(5 - 2\sqrt{3})t^2 + 3t^2} = \dfrac{2t^2}{(8 - 2\sqrt{3})t^2}$

$$= \dfrac{2}{8 - 2\sqrt{3}} = \dfrac{4 + \sqrt{3}}{13}.$$

41. $\mathbf{v} = \langle -9 - 4, 6 - 2\rangle = \langle -13, 4\rangle$; $\|\mathbf{v}\| = \sqrt{(-13)^2 + 4^2} = \sqrt{185}$

45. $\|2\mathbf{u}\| = \|2\langle -12, 5\rangle\| = \|\langle -24, 10\rangle\| = \sqrt{(-24)^2 + 10^2} = 26$;

$2\|\mathbf{u}\| = 2\|\langle -12, 5\rangle\| = 2\sqrt{(-12)^2 + 5^2} = 26$

49. The magnitude of $\mathbf{u}$ is $\|\langle -12, 5\rangle\| = \sqrt{(-12)^2 + 5^2} = 13$. The vector having the same

direction as $\mathbf{u}$ but with magnitude (length) 1 is $\dfrac{1}{\|\mathbf{u}\|}\mathbf{u} = \dfrac{1}{13}\langle -12, 5\rangle = \langle -\dfrac{12}{13}, \dfrac{5}{13}\rangle$.

53. Refer to page 418 of the text between Example 6 and Example 7. In this case,
$\mathbf{v} = \|\mathbf{v}\|\langle\cos\theta, \sin\theta\rangle = 18\langle\cos 225°, \sin 225°\rangle = \langle -9\sqrt{2}, -9\sqrt{2}\rangle$.

57. If we sketch $\mathbf{v}$ in standard position, the terminal point is $P(\sqrt{6} - \sqrt{2}, \sqrt{6} + \sqrt{2})$, which is

in the first quadrant. The distance from the origin to P is $r = \sqrt{(\sqrt{6} - \sqrt{2})^2 + (\sqrt{6} + \sqrt{2})^2}$
$= 4$, so $\|\mathbf{v}\| = 4$. The smallest positive angle that $\mathbf{v}$ makes with the x-axis is

$\theta = \tan^{-1}\left(\dfrac{\sqrt{6} + \sqrt{2}}{\sqrt{6} - \sqrt{2}}\right) = 75°$.

Keystrokes: $6\ \boxed{\sqrt{}}\ \boxed{+}\ 2\ \boxed{\sqrt{}}\ \boxed{=}\ \boxed{\div}\ \boxed{(}\ 6\ \boxed{\sqrt{}}\ \boxed{-}\ 2$

$\boxed{\sqrt{}}\ \boxed{)}\ \boxed{=}$

61. $6\left(\cos\dfrac{5\pi}{4} + i\sin\dfrac{5\pi}{4}\right) = 6\left(-\dfrac{\sqrt{2}}{2} + i\left(-\dfrac{\sqrt{2}}{2}\right)\right) = -3\sqrt{2} - 3\sqrt{3}i$.

65. $10\text{cis}165° = 10(\cos165° + i\sin165°)$. The exact value of $\cos165°$ can be determined as

$\cos(135° + 30°) = \cos135°\cos30° - \sin135°\sin30° = (-\frac{\sqrt{2}}{2})(\frac{\sqrt{3}}{2}) - (\frac{\sqrt{2}}{2})(\frac{1}{2})$

$= \frac{-\sqrt{6} - \sqrt{2}}{4}$. Similarly, the exact value of $\sin165°$ can be determined as

$\sin(135° + 30°) = \sin135°\cos30° + \cos135°\sin30° = (\frac{\sqrt{2}}{2})(\frac{\sqrt{3}}{2}) + (-\frac{\sqrt{2}}{2})(\frac{1}{2})$

$= \frac{\sqrt{6} - \sqrt{2}}{4}$. Thus, $10(\cos165° + i\sin165°) = 10(\frac{-\sqrt{6} - \sqrt{2}}{4} + i \frac{\sqrt{6} - \sqrt{2}}{4})$

$= \frac{-5(\sqrt{6} + \sqrt{2})}{2} + \frac{5(\sqrt{6} - \sqrt{2})}{2}i$

69. Consider the graph of $z = -20 = -20 + 0i$; it lies on the real axis 20 units to the left of the origin. Clearly $r = 20$ and $\theta = 180°$. Thus $-20 = 20(\cos180° + i\sin180°)$ or $20\text{cis}180°$.

73. Refer to the box "Multiplication and Division in Trigonometric Form" on page 427.
$z_1z_2 = (9\text{cis}105°)(3\text{cis}60°) = (9)(3)\text{cis}(105° + 60°) = 27\text{cis}165°$ or $27(\cos165° + i\sin165°)$;
$\frac{z_1}{z_2} = \frac{9\text{cis}105°}{3\text{cis}60°} = \frac{9}{3}\text{cis}(105° - 60°) = 3\text{cis}45°$ or $3(\cos45° + i\sin45°)$.

77. $(4\text{cis}36°)^5 = 4^5\text{cis}(5(36°)) = 1024\text{cis}180°$. In rectangular form this is
$1024(\cos180° + i\sin180°) = 1024(-1 + i(0)) = -1024$

81. Refer to the box on page 434. The four roots are of the form $\sqrt[4]{16}\ \text{cis}(\frac{120° + 360°k}{4})$
$= 2\text{cis}(30° + 90°k)$. Letting $k = 0, 1, 2,$ and 3 gives all four roots:
$2\text{cis}(30° + 90°(0)) = 2\text{cis}30°$; $2\text{cis}(30° + 90°(1)) = 2\text{cis}120°$; $3\text{cis}(30° + 90°(2))$
$= 2\text{cis}210°$; and $2\text{cis}(30° + 90°(3)) = 2\text{cis}300°$.

85. $x^3 + 64 = 0 \Rightarrow x = -64$. Solving this equation is equivalent to finding all the cube roots of -64. The trigonometric form of -64 is $64\text{cis}180°$; so the three roots

have the form $\sqrt[3]{64}\ \text{cis}(\frac{180° + 360°k}{3})$

$= 4\text{cis}(60° + 120°k)$. Letting $k = 0, 1,$ and 2 gives all three roots:

$4\text{cis}(60° + 120°(0)) = 4\text{cis}60°$; $4\text{cis}(60° + 120°(1))$
$= 4\text{cis}180°$; and $4\text{cis}(60° + 120°(2)) = 4\text{cis}300°$.

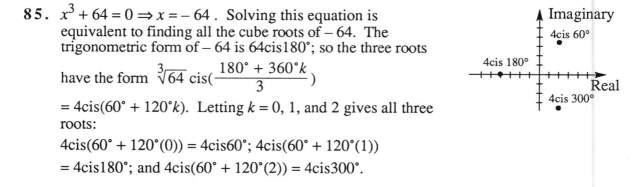

89. $|z + 3| < 5 \Leftrightarrow |z - (-3 + 0i)| < 5$ represents all real numbers z that are less than 5 units from $-3 + 0i$ (in the complex plane).

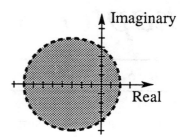

Chapter 8

Section 8.1

---

1. a) Using the point-slope form : $y - 5 = \frac{3}{2}(x + 2) \Rightarrow y - 5 = \frac{3}{2}x + 3 \Rightarrow y = \frac{3}{2}x + 8$

  b) Using the result of a), $y = \frac{3}{2}x + 8 \Rightarrow 2y = 3x + 16 \Rightarrow 3x - 2y + 16 = 0$

5. a) The line has slope $m = \tan\frac{\pi}{4} = 1$. Using the point-slope form :

  $y - 4 = 1(x - 1) \Rightarrow y = x + 3$

  b) Using the result of a), $y = x + 3 \Rightarrow x - y + 3 = 0$

9. The slope of the line is 4, so the angle of inclination is $\tan^{-1} 4 \approx 75.96°$.

13. The given equation in the form $y = mx + b$ is $y = 2x + 5$, so the slope of the line is 2. The angle of inclination is $\tan^{-1} 2 \approx 63.43°$.

17. The slope of the line passing through (4, 4) and (–1, 3) is $\frac{4 - 3}{4 - (-1)} = \frac{1}{5}$. The angle of inclination is $\tan^{-1}\frac{1}{5} \approx 11.31°$.

21. The given equations in he form $y = mx + b$ are $y = 2x + 3$, and $y = -\frac{1}{4}x + \frac{7}{4}$, so the respective slopes are 2 and $-\frac{1}{4}$. The angle θ between the lines is given by

$$\tan\theta = \left| \frac{(-\frac{1}{4}) - 2}{1 + (-\frac{1}{4})(2)} \right| = \left| \frac{-9}{2} \right| = \frac{9}{2}. \text{ Thus, } \theta = \tan^{-1}\frac{9}{2} \approx 77.47°.$$

25. The line determined by the first equation has a slope of $\frac{1}{3}$, but the second equation determines a vertical line and therefore does not have a slope . To get around this, consider two lines perpendicular to the respective given lines and the angle θ between them. The angle between the given lines is the same as the angle θ. The slopes of the respective perpendiculars are –3 and 0. Thus

$$\tan\theta = \left| \frac{(-3) - 0}{1 + (-3)(0)} \right| = 3. \text{ Thus, } \theta = \tan^{-1} 3 \approx 71.57°.$$

29. Refer to the (first) formula at the bottom of page 448 in the text. We have $m = \dfrac{3}{4}$, $b = -1$,

$h = 5$, and $k = -3$: $d = \dfrac{|\,mh + b - k\,|}{\sqrt{1 + m^2}} = \dfrac{|\,(\frac{3}{4})5 + (-1) - (-3)\,|}{\sqrt{1 + (3/4)^2}} = \dfrac{23/4}{\sqrt{25/16}} = \dfrac{23}{5}.$

33. Refer to the second formula at the bottom of page 448 in the text. We have $D = 3$, $E = 2$,

$F = -6$, $h = 0$, and $k = 0$: $d = \dfrac{|\,Dh + Ek + F\,|}{\sqrt{D^2 + E^2}} = \dfrac{|\,(3)0 + (2)0 + (-6)\,|}{\sqrt{3^2 + 2^2}} = \dfrac{6}{\sqrt{13}}.$

37. Letting b take on the values $-1, -\dfrac{1}{2}, 0,$

$\dfrac{1}{2}$, we have $y = x - 1$, $y = x - \dfrac{1}{2}$, $y = x$,

and $y = x + \dfrac{1}{2}$. The graphs of these four

equations are shown at right.

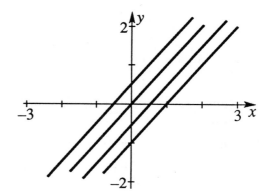

41. Refer to the point-slope form of a linear equation in Section 2.4. Since the equation of a line passing through (h, k) with slope m is $y - k = m(x - h)$, the given equation determines lines passing through $(-2, 2)$ with negative slopes, since $m < 0$. A few examples of this family of lines are shown on the right.

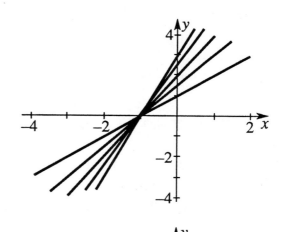

45. If we factor out c in the right side of the equation, we get $y = c(x + 1)$ Since the equation of a line passing through (h, k) with slope m is $y - k = m(x - h)$, the equation $y = c(x + 1)$ determines the family of all lines passing through $(-1, 0)$ with slopes ranging from 1 to 3, inclusive, since $1 \le c \le 3$.

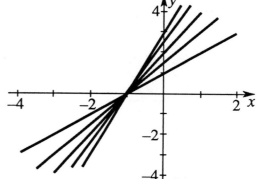

49. The given equation in point-slope form is $y = x + \dfrac{5}{2}$, so the slope of the given line is $m_1 = 1$. Let m_2 represent the slope of a line that makes an angle of $60°$ with the given line. Then $\tan\theta = |\dfrac{m_1 - m_2}{1 + m_1 m_2}| \Rightarrow \tan 60° = |\dfrac{1 - m_2}{1 + m_2}|$

$\Rightarrow \sqrt{3} = \dfrac{1 - m_2}{1 + m_2}$ or $-\sqrt{3} = \dfrac{1 - m_2}{1 + m_2}$. Solving the first of these two equations, we get

$\sqrt{3} = \dfrac{1 - m_2}{1 + m_2} \Rightarrow \sqrt{3} + \sqrt{3}\,m_2 = 1 - m_2 \Rightarrow \sqrt{3}\,m_2 + m_2 = 1 - \sqrt{3}$

$\Rightarrow m_2(\sqrt{3} + 1) = 1 - \sqrt{3} \Rightarrow m_2 = \dfrac{1 - \sqrt{3}}{\sqrt{3} + 1} = \dfrac{(1 - \sqrt{3})(\sqrt{3} - 1)}{(\sqrt{3} + 1)(\sqrt{3} - 1)} = \sqrt{3} - 2$.

Similarly, solving the equation $-\sqrt{3} = \dfrac{1 - m_2}{1 + m_2}$ leads to $m_2 = -\sqrt{3} - 2$.

53. Refer to Example 9, page 450 (in the text). The graph of the circle is centered at $(-3, 9)$ with radius $\sqrt{20}$. There are two lines tangent to the circle passing through $(4, 13)$. The equations for either of them can be written in the form $y = mx + b$ as follows :

$y - 13 = m(x - 4) \Rightarrow y = mx + (13 - 4m)$. The tangent lines are $\sqrt{20}$ units from the center $(-3, 9)$. Therefore

$d = \dfrac{|\,mh + b - k\,|}{\sqrt{1 + m^2}} = \dfrac{|\,m(-3) + (13 - 4m) - (9)\,|}{\sqrt{1 + m^2}} \Rightarrow \sqrt{20} = \dfrac{-7m + 4}{\sqrt{1 + m^2}}.$

Squaring both sides and solving for m we get

$20 = \dfrac{49m^2 - 56m + 16}{1 + m^2} \Rightarrow 20m^2 + 20 = 49m^2 - 56m + 16$

$\Rightarrow 29m^2 - 56m - 4 = 0 \Rightarrow (29m + 2)(m - 2) = 0 \Rightarrow m = -\dfrac{2}{29}$ or $m = 2$.

57. There are two lines that are 3 units from $(4, -4)$ and passing through $(2, 6)$. The equations for either of them can be written in the form $y = mx + b$ as follows :

$y - 6 = m(x - 2) \Rightarrow y = mx + (6 - 2m)$. These lines are 3 units from the $(h, k) = (4, -4)$.

Therefore $d = \dfrac{|\,mh + b - k\,|}{\sqrt{1 + m^2}} = \dfrac{|\,m(4) + (6 - 2m) - (-4)\,|}{\sqrt{1 + m^2}} \Rightarrow 3 = \dfrac{2m + 10}{\sqrt{1 + m^2}}$. Squaring

both sides and solving for m we get $9 = \dfrac{4m^2 + 40m + 100}{1 + m^2}$

$\Rightarrow 9m^2 + 9 = 4m^2 + 40m + 100 \Rightarrow 5m^2 - 40m - 91 = 0 \Rightarrow m = \dfrac{40 \pm \sqrt{1600 - 4(5)(-91)}}{2(5)}$

$\Rightarrow m = \dfrac{20 \pm 3\sqrt{95}}{5}$. Thus, the equations of the lines are $y - 6 = m(x - 2)$

$\Rightarrow y - 6 = \dfrac{20 + 3\sqrt{95}}{5}(x - 2)$ or $y - 6 = \dfrac{20 - 3\sqrt{95}}{5}(x - 2)$

61. The equation $y = mx + b$ in general form is $mx - y + b = 0$, so $D = m$, $E = -1$, and $F = b$.

Therefore $\dfrac{|Dh + Ek + F|}{\sqrt{D^2 + E^2}} = \dfrac{|mh + (-1)k + b|}{\sqrt{m^2 + (-1)^2}} = \dfrac{|mh + b - k|}{\sqrt{1 + m^2}}$.

Section 8.2

---

1. If we graph the vertex and focus, we see that the parabola opens to the right and that the distance from the vertex to the focus is $c = 2$ (See the summary on page 455 of the text.). The equation of the parabola is
$$y^2 = 4cx \implies y^2 = 4(2)x \implies y^2 = 8x.$$

The directrix is $x = -c \implies x = -2$. The length of the (vertical) focal chord is $|4c| = 8$, so count four units up from the focus, and count four units down from the focus to get the points $(2, 4)$ and $(2, -4)$ on the parabola.

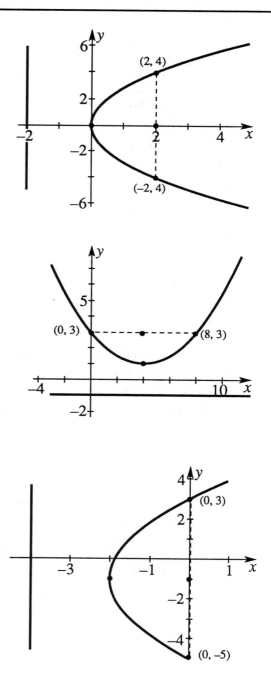

5. If we graph the vertex and focus, we see that the parabola opens up and that the distance from the vertex to the focus is $c = 2$. The equation of the parabola is
$$(x - h)^2 = 4c(y - k) \implies (x - 4)^2 = 4(2)(y - 1)$$
$\implies (x - 4)^2 = 8(y - 1)$. Since the distance from the vertex to the directrix is $c = 2$, the directrix is the line $y = -1$. The length of the (horizontal) focal chord is $|4c| = 8$, so count four units right from the focus, and count four units left from the focus to get the points $(8, 3)$ and $(0, 3)$ on the parabola.

9. If we graph the vertex and focus, we see that the parabola opens right and that the distance from the vertex to the focus is $c = 2$. The equation of the parabola is $(y - k)^2 = 4c(x - h)$
$\implies (y - (-1))^2 = 4(2)(x - (-2))$
$\implies (y + 1)^2 = 8(x + 2)$. Since the distance from the vertex to the directrix is $c = 2$, the directrix is the line $x = -4$. The length of the (vertical) focal chord is $|4c| = 8$, so count four units up from the focus, and count four units down from the focus to get the points $(0, 3)$ and $(0, -5)$ on the parabola.

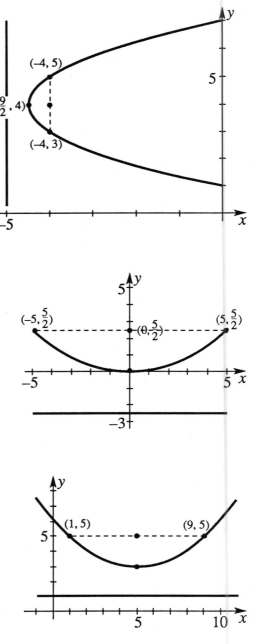

13. If we graph the directrix and focus, we see that the parabola opens right and that the vertex is $(-\frac{9}{2}, 4)$. The distance from the vertex to the focus is $c = \frac{1}{2}$. The equation of the parabola is

$$(y - k)^2 = 4c(x - h)$$

$$\Rightarrow (y - 4)^2 = 4(\tfrac{9}{2})(x - -\tfrac{9}{2})$$

$$\Rightarrow (y - 4)^2 = 2(x + \tfrac{9}{2}).$$ The length of the (vertical) focal chord is $|4c| = 2$, so count one unit up from the focus, and count one unit down from the focus to get the points $(-4, 5)$ and $(-4, 3)$ on the parabola.

17. Comparing the given equation to $x^2 = 4cy$, we see that $4c = 10$. Solving for c, we find the distance from the focus to the vertex $(0, 0)$ is $c = \frac{5}{2}$. Thus the focus is at $(0, \frac{5}{2})$. The length of the (horizontal) focal chord is $|4c| = 10$, so count five units right from the focus, and count five units left from the focus to get the points $(5, \frac{5}{2})$ and $(-5, \frac{5}{2})$ on the parabola.

21. The equation is quadratic in x, so we complete the square with the terms involving x:

$$x^2 - 10x - 8y + 49 = 0$$

$$\Rightarrow x^2 - 10x = 8y - 49 \Rightarrow x^2 - 10x + 25$$

$$= 8y - 49 + 25 \Rightarrow (x - 5)^2 = 8(y - 3).$$ This is the parabola $x^2 = 8y$ shifted 5 units right and 3 units up. The vertex is at $(5, 3)$, and the focus is $c = 2$ units above. The (horizontal) directrix is 2 units below $(5, 3)$ as well. The length of the (horizontal) focal chord is $|4c| = 8$, so count four units right from the focus, and count four units left from the focus to get the points $(9, 5)$ and $(1, 5)$ on the parabola.

25. The equation is quadratic in y, so we complete the square with the terms involving y:
$y^2 + 8y - 4x + 16 = 0$
$\Rightarrow y^2 + 8y = 4x - 16 \Rightarrow y^2 + 8y + 16$
$= 4x - 16 + 16 \Rightarrow (y + 4)^2 = 4x$. This is the parabola $y^2 = 4x$ shifted 4 units down. The vertex is at $(0, -4)$, and the focus is $c = 1$ unit to the right. The (vertical) directrix is 1 unit to the left of $(0, -4)$ as well. The length of the (vertical) focal chord is $|4c| = 4$, so count two units above from the focus, and count two units below the focus to get the points $(1, -2)$ and $(1, -6)$ on the parabola.

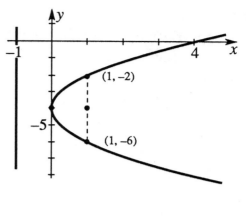

29. The equation is quadratic in x, so we complete the square with the terms involving x:
$3x^2 - 12x - 8y = 4 \Rightarrow 3(x^2 - 4x) = 8y + 4$
$\Rightarrow 3(x^2 - 4x + 4) = 8y + 4 + 12$
$\Rightarrow 3(x - 2)^2 = 8(y + 2) \Rightarrow (x - 2)^2 = \frac{8}{3}(y + 2)$.

This is the parabola $x^2 = \frac{8}{3}y$ shifted 2 units right and 2 units down. The vertex is at $(2, -2)$, and the focus is $c = \frac{2}{3}$ units above. The (horizontal) directrix is $\frac{2}{3}$ units below $(2, -2)$ as well.

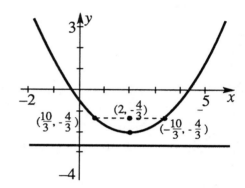

The length of the (horizontal) focal chord is $|4c| = \frac{8}{3}$, so count $\frac{4}{3}$ units right from the focus, and count $\frac{4}{3}$ units left from the focus to get the points $(\frac{10}{3}, -\frac{4}{3})$ and $(\frac{2}{3}, -\frac{4}{3})$ on the parabola.

33. Since the directrix is vertical, the parabola opens left or right. Since the length of the focal chord is 6, $|4c| = 6 \Rightarrow c = \pm\frac{3}{2}$. In the case of $c = \frac{3}{2}$, the parabola opens to the right and the vertex is $\frac{3}{2}$ units to the left of the focus; that is the vertex is at $(-\frac{1}{2}, 2)$. In the case of $c = -\frac{3}{2}$, the parabola opens to the left and the vertex is $\frac{3}{2}$ units to the right of the focus; so the vertex is at $(\frac{5}{2}, 2)$. Substituting each case into $(y - k)^2 = 4c(x - h)$, we have
$(y - 2)^2 = 6(x + \frac{1}{2})$ or $(y - 2)^2 = -6(x - \frac{5}{2})$.

37. Since the directrix is vertical, the parabola opens left or right, so we know the equation can be written in the form $(y - k)^2 = 4c(x - h)$. Furthermore, the vertex is $(h, k) = (2, 0)$, so the equation is $(y - 0)^2 = 4c(x - 2) \Rightarrow y^2 = 4c(x - 2)$. We are given that $(1, 2\sqrt{2})$ satisfies this equation: $(2\sqrt{2})^2 = 4c(1 - 2) \Rightarrow 8 = -4c \Rightarrow 4c = -8$. Thus the equation is $y^2 = -8(x - 2)$.

41. The angle between the the tangent line and the line from the focus to P is the same as the angle of inclination of the tangent line. (See Example 7 on page 458 and the discussion that follows.) The angle is $\tan^{-1} \dfrac{3}{2} \approx 56.31°$.

45. If we place the parabola on a coordinate system so that the vertex is at the origin, the equation of the parabola can be written in the form $x^2 = 4cy$. Since $w = 6$ and $h = 4$, the parabola contains the point $(3, 4)$. Substituting this into the equation and solving for c gives $3^2 = 4c(4) \Rightarrow c = \dfrac{9}{16}$. The bulb should be placed $\dfrac{9}{16}$ units above the vertex.

49. If we place the parabola on a coordinate system so that the vertex is at the origin, the equation of the parabola can be written in the form $x^2 = 4cy$. Since the cable is 30 feet above the road surface at a point 40 feet from the vertex, the parabola contains the point $(40, 30)$. Substituting this into the equation and solving for c gives $40^2 = 4c(30)$ $\Rightarrow 4c = \dfrac{160}{3}$. The equation of the cable is $x^2 = \dfrac{160}{3} y$. Since the towers are 270 feet apart, substitute 135 for x in the equation and solve for y: $135^2 = \dfrac{160}{3} y \Rightarrow y \approx 342$ feet.

Section 8.3

1. The graph is an ellipse. To find the x-intercepts set $y = 0 \Rightarrow \dfrac{x^2}{49} + \dfrac{0^2}{9} = 1$

$\Rightarrow x = \pm 7$. If we set $x = 0$ and solve for y, we obtain the y-intercepts:

$\dfrac{0^2}{49} + \dfrac{y^2}{9} = 1 \Rightarrow y = \pm 3$. The points $(7, 0)$, $(-7, 0)$, $(0, 3)$, and $(0, -3)$ are enough to get a sketch of the graph.

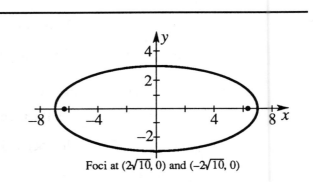

Foci at $(2\sqrt{10}, 0)$ and $(-2\sqrt{10}, 0)$

The major axis is horizontal. The foci are $c = \sqrt{a^2 - b^2} = \sqrt{49 - 9} = 2\sqrt{10}$ units from the center $(0, 0)$ along the major axis.

5. Divide each side of the equation by 9: $9x^2 + y^2 = 9$

$\Rightarrow \dfrac{x^2}{1} + \dfrac{y^2}{9} = 1$. To find the x-intercepts set $y = 0$:

$\Rightarrow \dfrac{x^2}{1} + \dfrac{0^2}{9} = 1 \Rightarrow x = \pm 1$. If we set $x = 0$ and

solve for y, we obtain the y-intercepts:

$\dfrac{0^2}{1} + \dfrac{y^2}{9} = 1 \Rightarrow y = \pm 3$. The points $(1, 0)$,

$(-1, 0)$, $(0, 3)$, and $(0, -3)$ are enough to get a
sketch of the graph. The major axis is vertical. The

foci are $c = \sqrt{a^2 - b^2} = \sqrt{9 - 1} = 2\sqrt{2}$ units
from the center $(0, 0)$ along the major axis.

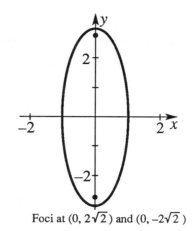

Foci at $(0, 2\sqrt{2})$ and $(0, -2\sqrt{2})$

9. The graph of $\dfrac{(x + 3)^2}{25} + \dfrac{(y - 5)^2}{16} = 1$ is the

ellipse $\dfrac{x^2}{25} + \dfrac{y^2}{16} = 1$ centered at $(-3, 5)$. The

graph of $\dfrac{x^2}{25} + \dfrac{y^2}{16} = 1$ has intercepts $(5, 0)$,

$(-5, 0)$, $(0, 4)$, and $(0, -4)$. The major axis is
horizontal. The foci are

$c = \sqrt{a^2 - b^2} = \sqrt{25 - 16} = 3$ units from the
center along the major axis. Shifting this ellipse
so that the center is at $(-3, 5)$, the
corresponding vertices are $(2, 5)$, $(-8, 5)$,
$(-3, 9)$, and $(-3, 1)$. The corresponding foci
are at $(0, 5)$ and $(-6, 5)$.

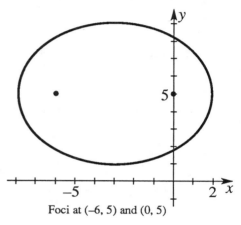

Foci at $(-6, 5)$ and $(0, 5)$

13. Since $2x^2$ is nonnegative and $5y^2$ is nonnegative, $2x^2 + 5y^2$ is nonnegative. Therefore,
there are no real values x and y that satisfy $2x^2 + 5y^2 = -20$.

17. Note that y is nonnegative. Squaring both
sides gives $y^2 = 1 - 4x^2$

$\Rightarrow 4x^2 + y^2 = 1 \Rightarrow \dfrac{x^2}{1/4} + y^2 = 1$. The

graph is an ellipse with vertices $(0, 1)$,

$(0, -1)$, $(\frac{1}{2}, 0)$, and $(-\frac{1}{2}, 0)$. The major

axis is vertical. The foci are

$c = \sqrt{a^2 - b^2} = \sqrt{1 - 1/4} = \dfrac{\sqrt{3}}{2}$ units from

the center along the major axis. Since y
must be nonnegative, the graph of

$y = \sqrt{1 - 4x^2}$ is only the upper half of the
ellipse.

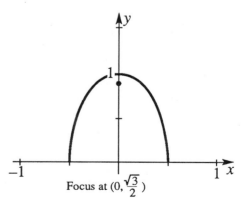

Focus at $(0, \frac{\sqrt{3}}{2})$

21. First we will complete the square in both variables.

$9x^2 + 4y^2 - 54x - 8y + 49 = 0$

$\Rightarrow 9x^2 - 54x + 4y^2 - 8y = -49$

$\Rightarrow 9(x^2 - 6x) + 4(y^2 - 2y) = -49$

$\Rightarrow 9(x^2 - 6x + 9) + 4(y^2 - 2y + 1) = -49 + 81 + 4$

$\Rightarrow 9(x - 3)^2 + 4(y - 1)^2 = 36$

$\Rightarrow \dfrac{(x-3)^2}{4} + \dfrac{(y-1)^2}{9} = 1$

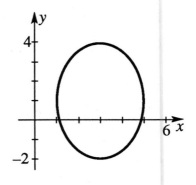

The graph is the ellipse $\dfrac{x^2}{4} + \dfrac{y^2}{9} = 1$ centered at $(3, 1)$.

25. The equation determines an ellipse with $a^2 = 49 \Rightarrow a = 7$. Thus $d(P, F_1) + d(P, F_2) = 2a = 14$.

29. The equation determines an ellipse with $a^2 = 64 \Rightarrow a = 8$. Thus $d(P, F_1) + d(P, F_2) = 2a = 16$.

33. The center of the ellipse is the midpoint of major axis, which is $\left(\dfrac{3+7}{2}, \dfrac{2+2}{2}\right) = (5, 2)$. The length of the major axis is $2a = 4 \Rightarrow a = 2$. The eccentricity is $\dfrac{c}{a} = \dfrac{c}{2}$ so $\dfrac{c}{2} = \dfrac{1}{2} \Rightarrow c = 1$. Since $b^2 = a^2 - c^2 \Rightarrow b^2 = 4 - 1 = 3$. The equation of the ellipse is $\dfrac{(x-h)^2}{a^2} + \dfrac{(y-k)^2}{b^2} = 1 \Rightarrow \dfrac{(x-5)^2}{4} + \dfrac{(y-2)^2}{3} = 1$.

37. Setting $y = 0$ we get the x-intercepts: $\dfrac{x^2}{16} - \dfrac{0^2}{9} = 1 \Rightarrow x = \pm 4$. Thus, the hyperbola opens along the x-axis. The asymptotes are $y = \pm\dfrac{b}{a}x = \pm\dfrac{3}{4}x$. The foci are $c = \sqrt{a^2 + b^2} = \sqrt{16 + 9} = 5$ units from the center $(0, 0)$ along the x-axis.

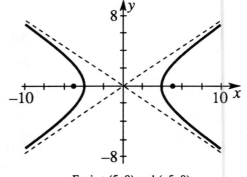

Foci at $(5, 0)$ and $(-5, 0)$

41. Dividing each side by 2, we get

$\frac{x^2}{2} - \frac{y^2}{2} = 1$. Setting $y = 0$ we get the x-

intercepts: $\frac{x^2}{2} - \frac{0^2}{2} = 1 \Rightarrow x = \pm \sqrt{2}$.

Thus, the hyperbola opens along the x-axis. The asymptotes are

$y = \pm \frac{b}{a} x = \pm \frac{\sqrt{2}}{\sqrt{2}} x = \pm x$. The foci are

$c = \sqrt{a^2 + b^2} = \sqrt{2 + 2} = 2$ units from the center $(0, 0)$ along the x-axis.

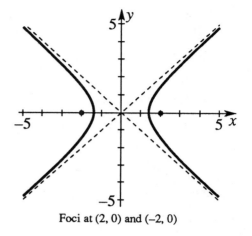

Foci at $(2, 0)$ and $(-2, 0)$

45. Dividing each side by -9, we get

$\frac{(y + 1)^2}{9} - \frac{x^2}{9} = 1$. The graph of this is the

same as that of $\frac{y^2}{9} - \frac{x^2}{9} = 1$ shifted 1 unit

down. The graph of

$\frac{y^2}{9} - \frac{x^2}{9} = 1$ has intercepts $(0, \pm 3)$. Therefore,

the hyperbola opens along the y-axis. The

asymptotes for $\frac{y^2}{9} - \frac{x^2}{9} = 1$ are

$y = \pm \frac{a}{b} x = \pm \frac{3}{3} x \Rightarrow y = \pm x$.

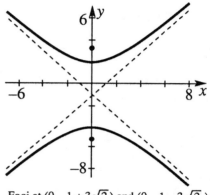

Foci at $(0, -1 + 3\sqrt{2})$ and $(0, -1 - 3\sqrt{2})$

The foci are $c = \sqrt{a^2 + b^2} = \sqrt{9 + 9} = 3\sqrt{2}$ units from the center along the y-axis. Finally, translate the graph (including asymptotes and foci) 1 unit down.

49. Note that y is positive. Squaring both sides gives $y^2 = 1 + 16x^2$

$\Rightarrow y^2 - 16x^2 = 1 \Rightarrow y^2 - \frac{x^2}{1/16} = 1$. The

graph is a hyperbola with intercepts $(0, 1)$, and $(0, -1)$. Thus, the hyperbola opens along the y-axis. The asymptotes are

$y = \pm \frac{a}{b} x = \pm \frac{1}{1/4} x = \pm 4x$. The foci are c

$= \sqrt{a^2 + b^2} = \sqrt{1 + \frac{1}{16}} = \frac{\sqrt{17}}{4}$ units

from the center $(0, 0)$ along the y-axis.

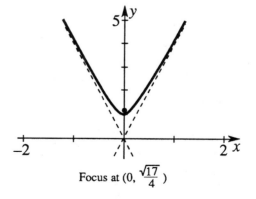

Focus at $(0, \frac{\sqrt{17}}{4})$

Since y must be positive, the graph of $y = \sqrt{1 + 16x^2}$ is only the upper half of the hyperbola.

53. First we will complete the square:

$$4x^2 - 25y^2 + 200y - 500 = 0$$
$$\Rightarrow 4x^2 - 25(y^2 - 8y) = 500$$
$$\Rightarrow 4x^2 - 25(y^2 - 8y + 16) = 500 - 400$$
$$\Rightarrow 4x^2 - 25(y - 4)^2 = 100$$
$$\Rightarrow \frac{x^2}{25} - \frac{(y-4)^2}{4} = 1$$

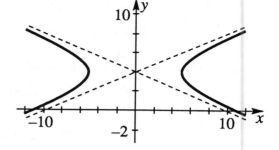

The graph is the hyperbola $\frac{x^2}{25} - \frac{y^2}{4} = 1$ centered at $(0, 4)$. The slopes of the asymptotes are $\pm\frac{b}{a} = \pm\frac{2}{5}$.

57. First we complete the square.

$$4x^2 - y^2 - 8x - 6y - 3 = 0$$
$$\Rightarrow 4x^2 - 8x - y^2 - 6y = 3$$
$$\Rightarrow 4(x^2 - 2x) - (y^2 + 6y) = 3$$
$$\Rightarrow 4(x^2 - 2x + 1) - (y^2 + 6y + 9) = 3 + 4 - 9$$
$$\Rightarrow 4(x - 1)^2 - (y + 3)^2 = -2$$
$$\Rightarrow \frac{(y+3)^2}{2} - \frac{(x-1)^2}{1/2} = 1. \text{ The graph is the}$$

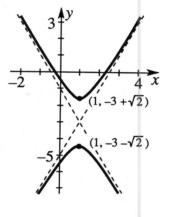

hyperbola $\frac{y^2}{2} - \frac{x^2}{1/2} = 1$ centered at $(1, -3)$.

The slopes of the asymptotes are $\pm\frac{a}{b} = \pm\frac{\sqrt{2}}{1/\sqrt{2}} = \pm 2$.

61. The center of the hyperbola is the midpoint of the segment joining the foci, so the center is

$(\frac{3+3}{2}, \frac{2+-6}{2}) = (3, -2)$. The value of c is the distance from the center to a focus

$\Rightarrow c = 4$. Since the hyperbola opens up and down, the equation has the form

$$\frac{(y-k)^2}{a^2} - \frac{(x-h)^2}{b^2} = 1 \Rightarrow \frac{(y+2)^2}{a^2} - \frac{(x-3)^2}{b^2} = 1. \text{ Since } (3, 1) \text{ is on the graph,}$$

$$\frac{(1+2)^2}{a^2} + \frac{(3-3)^2}{b^2} = 1 \Rightarrow \frac{3^2}{a^2} = 1 \Rightarrow a^2 = 9. \text{ Now we can determine } b^2 :$$

$b^2 = c^2 - a^2 = 16 - 9 = 7$. Therefore the equation is

$$\frac{(x-h)^2}{a^2} + \frac{(y-k)^2}{b^2} = 1 \Rightarrow \frac{(y+2)^2}{9} - \frac{(x-3)^2}{7} = 1.$$

65. The center of the hyperbola is the intersection of the asymptotes, so by carefully graphing the asymptotes we get the center is $(-3, 1)$. If we plot the point $(3, 1)$, which is on the hyperbola, we see that it must open left and right. Therefore the equation has the form

$$\frac{(x-h)^2}{a^2} - \frac{(y-k)^2}{b^2} = 1 \Rightarrow \frac{(x+3)^2}{a^2} - \frac{(y-1)^2}{b^2} = 1. \text{ Since } (3, 1) \text{ is on the graph,}$$

$$\frac{(3+3)^2}{a^2} + \frac{(1-1)^2}{b^2} = 1 \Rightarrow a^2 = 36. \text{ The slopes of the asymptotes are } \pm\frac{b}{a} = \pm\frac{2}{3},$$

and since $a = 6$ we have $\frac{b}{6} = \frac{2}{3} \Rightarrow b = 4$. Therefore the equation is

$$\frac{(x+3)^2}{a^2} - \frac{(y-1)^2}{b^2} = 1 \Rightarrow \frac{(x+3)^2}{36} - \frac{(y-1)^2}{16} = 1.$$

69. The graph is a hyperbola since the equation is quadratic in both variables and the two quadratic terms have opposite signs. Completing the square in each variable gives

$$9y^2 - 18y - 4x^2 - 8x = 4$$
$$\Rightarrow 9(y^2 - 2y) - 4(x^2 + 2x) = 4$$
$$\Rightarrow 9(y^2 - 2y + 1) - 4(x^2 + 2x + 1) = 4 + 9 - 4$$
$$\Rightarrow 9(y-1)^2 - 4(x+1)^2 = 9$$
$$\Rightarrow \frac{(y-1)^2}{1} - \frac{(x+1)^2}{9/4} = 1$$

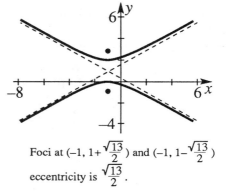

Foci at $(-1, 1+\frac{\sqrt{13}}{2})$ and $(-1, 1-\frac{\sqrt{13}}{2})$

eccentricity is $\frac{\sqrt{13}}{2}$.

The graph is the hyperbola $\frac{y^2}{1} - \frac{x^2}{9/4} = 1$ centered at $(-1, 1)$. The hyperbola crosses the vertical axis of symmetry $a = 1$ unit above and below the center. The slopes of the asymptotes are $\pm\frac{a}{b} = \pm\frac{1}{3/2} = \pm\frac{2}{3}$. The foci are $c = \sqrt{a^2 + b^2} = \sqrt{1 + \frac{9}{4}} = \frac{\sqrt{13}}{2}$ units above and below the center $(-1, 1)$. The eccentricity is $\frac{c}{a} = \frac{\sqrt{13}/2}{1} = \frac{\sqrt{13}}{2}$.

73. Letting $k = 0$, we get $x^2 + 27y^2 = 0$. The only solution to this equation is $(x, y) = (0, 0)$. Letting $k = 3$, we get

$$x^2 + 27y^2 = 3 \Rightarrow \frac{x^2}{3} + \frac{y^2}{1/9} = 1. \text{ The}$$

graph of this equation is an ellipse with

intercepts $(\pm\sqrt{3}, 0)$ and $(0, \pm\frac{1}{3})$. Letting $k = 9$, we get $x^2 + 27y^2 = 9 \Rightarrow \frac{x^2}{9} + \frac{y^2}{1/3} = 1$.

The graph of this equation is an ellipse with intercepts $(\pm 3, 0)$ and $(0, \pm\frac{1}{\sqrt{3}})$.

77. Let the coordinates of P be (x, y). Then a perpendicular segment from P to the y-axis will have length x, a perpendicular segment from P to the x-axis will have length y (see figure on the right). Two similar right triangles are formed: a smaller right triangle having legs x, $\sqrt{4 - x^2}$, and hypotenuse 2 ; and a larger

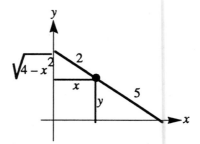

right triangle having corresponding sides $\sqrt{25 - y^2}$, y, and 5. The following proportion holds $\dfrac{\sqrt{4 - x^2}}{y} = \dfrac{2}{5}$. Squaring both sides and simplifying gives $\dfrac{x^2}{4} + \dfrac{y^2}{25} = 1$.

Section 8.4

1. The graph is an ellipse centered at $(0, 0)$, with $a = 7$, $b = 3$, and $c = \sqrt{a^2 - b^2} = 2\sqrt{10}$. The eccentricity is $e = \dfrac{c}{a} = \dfrac{2\sqrt{10}}{7}$. The directrices are given by $x = \pm\dfrac{a}{e} = \pm\dfrac{7}{2\sqrt{10}/7}$
$= \pm\dfrac{49\sqrt{10}}{20}$.

5. $x^2 + 4y^2 + 6x - 16y - 11 = 0 \Rightarrow x^2 + 6x + 4(y^2 - 4y) = 11$
$\Rightarrow (x^2 + 6x + 9) + 4(y^2 - 4y + 4) = 11 + 9 + 16 \Rightarrow (x + 3)^2 + 4(y - 2)^2 = 36$
$\Rightarrow \dfrac{(x + 3)^2}{36} + \dfrac{(y - 2)^2}{9} = 1$

The graph is an ellipse centered at $(-3, 2)$ with $a = 6$, $b = 3$, and $c = \sqrt{a^2 - b^2} = 3\sqrt{3}$. The eccentricity is $e = \dfrac{c}{a} = \dfrac{3\sqrt{3}}{6} = \dfrac{\sqrt{3}}{2}$. The directrices are $\dfrac{a}{e} = \dfrac{6}{\sqrt{3}/2} = 4\sqrt{3}$ units from the center. Thus, the equation of the directrices are $x = -3 + 4\sqrt{3}$, $x = -3 - 4\sqrt{3}$.

9. $x^2 - 4x - 4y + 8 = 0 \Rightarrow x^2 - 4x = 4y - 8 \Rightarrow x^2 - 4x + 4 = 4y - 8 + 4$
$\Rightarrow (x - 2)^2 = 4(y - 1)$. The graph is a parabola opening upward with vertex at $(2, 1)$. Comparing the last equation to the form $(x - h)^2 = 4c(y - k) \Rightarrow 4c = 4 \Rightarrow c = 1$. The eccentricity of any parabola is 1. The directrix is $c = 1$ unit below the vertex; therefore the equation of the directrix is $y = 0$.

13. Sketch the directrices and foci. The hyperbola is centered at the origin, and opens to the left and right. The foci are 10 units from the center, so $c = 10$. The directrices are 9 units from the center, so $\dfrac{a}{e} = 9 \Rightarrow \dfrac{a^2}{c} = 9 \Rightarrow a^2 = 90$. Since $b^2 = c^2 - a^2$, $b^2 = 10$. Thus, an equation of the hyperbola is $\dfrac{x^2}{90} - \dfrac{y^2}{10} = 1$.

17. We are given that $e = \frac{1}{4}$, so the set of points describes an ellipse. Let (x, y) be any point on the ellipse. Using the distance formula, we have $\sqrt{(x-2)^2 + y^2} = \frac{1}{4}|5 - x|$

$\Rightarrow (x-2)^2 + y^2 = \frac{1}{16}(5-x)^2 \Rightarrow x^2 - 4x + 4 + y^2 = \frac{1}{16}(x^2 - 10x + 25)$

$\Rightarrow 16x^2 - 64x + 64 + 16y^2 = x^2 - 10x + 25 \Rightarrow 15x^2 + 16y^2 - 54x + 39 = 0.$

21. $x^2 + 3y^2 = 36 \Rightarrow \frac{x^2}{36} + \frac{y^2}{12} = 1$. Therefore, $c = \sqrt{a^2 - b^2} = \sqrt{36 - 12} = 2\sqrt{6}$, and so the foci are at $(\pm 2\sqrt{6}, 0)$. The slope of the tangent line is $m_1 = -\frac{1}{3}$ and the slope of the segment from $(2\sqrt{6}, 0)$ to $(3, 3)$ is $m_2 = \dfrac{3}{3 - 2\sqrt{6}} = -\dfrac{3 + 2\sqrt{6}}{5}$. The angle θ between these lines can be determined by

$$\tan \theta = \left| \frac{m_1 - m_2}{1 + m_1 m_2} \right| = \left| \frac{(-\frac{1}{3}) - (-\frac{3 + 2\sqrt{6}}{5})}{1 + (-\frac{1}{3})(-\frac{3 + 2\sqrt{6}}{5})} \right| = \frac{2 + 3\sqrt{6}}{9 + \sqrt{6}}$$

$$\Rightarrow \theta = \tan^{-1}\left(\frac{2 + 3\sqrt{6}}{9 + \sqrt{6}}\right) \approx 39.23°.$$

Section 8.5

1. Comparing $x^2 - 2xy + y^2 + 5x + 2y - 45 = 0$ to the general second-degree equation $Ax^2 + Bxy + Cy^2 + Dx + Ey + F = 0$, we have $A = 1$, $B = -2$, and $C = 1$. Since $B^2 - 4AC = (-2)^2 - 4(1)(1) = 0$, the graph of the equation is a (rotated) parabola.

5. Comparing $x^2 + 10xy + 4y^2 + 0x + 2y - 16 = 0$ to the general second-degree equation $Ax^2 + Bxy + Cy^2 + Dx + Ey + F = 0$, we have $A = 1$, $B = 10$, and $C = 4$. Since $B^2 - 4AC = (10)^2 - 4(1)(4) > 0$, the graph of the equation is a (rotated) hyperbola.

9. Comparing $4x^2 + \sqrt{3}xy + 3y^2 + 11x + 2y - 35 = 0$ to the general second-degree equation $Ax^2 + Bxy + Cy^2 + Dx + Ey + F = 0$, we have $A = 4$, $B = \sqrt{3}$, and $C = 3$. The angle θ of rotation can be determined by $\cot 2\theta = \dfrac{A - C}{B} = \dfrac{4 - 3}{\sqrt{3}} = \dfrac{1}{\sqrt{3}}$. The smallest positive angle with cotangent $\dfrac{1}{\sqrt{3}}$ is $60° \Rightarrow 2\theta = 60° \Rightarrow \theta = 30°$.

13. To find the angle of rotation, solve $\cot 2\theta = \dfrac{A-C}{B} = \dfrac{29-29}{42} = 0 \Rightarrow 2\theta = 90°$

$\Rightarrow \theta = 45°$. The rotation of axes formulas are $x = x'\cos\theta - y'\sin\theta = \dfrac{\sqrt{2}}{2}x' - \dfrac{\sqrt{2}}{2}y'$

and $y = x'\sin\theta + y'\cos\theta = \dfrac{\sqrt{2}}{2}x' + \dfrac{\sqrt{2}}{2}y'$. Substituting for x and y into the equation

$29x^2 + 42xy + 29y^2 = 200$ we have

$29\left(\dfrac{\sqrt{2}}{2}x' - \dfrac{\sqrt{2}}{2}y'\right)^2 + 42\left(\dfrac{\sqrt{2}}{2}x' - \dfrac{\sqrt{2}}{2}y'\right)\left(\dfrac{\sqrt{2}}{2}x' + \dfrac{\sqrt{2}}{2}y'\right) + 29\left(\dfrac{\sqrt{2}}{2}x' + \dfrac{\sqrt{2}}{2}y'\right)^2 = 200$

$\Rightarrow 29(\tfrac{1}{2})(x'-y')^2 + 42(\tfrac{1}{2})(x'-y')(x'+y') + 29(\tfrac{1}{2})(x'+y')^2 = 200$

$\Rightarrow (\tfrac{29}{2})(x'^2 - 2x'y' + y'^2) + 21(x'^2 - y'^2) + (\tfrac{29}{2})(x'^2 + 2x'y' + y'^2) = 200$

$\Rightarrow 50x'^2 + 8y'^2 = 200$

$\Rightarrow \dfrac{x'^2}{4} + \dfrac{y'^2}{25} = 1$. The graph is the ellipse
with intercepts (2,0), (–2, 0), (0, 5), and
(0, –5) rotated 45°.

17. To find the angle of rotation, solve $\cot 2\theta = \dfrac{A-C}{B} = \dfrac{11-1}{10\sqrt{3}} = \dfrac{1}{\sqrt{3}} \Rightarrow 2\theta = 60°$

$\Rightarrow \theta = 30°$. The rotation of axes formulas are $x = x'\cos\theta - y'\sin\theta = \dfrac{\sqrt{3}}{2}x' - \dfrac{1}{2}y'$

and $y = x'\sin\theta + y'\cos\theta = \dfrac{1}{2}x' + \dfrac{\sqrt{3}}{2}y'$. Substituting for x and y into the equation

$11x^2 + 10\sqrt{3}xy + y^2 = 4$ we have

$11\left(\dfrac{\sqrt{3}}{2}x' - \dfrac{1}{2}y'\right)^2 + 10\sqrt{3}\left(\dfrac{\sqrt{3}}{2}x' - \dfrac{1}{2}y'\right)\left(\dfrac{1}{2}x' + \dfrac{\sqrt{3}}{2}y'\right) + \left(\dfrac{1}{2}x' + \dfrac{\sqrt{3}}{2}y'\right)^2 = 4$

$\Rightarrow 11(\tfrac{1}{4})(\sqrt{3}x' - y')^2 + 10\sqrt{3}(\tfrac{1}{4})(\sqrt{3}x' - y')(x' + \sqrt{3}y') + (\tfrac{1}{4})(x' + \sqrt{3}y')^2 = 4$

$\Rightarrow (\tfrac{11}{4})(3x'^2 - 2\sqrt{3}x'y' + y'^2) + (\tfrac{5\sqrt{3}}{2})(\sqrt{3}x'^2 + 2x'y' - \sqrt{3}y'^2)$

$\qquad\qquad\qquad\qquad + (\tfrac{1}{4})(x'^2 + 2\sqrt{3}x'y' + 3y'^2) = 4$

$\Rightarrow 16x'^2 - 4y'^2 = 4$

$\Rightarrow \dfrac{x'^2}{1/4} - \dfrac{y'^2}{1} = 1$. The graph is the

hyperbola with intercepts $(\frac{1}{2}, 0)$, $(-\frac{1}{2}, 0)$,

and asymptotes $y = \pm 4x$ rotated 30°.

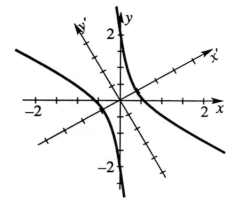

21. Comparing $xy + 2x - 4y = 0$ to the general second-degree equation

$Ax^2 + Bxy + Cy^2 + Dx + Ey + F = 0$, we have $A = 0$, $B = 1$, and $C = 0$. Since

$B^2 - 4AC = 1^2 - 4(0)(0) > 0$, the graph of the equation is a (rotated) hyperbola. Solving

for y we get $xy + 2x - 4y = 0 \Rightarrow xy - 4y = -2x \Rightarrow y(x - 4) = -2x \Rightarrow y = -\dfrac{2}{x-4}$.

This is a rational function, and we can use the
graphing techniques developed in Section 3.4.
Setting the numerator equal to zero, we get the
y-intercept when $-2x = 0 \Rightarrow x = 0$. Setting the
denominator equal to zero, we get the vertical
asymptote when $x - 4 = 0 \Rightarrow x = 4$. The line
$y = -2$ is a horizontal asymptote.

25. Comparing $5x^2 - 4xy + y^2 - 4 = 0$ to the general second-degree equation

$Ax^2 + Bxy + Cy^2 + Dx + Ey + F = 0$, we have $A = 5$, $B = -4$, and $C = 1$. Since

$B^2 - 4AC = (-4)^2 - 4(5)(1) < 0$, the graph of the equation is an (rotated) ellipse. To solve

for y we first write the equation as a quadratic equation in y, and then use the quadratic

formula: $y^2 - 4xy + (5x^2 - 4) = 0 \Rightarrow y = \dfrac{-b \pm \sqrt{b^2 - 4ac}}{2a}$

$\Rightarrow y = \dfrac{-(-4x) \pm \sqrt{(-4x)^2 - 4(1)(5x^2 - 4)}}{2(1)}$

$$\Rightarrow y = \frac{4x \pm 2\sqrt{4 - x^2}}{2}$$

$\Rightarrow y = 2x \pm \sqrt{4 - x^2}$. If we first graph $y = 2x$

and $y = \sqrt{4 - x^2}$ on the same set of axes,
then we can use the technique of adding y-coordinates to graph the ellipse

$$y = 2x \pm \sqrt{4 - x^2} \; .$$

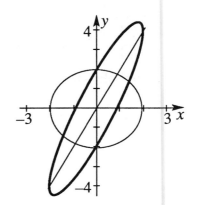

29. Substituting $(x'\cos\theta - y'\sin\theta)$ for x and $(x'\sin\theta + y'\cos\theta)$ for y into
$Ax^2 + Bxy + Cy^2 + Dx + Ey + F = 0$, yields
$A(x'\cos\theta - y'\sin\theta)^2 + B(x'\cos\theta - y'\sin\theta)(x'\sin\theta + y'\cos\theta)$
$\qquad + C(x'\sin\theta + y'\cos\theta)^2 + D(x'\cos\theta - y'\sin\theta) + E(x'\sin\theta + y'\cos\theta) + F = 0$.
Expanding and simplifying, we get $A'x^2 + B'xy + C'y^2 + D'x + E'y + F' = 0$, where
$A' = A\cos^2\theta + B\cos\theta\sin\theta + C\sin^2\theta$, $B' = B(\cos^2\theta - \sin^2\theta) + 2(C - A)\cos\theta\sin\theta$, and
$C' = A\sin^2\theta - B\cos\theta\sin\theta + C\cos^2\theta$. Expanding B'^2 we get
$B'^2 = [B(\cos^2\theta - \sin^2\theta) + 2(C - A)\cos\theta\sin\theta]^2$
$= B^2[\cos^4\theta - 2\cos^2\theta\sin^2\theta + \sin^4\theta] + 4B(C - A)[\cos^3\theta\sin\theta - \cos\theta\sin^3\theta]$
$\qquad\qquad\qquad\qquad\qquad\qquad + [C - A]^2(4\cos^2\theta\sin^2\theta)$

Expanding $4A'C'$, we get
$4A'C' = 4(A^2 - B^2 + C^2)\cos^2\theta\sin^2\theta + 4(BC - AB)\cos^3\theta\sin\theta + 4AC\cos^4\theta + 4AC\sin^4\theta$
$\qquad\qquad\qquad\qquad\qquad\qquad + 4(AB - BC)\cos\theta\sin^3\theta$.

Subtracting these expressions and simplifying gives
$B'^2 - 4A'C' = (B^2 - 4AC)(\cos^4\theta + \sin^4\theta) + (2B^2 - 8AC)\cos^2\theta\sin^2\theta$
$= (B^2 - 4AC)(\cos^4\theta + 2\cos^2\theta\sin^2\theta + \sin^4\theta) = (B^2 - 4AC)(\cos^2\theta + \sin^2\theta)^2 = B^2 - 4AC$.

Miscellaneous Exercises for Chapter 8

1. a) Using the point-slope form : $y - 7 = \frac{4}{3}(x + 5) \Rightarrow y - 7 = \frac{4}{3}x + \frac{20}{3} \Rightarrow y = \frac{4}{3}x + \frac{41}{3}$
    b) Using the result of a), $y = \frac{4}{3}x + \frac{41}{3} \Rightarrow 3y = 4x + 41 \Rightarrow 4x - 3y + 41 = 0$

5. The slope of the line is $\frac{\sqrt{3}}{3}$, so the angle of inclination is $\tan^{-1}\frac{\sqrt{3}}{3} = 30°$.

9. The angle between the lines is $83° - 22° = 61°$.

13. If we write the equation in the form
$y = mx + b$, we get $y = \frac{5}{2}x - \frac{c}{2}$. Therefore
the family consists of all lines with slope $\frac{5}{2}$.

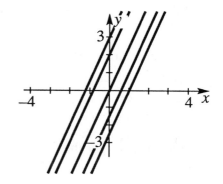

17. The slope of the line $5x - 12y = 1$ is $\frac{5}{12}$. Any line that is (a constant) 5 units from this given line must be parallel to $5x - 12y = 1$. Therefore the lines that we seek must have a slope of $\frac{5}{12}$, and it follows that the general form of their equations can be written as $5x - 12y + F = 0$. From Section 8.1 we know that the distance from a point (h, k) to the line $Dh + Ek + F = 0$ can be determined by $d = \dfrac{|Dh + Ek + F|}{\sqrt{D^2 + E^2}}$. A point on the line $5x - 12y = 1$ is $(h, k) = (5, 2)$. Since the distance from this point to the line $5x - 12y + F = 0$ must be 5, we have $5 = \dfrac{|(5)(5) + (-12)(2) + F|}{\sqrt{5^2 + (-12)^2}} \Rightarrow 65 = |F + 1|$. Solving for F we have $F + 1 = 65$ or $F + 1 = -65 \Rightarrow F = 64$ or $F = -66$. Substituting into $5x - 12y + F = 0$, we get the equations $5x - 12y + 64 = 0$ or $5x - 12y - 66 = 0$.

21. The equation is quadratic in x, so we complete the square with the terms involving x: $x^2 + 8x - 10y + 6 = 0$
$\Rightarrow x^2 + 8x = 10y - 6 \Rightarrow x^2 + 8x + 16 = 10y - 6 + 16 \Rightarrow (x + 4)^2 = 10(y + 1)$.
This is the parabola $x^2 = 10y$ shifted 4 units left and 1 unit down. The vertex is at $(-4, -1)$, and the focus is $c = \frac{5}{2}$ units above. The (horizontal) directrix

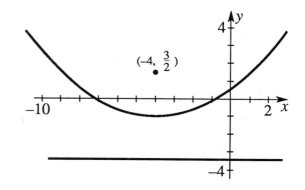

is $c = \frac{5}{2}$ units below $(-4, -1)$ as well. The length of the (horizontal) focal chord is $|4c| = 10$, so count five units right from the focus, and count five units left from the focus to get the points $(-9, 3/2)$ and $(1, 3/2)$ on the parabola.

25. If we graph the directrix and focus, we see that the parabola opens left and that the vertex is (3, 3). The distance from the vertex to the focus is $c = 1$. The equation of the parabola is

$(y - k)^2 = 4c(x - h) \Rightarrow (y - 3)^2 = 4(-1)(x - 3)$

$\Rightarrow (y - 3)^2 = -4(x - 3)$. The length of the (vertical) focal chord is $|4c| = 4$, so count two units up from the focus, and count two units down from the focus to get the points (2, 5) and (2, 1) on the parabola.

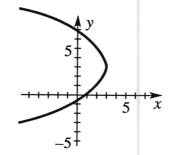

29. Divide each side of the equation by 9:

$x^2 + 9y^2 = 9 \Rightarrow \dfrac{x^2}{9} + \dfrac{y^2}{1} = 1$. To find the x-

intercepts set $y = 0 \Rightarrow \dfrac{x^2}{9} + \dfrac{0^2}{1} = 1$

$\Rightarrow x = \pm 3$. If we set $x = 0$ and solve for y, we obtain the y-intercepts:

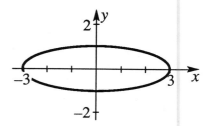

$\dfrac{0^2}{9} + \dfrac{y^2}{1} = 1 \Rightarrow y = \pm 1$. The points (3, 0), (−3, 0), (0, 1), and (0, −1) are enough to get

a sketch of the graph. The major axis is horizontal. The foci are $c = \sqrt{a^2 - b^2} = \sqrt{9 - 1}$ = $2\sqrt{2}$ units from the center (0, 0) along the major axis.

33. First we will complete the square in both variables:

$4x^2 + 3y^2 - 16x + 30y + 79 = 0$

$\Rightarrow 4x^2 - 16x + 3y^2 + 30y = -79$

$\Rightarrow 4(x^2 - 4x) + 3(y^2 + 10y) = -79$

$\Rightarrow 4(x^2 - 4x + 4) + 3(y^2 + 10y + 25) = -79 + 16 + 75$

$\Rightarrow 4(x - 2)^2 + 3(y + 5)^2 = 12$

$\Rightarrow \dfrac{(x - 2)^2}{3} + \dfrac{(y + 5)^2}{4} = 1$

The graph is the ellipse $\dfrac{x^2}{3} + \dfrac{y^2}{4} = 1$ centered at (2, −5).

37. If we isolate y we get $9y^2 = 36x^2 \Rightarrow y = 2x$ or $y = -2x$.

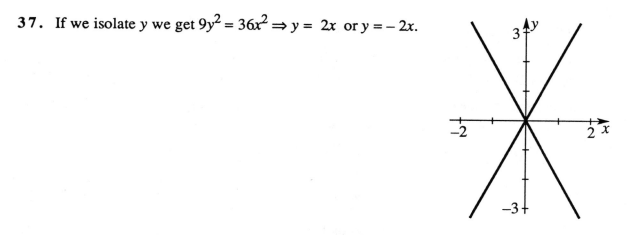

41. If we divide both sides of the equation by 8, we have $\dfrac{x^2}{4} + \dfrac{y^2}{8/25} = 1$
$\Rightarrow a = 2$. Therefore, $d(P, F_1) + d(P, F_2) = 2a = 4$.

45. The center of the ellipse is the midpoint of major axis, which is $\left(\dfrac{-2 + -2}{2}, \dfrac{6 + -0}{2}\right)$
$= (-2, 3)$. The length of the major axis is $2a = 6 \Rightarrow a = 3$. The eccentricity is $\dfrac{c}{a} = \dfrac{c}{3}$
so $\dfrac{c}{3} = \dfrac{2}{3} \Rightarrow c = 2$. Since $b^2 = a^2 - c^2 \Rightarrow b^2 = 9 - 4 = 5$. The equation of the ellipse is
$\dfrac{(x - h)^2}{a^2} + \dfrac{(y - k)^2}{b^2} = 1 \Rightarrow \dfrac{(x + 2)^2}{5} + \dfrac{(y - 3)^2}{9} = 1$.

49. Refer to the solution to Problem 29 (this section) of the manual. Since $a = 3$ and $c = 2\sqrt{2}$,
$e = \dfrac{c}{a} = \dfrac{2\sqrt{2}}{3}$. The equations of the directrices are $x = \pm\dfrac{a}{e} = \pm\dfrac{3}{2\sqrt{2}/3} = \pm\dfrac{9\sqrt{2}}{4}$.

53. Comparing $7x^2 - 2xy - 5y^2 + 12x - 56 = 0$ to the general second-degree equation
$Ax^2 + Bxy + Cy^2 + Dx + Ey + F = 0$, we have $A = 7$, $B = -2$, and $C = -5$. Since
$B^2 - 4AC = (-2)^2 - 4(7)(-5) > 0$, the graph of the equation is a (rotated) hyperbola.

57. Comparing $5x^2 + 2xy + 5y^2 - 2x + 14y = 60$ to the general second-degree equation
$Ax^2 + Bxy + Cy^2 + Dx + Ey + F = 0$, we have $A = 5$, $B = 2$, and $C = 5$. The angle θ of
rotation can be determined by $\cot 2\theta = \dfrac{A - C}{B} = \dfrac{5 - 5}{2} = 0$. The smallest positive
angle with cotangent 0 is $90° \Rightarrow 2\theta = 90° \Rightarrow \theta = 45°$.

61. To find the angle of rotation, solve $\cot 2\theta = \dfrac{A-C}{B} = \dfrac{5-8}{-4} = \dfrac{3}{4}$. Instead of solving for θ, we are really interested in $\cos\theta$ and $\sin\theta$. Assuming $0° < 2\theta < 180°$, since $\cot 2\theta$ is positive, 2θ must be in Quadrant I. If we sketch 2θ in standard position the point $(3, 4)$ is on the terminal side $\Rightarrow \cos 2\theta = \dfrac{3}{5}$ and $\sin 2\theta = \dfrac{4}{5}$. According to the half-angle identities,

$$\cos\theta = \sqrt{\frac{1 + \cos 2\theta}{2}} = \sqrt{\frac{1 + 3/5}{2}} = \frac{2}{\sqrt{5}}, \text{ and } \sin\theta = \sqrt{\frac{1 - \cos 2\theta}{2}}$$

$$= \sqrt{\frac{1 - 3/5}{2}} = \frac{1}{\sqrt{5}}. \quad \text{The rotation of axes formulas are } x = x'\cos\theta - y'\sin\theta$$

$$= \frac{2}{\sqrt{5}}x' - \frac{1}{\sqrt{5}}y' \text{ and } y = x'\sin\theta + y'\cos\theta = \frac{1}{\sqrt{5}}x' + \frac{2}{\sqrt{5}}y'. \text{ Substituting for } x \text{ and } y \text{ into}$$

the equation $5x^2 - 4xy + 8y^2 = 36$ we have

$$5\left(\frac{2}{\sqrt{5}}x' - \frac{1}{\sqrt{5}}y'\right)^2 - 4\left(\frac{2}{\sqrt{5}}x' - \frac{1}{\sqrt{5}}y'\right)\left(\frac{1}{\sqrt{5}}x' + \frac{2}{\sqrt{5}}y'\right) + 8\left(\frac{1}{\sqrt{5}}x' + \frac{2}{\sqrt{5}}y'\right)^2 = 36$$

$$\Rightarrow 5(\tfrac{1}{5})(2x' - y')^2 - 4(\tfrac{1}{5})(2x' - y')(x' + 2y') + 8(\tfrac{1}{5})(x' + 2y')^2 = 36$$

$$\Rightarrow (4x'^2 - 4x'y' + y'^2) - \tfrac{4}{5}(2x'^2 + 3x'y' - 2y'^2) + \tfrac{8}{5}(x'^2 + 4x'y' + 4y'^2) = 36$$

$\Rightarrow 4x'^2 + 9y'^2 = 36$. Thus, we rotate the axes through an angle θ such that

$\tan\theta = \dfrac{\sin\theta}{\cos\theta} = \dfrac{1/\sqrt{5}}{2/\sqrt{5}} = \dfrac{1}{2} \approx 26.6°$. On the rotated $x'y'$-axes sketch the ellipse with vertices $(3, 0)$, $(-3, 0)$, $(0, 2)$, and $(0, -2)$.

65. The sum of the distances from $(-6, 0)$ to $(0, 3)$ and from $(1, 1)$ to $(0, 3)$ is

$\sqrt{(0 - (-6))^2 + (3 - 0)^2} + \sqrt{(0 - 1)^2 + (3 - 1)^2} = 4\sqrt{5}$. Let (x, y) be any point on the ellipse. By definition, the sum of the distances from $(-6, 0)$ to (x, y) and from $(1, 1)$ to (x, y) is $4\sqrt{5}$. Therefore,

$\sqrt{(x - (-6))^2 + (y - 0)^2} + \sqrt{(x - 1)^2 + (y - 1)^2} = 4\sqrt{5}$

$\Rightarrow \sqrt{(x - 1)^2 + (y - 1)^2} = 4\sqrt{5} - \sqrt{(x + 6)^2 + y^2}$

$\Rightarrow \sqrt{(x^2 - 2x + 1) + (y^2 - 2y + 1)} = 4\sqrt{5} - \sqrt{(x^2 + 12x + 36) + y^2}$.   Squaring both sides, we get

$(x^2 - 2x + 1) + (y^2 - 2y + 1) = 80 - 8\sqrt{5}\sqrt{x^2 + 12x + 36 + y^2} + x^2 + 12x + 36 + y^2$

$\Rightarrow x^2 + y^2 - 2x - 2y + 2 = 116 - 8\sqrt{5}\sqrt{x^2 + 12x + 36 + y^2} + x^2 + 12x + y^2$

$\Rightarrow 8\sqrt{5}\sqrt{x^2 + 12x + 36 + y^2} = 114 + 14x + 2y$

$\Rightarrow 4\sqrt{5}\sqrt{x^2 + 12x + 36 + y^2} = 57 + 7x + y$.   Squaring both sides again, we get

$80(x^2 + 12x + 36 + y^2) = (57 + 7x + y)^2$

$\Rightarrow 80x^2 + 960x + 2880 + 80y^2 = 49x^2 + 14xy + y^2 + 798x + 114y + 3249$

$\Rightarrow 31x^2 - 14xy + 79y^2 + 162x - 114y - 369 = 0$.

69. $(4x^2 + y^2 - 16)(x - 2y + 5) = 0$ if and only if
$4x^2 + y^2 - 16 = 0$ or

$x - 2y + 5 = 0$. Therefore we graph the ellipse
$4x^2 + y^2 - 16 = 0$ and the line $x - 2y + 5 = 0$
on the same axes.

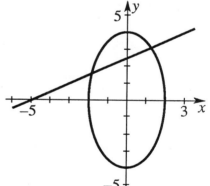

Chapter 9

Section 9.1

1. The point A is two units from the origin on the terminal side of $\frac{\pi}{4}$ in standard position. The point B is two units from the origin on the ray opposite the terminal side of $\frac{\pi}{4}$ in standard position.

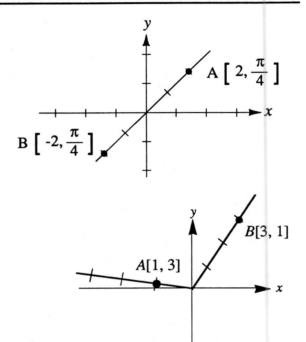

5. The point A is one unit from the origin on the terminal side of 3 in standard position. (This is a radian measure. This angle is about 172°.) The point B is three units from the origin on the terminal side of 1 in standard position. (This angle is about 57°.)

9. Using the equations on page 502, we get

$$x = r \cos \theta = -2 \cos \tfrac{\pi}{6} = -2\left(\frac{\sqrt{3}}{2}\right) = -\sqrt{3} \quad \text{and} \quad y = r \sin \theta = -2 \sin \tfrac{\pi}{6} = -2\left(\frac{1}{2}\right) = -1$$

The rectangular coordinates of the point are $(-\sqrt{3}, -1)$.

13. The polar angle for such a representation must be coterminal to or opposite to $\frac{\pi}{4}$. The only angle in this range is $\frac{9\pi}{4}$. Because $\frac{9\pi}{4}$ is coterminal to $\frac{\pi}{4}$, we choose $r = 2$ instead of -2. Thus our answer is $\left[2, \frac{9\pi}{4}\right]$.

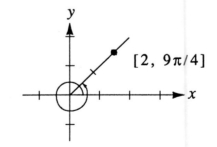

17. Because $r \geq 0$, the polar angle for such a representation must be coterminal with π. Those coterminal to π in the interval $[0, 6\pi]$ are π, 3π, and 5π. Thus our answers are $[2, \pi]$, $[2, 3\pi]$, and $[2, 5\pi]$.

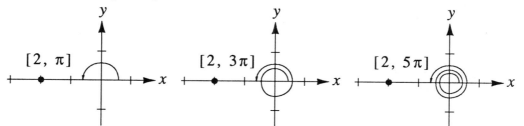

21. The point is on the negative y-axis six units from the origin. The two angles that are coterminal with the negative y-axis are $-\dfrac{\pi}{2}$ and $\dfrac{3\pi}{2}$. Thus our answers are $[6, -\dfrac{\pi}{2}]$ and $[6, \dfrac{3\pi}{2}]$.

25. We substitute $x = r\cos\theta$ and $y = r\sin\theta$ into the rectangular equation:
$$x^2 + y^2 = 6x$$
$$(r\cos\theta)^2 + (r\sin\theta)^2 = 6(r\cos\theta)$$
$$r^2\left(\cos^2\theta + \sin^2\theta\right) = 6r\cos\theta \qquad \text{(simplifying and factoring)}$$
$$r^2 = 6r\cos\theta \qquad \text{(Pythagorean Identity)}$$
$$r^2 - 6r\cos\theta = 0$$
$$r(r - 6\cos\theta) = 0$$
This yields either $r = 0$ or $r = 6\cos\theta$. The graph of $r = 0$ is the origin only. The graph of $r = 6\cos\theta$ passes through the origin since $[0, \dfrac{\pi}{2}]$ makes this equation true. This means that the polar equation we seek is $r = 6\cos\theta$ only.

29. First we solve the equation for y:
$$x^2 + 4xy + 4y^2 = 0 \quad \Rightarrow \quad (x + 2y)^2 = 0 \quad \Rightarrow \quad x + 2y = 0 \quad \Rightarrow \quad y = -\tfrac{1}{2}x$$
Next, we substitute $x = r\cos\theta$ and $y = r\sin\theta$ into this rectangular equation:
$$y = -\tfrac{1}{2}x \quad \Rightarrow \quad r\sin\theta = -\tfrac{1}{2}r\cos\theta \quad \Rightarrow \quad \frac{\sin\theta}{\cos\theta} = -\tfrac{1}{2} \quad \Rightarrow \quad \tan\theta = -\tfrac{1}{2}$$
The polar equation we seek is $\tan\theta = -\dfrac{1}{2}$.

33. First we rewrite the equation as $r(1 - \cos\theta) = 4$, or $r - r\cos\theta = 4$. Then substituting we get

$$\sqrt{x^2 + y^2} - x = 4$$

$$\sqrt{x^2 + y^2} = x + 4$$

$$x^2 + y^2 = (x + 4)^2$$

(squaring both sides)

$$x^2 + y^2 = x^2 + 8x + 16$$

$$\tfrac{1}{8}y^2 - 2 = x$$

(solving for x)

The graph of this equation is a parabola opening to the right.

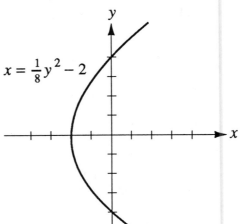

37. Using the equations on page 502, we get $x = r\cos\theta = 2\cos 5 = 0.57$ and $x = r\sin\theta = 2\sin 5 = -1.92$. The rectangular coordinates are $(0.57, -1.92)$.

Keystrokes: 2 $\boxed{\text{X}}$ 5 $\boxed{\cos}$ $\boxed{=}$ and 2 $\boxed{\text{X}}$ 5 $\boxed{\sin}$ $\boxed{=}$

41. The graph of $\theta = \frac{\pi}{3}$ is the line that includes the terminal side of $\frac{\pi}{3}$ and the ray opposite it. Of these points, the points on the ray opposite the terminal side of $\frac{\pi}{3}$ (including the origin) are those in which $r \le 0$.

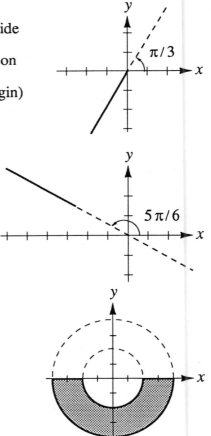

45. The graph of $\theta = \frac{5\pi}{6}$ is the line that includes the terminal side of $\frac{5\pi}{6}$ and the ray opposite it. Of these points, the points on the terminal side of $\frac{5\pi}{6}$ that are at least 3 units from the origin are those described by the inequalities.

49. The points for which $0 \le \theta \le \pi$ and $r \le 0$ are in Quadrants III and IV. The points for which $-4 \le r \le -2$ are those that are inside the circle of radius 4 and outside the circle of radius 2 (both centered at the origin). This gives us the points shown in the figure.

53. Let (x_a, y_a) and (x_b, y_b) be the rectangular coordinates of A and B respectively. The distance formula from Chapter 2 gives

$$d(A, B) = \sqrt{(x_a - x_b)^2 + (y_a - y_b)^2}$$

$$= \sqrt{(a \cos \alpha - b \cos \beta)^2 + (a \sin \alpha - b \sin \beta)^2}$$

$$= \sqrt{a^2 \cos^2 \alpha - 2ab \cos \alpha \cos \beta + b^2 \cos^2 \beta + a^2 \sin^2 \alpha - 2ab \sin \alpha \sin \beta + b^2 \sin^2 \beta}$$

$$= \sqrt{a^2 \left(\cos^2 \alpha + \sin^2 \alpha\right) + b^2 \left(\cos^2 \beta + \sin^2 \beta\right) - 2ab \left(\cos \alpha \cos \beta - \sin \alpha \sin \beta\right)}$$

$$= \sqrt{a^2 + b^2 - 2ab \cos(\alpha - \beta)}$$

Section 9.2

1. We substitute $x = r\cos\theta$ in the equation to get the equation $x = 4$. This is the vertical line with x-intercept 4.

5. From the catalog of circles on page 520, the graph is a circle in Quadrants III and IV that is tangent to the x-axis at the origin and has a diameter of 6.

9. First, note that

$$r\sec\theta = 4 \Rightarrow r\left(\frac{1}{\cos\theta}\right) = 4 \Rightarrow r = 4\cos\theta$$

From the catalog of circles on page 520, the graph is a circle in Quadrants I and IV that is tangent to the y-axis at the origin and has a diameter of 4.

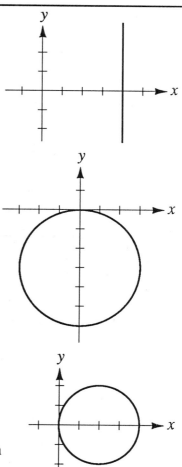

13. The symmetry tests:

x − axis test: $r = 2(1 - \sin(-\theta))$

$$r = 2(1 + \sin\theta)$$

No information

y − axis test: $r = 2(1 - \sin(\pi - \theta))$

$$r = 2(1 - \sin\theta) \quad \text{Yes}$$

origin test: $-r = 2(1 - \sin\theta)$

$$r = -2(1 - \sin\theta)$$

No information

The only symmetry is with the y-axis, as predicted by the symmetry tests.

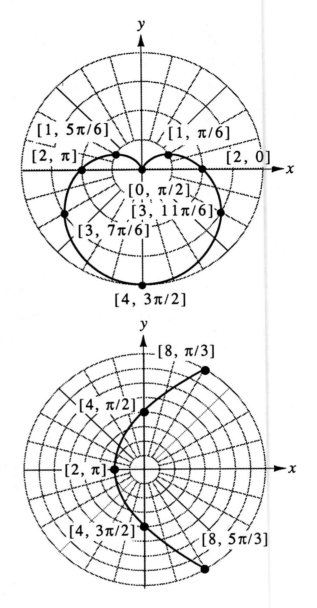

17. The symmetry tests:

x − axis test: $r = \dfrac{4}{1 - \cos(-\theta)}$

$$r = \dfrac{4}{1 - \cos\theta}$$

Yes

y − axis test: $r = \dfrac{4}{1 - \cos(\pi - \theta)}$

$$r = \dfrac{4}{1 + \cos\theta}$$

No information

origin test: $-r = \dfrac{4}{1 - \cos(-\theta)}$

$$-r = \dfrac{4}{1 + \cos\theta}$$

No information

The only symmetry is with the x-axis, as predicted by the symmetry tests.

21. The rectangular graph $y = 2 - 4\sin x$ has an amplitude of 4, a period 2π, and a vertical translation of 2. Transferring the values from the rectangular graph gives us the polar graph shown.

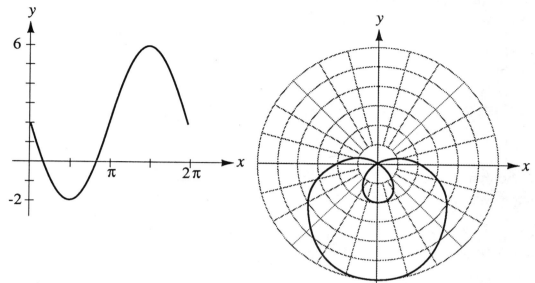

25. The rectangular graph $y = 4\cos 4x$ has an amplitude of 4, a period $\frac{\pi}{2}$, and no translations. Transferring the values from the rectangular graph gives us the polar graph shown.

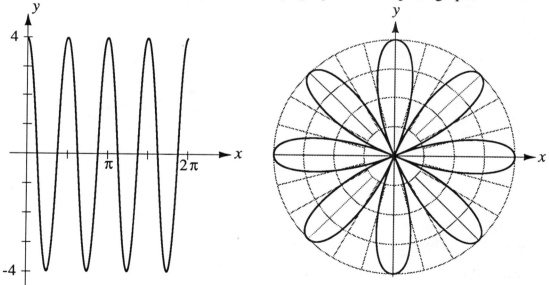

29. Because $r^2 \geq 0$, the only allowable values of θ are those that make $\sin 2\theta$ nonnegative. Examine the graph of $y = \sin 2x$:

It shows that $\sin 2\theta \geq 0$ for $0 \leq \theta \leq \dfrac{\pi}{2}$ and $\pi \leq \theta \leq \dfrac{3\pi}{2}$.

Since $(-r)^2 = 16 \sin 2\theta \Rightarrow r^2 = 16 \sin 2\theta$, the graph has symmetry with respect to the origin (The other tests fail). The graph is shown.

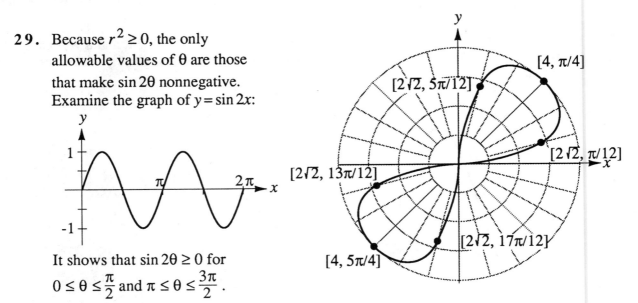

33. First, note that
$$18 \sin\theta \cos\theta = 9 \sin 2\theta.$$
The allowable values of θ are
$$0 \leq \theta \leq \frac{\pi}{2} \text{ and } \pi \leq \theta \leq \frac{3\pi}{2}$$
(see solution to Problem 29).
Since $(-r)^2 = 9 \sin 2\theta$
$$\Rightarrow r^2 = 9 \sin 2\theta,$$
the graph has symmetry with respect to the origin (The other tests fail). The graph is shown.

37. From the catalog of roses on page 518, the first equation has a polar graph is a three-leafed rose shown. The second equation is a rotation of the first. The angle of rotation is π.

41. This is best graphed by plotting a few points and connecting them with a smooth curve.

θ	r *(approx)*
$-\dfrac{\pi}{2}$	$-\dfrac{\pi}{2}$ (−1.6)
$-\pi$	$-\pi$ (−3.1)
$-\dfrac{3\pi}{2}$	$-\dfrac{3\pi}{2}$ (−4.7)
-2π	-2π (−6.3)
-3π	-3π (−9.4)

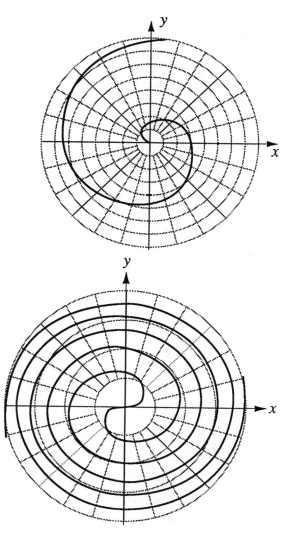

45. This is best graphed by plotting a few points and connecting them with a smooth curve. Note that for each value of θ, there are two values of r.

θ	r *(approx)*
0	0
$\dfrac{\pi}{2}$	$\pm\sqrt{\dfrac{\pi}{2}}$ (±1.3)
π	$\pm\sqrt{\pi}$ (±1.8)
$\dfrac{3\pi}{2}$	$\pm\sqrt{\dfrac{3\pi}{2}}$ (±2.2)
2π	$\pm\sqrt{2\pi}$ (±2.5)
3π	$\pm\sqrt{3\pi}$ (±3.1)

Section 9.3

1. We plot points for values of $t = 1, 2, 3, 4, 5$ and sketch the curve. The orientation is in the direction of increasing t. To find the rectangular equation, we solve the first equation for t and substitute this in the second equation:

$$x = t - 1 \;\Rightarrow\; t = x + 1$$
$$y = 4 - 2t = 4 - 2(x + 1)$$
$$\Rightarrow y = 2 - 2x$$

t	x	y
0	−1	4
1	0	2
2	1	0
3	2	−2
4	3	−4
5	4	−6

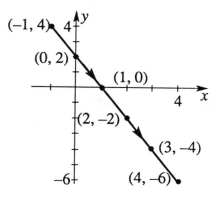

5. We plot points for values of $t = 0, 1, 2, 3, 4$ and sketch the curve. The orientation is in the direction of increasing t. To find the rectangular equation, we solve the second equation for t and substitute this in the first equation:

t	x	y
0	−2	0
1	0	1
2	2	$\sqrt{2}$
3	4	$\sqrt{3}$
4	6	2

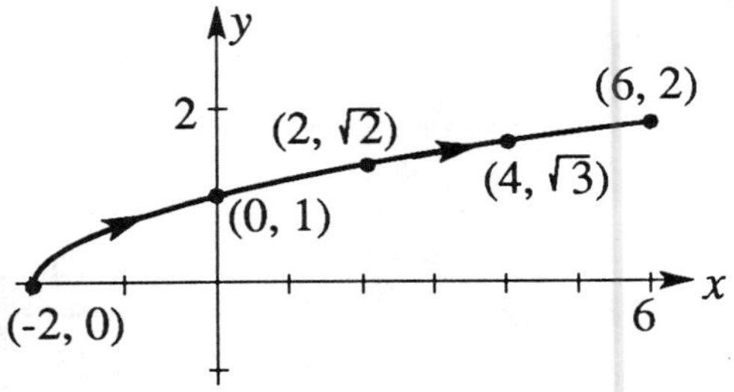

$y = \sqrt{t} \Rightarrow t = y^2$

$x = 2t - 2 = 2(y^2) - 2 \Rightarrow x + 2 = 2y^2$

9. We plot points for values of $t = 0, \frac{\pi}{3}, \frac{\pi}{2}, \frac{2\pi}{3}, \pi$ and sketch the curve. To find the rectangular equation, substitute $\cos t$ for x in the second equation:

$x = 2t - 2 \Rightarrow t = \frac{x+2}{2}$

$y = (\cos t)^2 \Rightarrow y = x^2$

t	x	y
0	1	1
$\frac{\pi}{3}$	$\frac{1}{2}$	$\frac{1}{4}$
$\frac{\pi}{2}$	0	0
$\frac{2\pi}{3}$	$-\frac{1}{2}$	$\frac{1}{4}$
π	−1	1

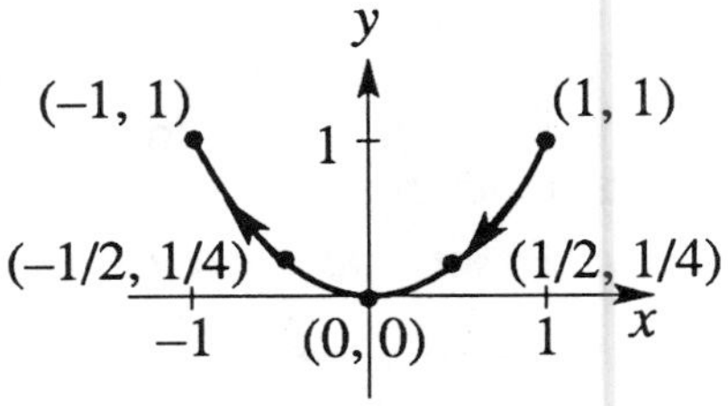

13. Using the table of parametric representations on page 527, we get

$$x = 1 + (9-1)t, \; y = 2 + (-4-2)t$$

or $\quad x = 1 + 8t, \; y = 2 - 6t$

The domain of t is $[0, 1]$.

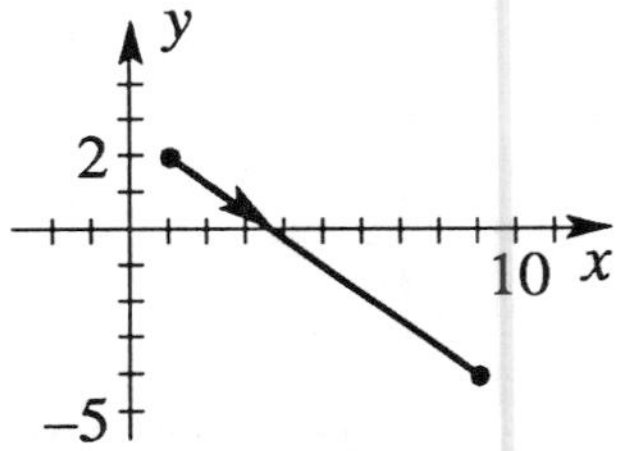

17. Using the table of parametric representations on page 527, we get

$$x = 3\cos t, \quad y = 3\sin t$$

The domain of t is $\left[0, \frac{\pi}{2}\right]$.

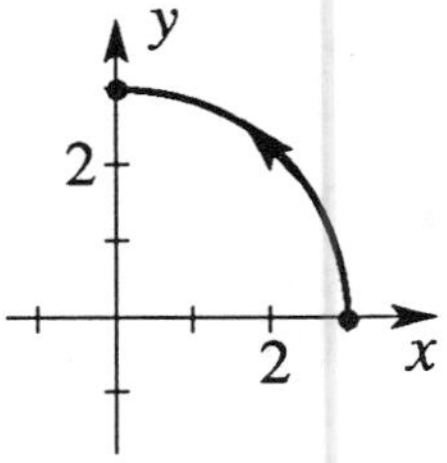

21. Using the table of parametric representations on page 527, we get

$$x = (2\cos t)\cos t, \quad y = (2\cos t)\sin t,$$

or $\quad x = 2\cos^2 t, \quad y = 2\cos t \sin t.$

The domain of t is $[0, \pi]$.

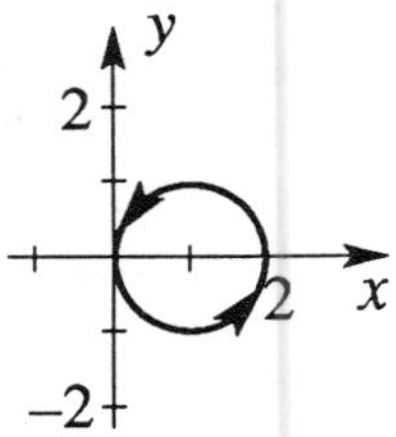

25. The secant and tangent functions are connected by the Pythagorean identity $\tan^2 t + 1 = \sec^2 t$. Substituting we get $y^2 + 1 = x^2$, or $x^2 - y^2 = 1$. This is a hyperbola with x-intercepts $(0, \pm 1)$ and asymptotes $y = \pm x$. Over the given domain, $\sec t$ varies from 1 to ∞ and $\tan t$ varies from 0 to $+\infty$. The graph with its orientation is shown.

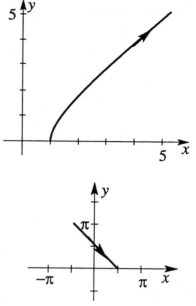

29. The arcsine and arccosine functions are connected by the identity $\arcsin t + \arccos t = \frac{\pi}{2}$ (page 368) Substituting we get $x + y = \frac{\pi}{2}$. This is a line with x-intercept $\left(\frac{\pi}{2}, 0\right)$ and y-intercept $\left(0, \frac{\pi}{2}\right)$. Over the given domain, $\arcsin t$ varies from $-\frac{\pi}{2}$ to $\frac{\pi}{2}$ and $\arccos t$ varies from 0 to π. The graph with its orientation is shown.

33. The value of x changes at a constant rate of $\frac{1}{2}$ unit/second, so x is a linear function of t with slope $\frac{1}{2}$. Assume that $x = \frac{1}{2} t + b$. We also know that when $t = 0$, $x = 0$, so $0 = \frac{1}{2} 0 + b \Rightarrow b = 0$. Thus $x = \frac{1}{2} t$. It follows that $y = 2x + 1 \Rightarrow y = 2\left(\frac{1}{2}t\right) + 1 \Rightarrow y = t + 1$.

Section 9.4

1. First, note that
$$r = \frac{4}{1 + \cos\theta} = \frac{4(1)}{1 + 1\cos\theta}$$
The eccentricity e is 1, so the graph of this polar equation is a parabola. The value of p is 4, so by the catalog on page 538, the directrix is $r = 4\sec\theta$.

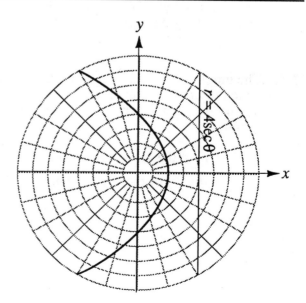

5. First, note that

$$r = \frac{2}{1 + \frac{2}{5}\cos\theta} = \frac{\frac{2}{5}(5)}{1 + \frac{2}{5}\cos\theta}$$

The eccentricity e is $\frac{2}{5}$, so the graph of this polar equation is an ellipse. The value of p is 5, so by the catalog on page 538, the directrix is $r = 5\sec\theta$.

9. We use a reciprocal identity to rewrite the equation in a familiar form:

$$r = \frac{12\left(\frac{1}{\cos\theta}\right)}{2\left(\frac{1}{\cos\theta}\right) + 6} = \frac{\frac{12}{\cos\theta}}{\frac{2}{\cos\theta} + 6} \cdot \frac{\cos\theta}{\cos\theta}$$

$$= \frac{12}{2 + 6\cos\theta} = \frac{6}{1 + 3\cos\theta}$$

So, $r = \dfrac{6}{1 + 3\cos\theta} = \dfrac{3(2)}{1 + 3\cos\theta}$

The eccentricity e is 3, so the graph of this equation is a hyperbola. The value of p is 2; by the catalog on page 538, the directrix is $r = 2\sec\theta$

13. The graph of the first equation is a parabola which (by the catalog on page 538) opens down. A few points are plotted to sketch the graph.

The graph of the second equation is a rotation (by $\frac{\pi}{4}$) of the first graph.

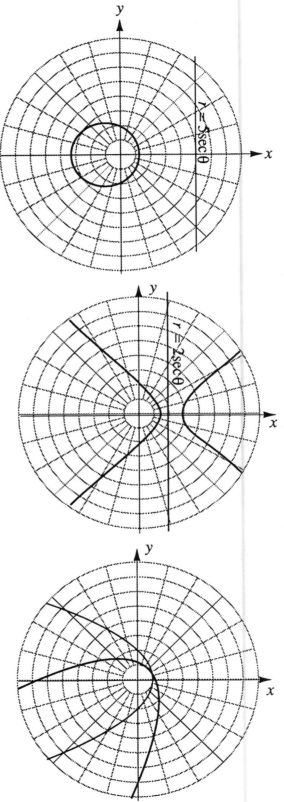

17. The graph of the first equation is a parabola which (by the catalog on page 538) opens down. A few points are plotted to sketch the graph.

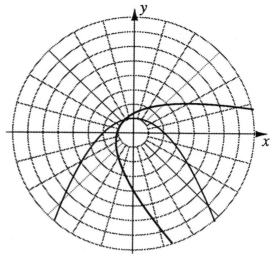

Because the second equation is a rotation of the first, it can be rewritten in the form

$$r = \frac{2}{1 + \sin(\theta - \alpha)},$$ or, using a subtraction identity for the sine,

$$r = \frac{2}{1 + \cos \alpha \sin \theta - \sin \alpha \cos \theta}.$$

Compare this form with the second equation.

It follows that $\cos \alpha = \frac{1}{2}$ and $\sin \alpha = \frac{\sqrt{3}}{2}$. This implies that $\alpha = \frac{\pi}{3}$. The equation can be written as $r = \dfrac{2}{1 + \sin\left(\theta - \frac{\pi}{3}\right)}$. The graph of this is a rotation (by $\frac{\pi}{3}$) of the first graph.

21. The conic section is an ellipse since the eccentricity is $\frac{1}{2}$. The polar form of the directrix

$y = -2$ is $r \sin \theta = -2$ or $r = -2 \csc \theta$. This implies that $p = 2$ and the ellipse has the polar equation

$$r = \frac{2\left(\frac{1}{2}\right)}{1 - \frac{1}{2}\sin \theta} \quad \text{or} \quad r = \frac{1}{1 - \frac{1}{2}\sin \theta}$$

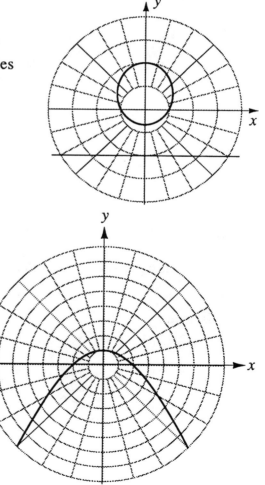

25. First, note that the equation can be written $x^2 = -4(1 + y)$. From Section 8.2, we know that the graph is a parabola opening down with a vertex of (0, 4), focus (0,0) and directrix $y = 2$ Thus, $e = 1$ and $p = 2$. From the catalog on page 538, The polar equation is

$$r = \frac{1(2)}{1 + 1 \sin \theta} \quad \text{or} \quad r = \frac{2}{1 + \sin \theta}$$

29. The conic section is an ellipse since the eccentricity is $\frac{1}{2}$. From the catalog on page 538, the equation is of the form

$$r = \frac{ep}{1 - e\sin\theta} \text{ or } r = \frac{\frac{1}{2}p}{1 - \frac{1}{2}\sin\theta}$$

It remains to find p, the distance from the focus at $[0, 0]$ to the directrix. From the figure 58 on page 481, this distance is

$$p = \frac{a}{e} - ae. \text{ With one focus at the origin}$$

and one focus at $[2, \frac{\pi}{2}]$, the distance between foci is $2c = 2$, so $c = 1$. Recall that $c = ae$, so $a = \frac{c}{e} = \frac{1}{(1/2)} = 2$. So,

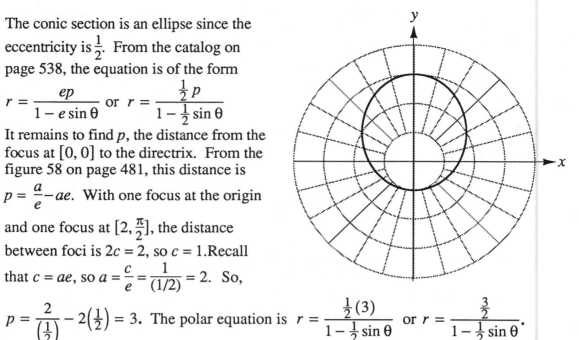

$$p = \frac{2}{\left(\frac{1}{2}\right)} - 2\left(\frac{1}{2}\right) = 3. \text{ The polar equation is } r = \frac{\frac{1}{2}(3)}{1 - \frac{1}{2}\sin\theta} \text{ or } r = \frac{\frac{3}{2}}{1 - \frac{1}{2}\sin\theta}.$$

33. We proceed in the manner of Example 4, part b (page 537), using the identity $\sin\theta = \cos\left(\theta - \frac{\pi}{2}\right)$. This identity can be verified using a subtraction identity:

$$\cos\left(\theta - \frac{\pi}{2}\right) = \cos\theta\cos\frac{\pi}{2} + \sin\theta\sin\frac{\pi}{2}$$
$$= \cos\theta(0) + \sin\theta(1)$$
$$= \sin\theta$$

So, $r = \dfrac{ep}{1 - e\sin\theta} = \dfrac{ep}{1 - e\cos\left(\theta - \frac{\pi}{2}\right)}$. The angle of rotation is $\frac{\pi}{2}$.

37. From the table on page 540, the semimajor axis of orbit and eccentricity of orbit of Icarus are 2.165 AU and 0.82712 respectively. In the manner of Example 7 (page 540), the orbit of Icarus is described by the polar equation $r = \dfrac{a\left(1 - e^2\right)}{1 - e\cos\theta} = \dfrac{2.165\left(1 - 0.82712^2\right)}{1 - 0.82712\cos\theta}$. So,

$$\left(\begin{array}{c}\text{distance at} \\ \text{perihelion}\end{array}\right) = \frac{2.165\left(1 - 0.82712^2\right)}{1 - 0.82712\cos\pi} \qquad \left(\begin{array}{c}\text{distance at} \\ \text{aphelion}\end{array}\right) = \frac{2.165\left(1 - 0.82712^2\right)}{1 - 0.82712\cos 0}$$

$$= \frac{2.165\left(1 - 0.82712^2\right)}{1 + 0.82712} \qquad\qquad\qquad = \frac{2.165\left(1 - 0.82712^2\right)}{1 - 0.82712}$$

$$= 0.3743 \qquad\qquad\qquad\qquad\qquad = 3.9557$$

Miscellaneous Exercises for Chapter 9

1. Using the equations on page 502, we get

$$x = r\cos\theta = 8\cos\tfrac{2\pi}{3} = 8\left(-\frac{1}{2}\right) = -4 \quad \text{and} \quad y = r\sin\theta = 8\sin\tfrac{2\pi}{3} = 8\left(\frac{\sqrt{3}}{2}\right) = 4\sqrt{3}$$

The rectangular coordinates of the point are $(-4, 4\sqrt{3})$.

5. Using the equations on page 502, we get

$$x = r\cos\theta = -3\cos\tfrac{5\pi}{4} = -3\left(-\frac{\sqrt{2}}{2}\right) = \frac{3\sqrt{2}}{2} \quad \text{and} \quad y = r\sin\theta = -3\sin\tfrac{5\pi}{4} = -3\left(-\frac{\sqrt{2}}{2}\right) = \frac{3\sqrt{2}}{2}$$

The rectangular coordinates of the point are $(-4, 4\sqrt{3})$.

9. The point is on the negative y-axis eight units from the origin. The angle in the interval $[0, 2\pi]$ opposite the negative y-axis is $\frac{\pi}{2}$. Thus our answer is $[-8, \frac{\pi}{2}]$.

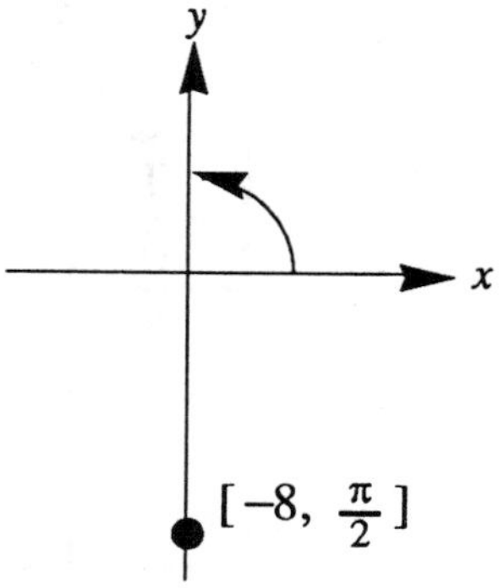

13. The graph of $\theta = \frac{\pi}{2}$ is the line that includes the terminal side of $\frac{\pi}{2}$ and the ray opposite it. (This line is the y-axis.) Of these points, the points on the nonnegative y-axis (including the origin) are those in which $r \geq 0$.

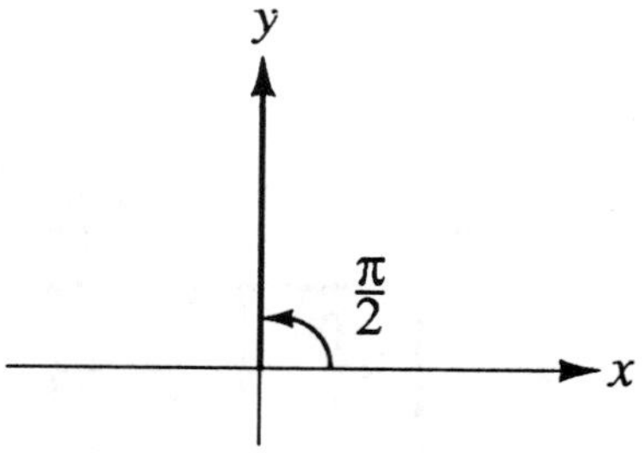

17. The graph of $\theta = \frac{5\pi}{3}$ is the line that includes the terminal side of $\frac{5\pi}{3}$ and the ray opposite it. Those points for which $r \geq -3$ are shown in the figure.

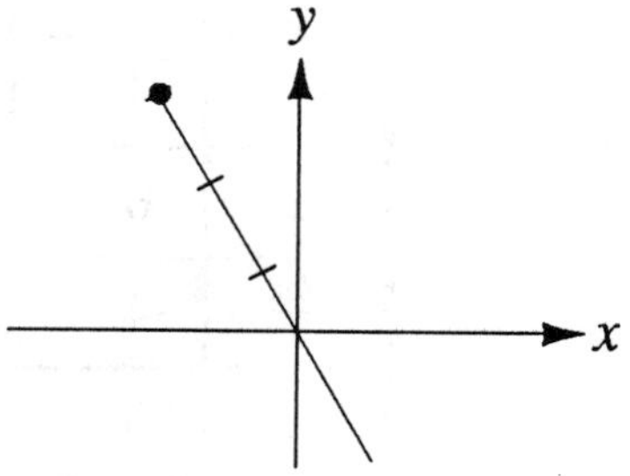

21. We substitute $y = r\sin\theta$ in the equation to get the equation $y = 6$. This is the horizontal line with y-intercept 6.

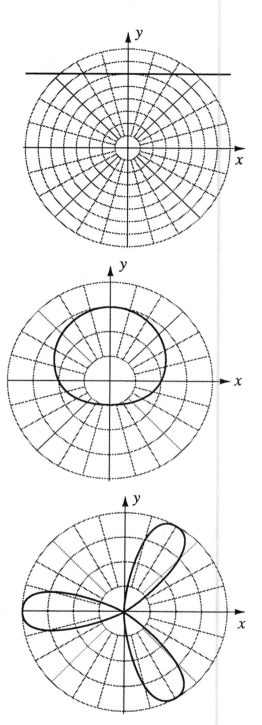

25. From the catalog on page 516, the graph of the equation is a limaçon.

θ	r
0	2
$\dfrac{\pi}{2}$	3
π	2
$\dfrac{3\pi}{2}$	1

29. From the catalog on page 518, the graph of the equation is a three–leafed rose.

θ	r
0	-4
$\dfrac{\pi}{6}$	0
$\dfrac{\pi}{3}$	4
$\dfrac{\pi}{2}$	0
$\dfrac{2\pi}{3}$	-4
$\dfrac{5\pi}{6}$	0
π	4

33. The graph in question is a rotation by $\frac{\pi}{4}$ of the graph of $r = 3 + 6\sin\theta$. The graph of $r = 3 + 6\sin\theta$ is a limaçon with an inner loop as shown in the figure at right. This graph is rotated $\frac{\pi}{4}$ to arrive at the desired graph

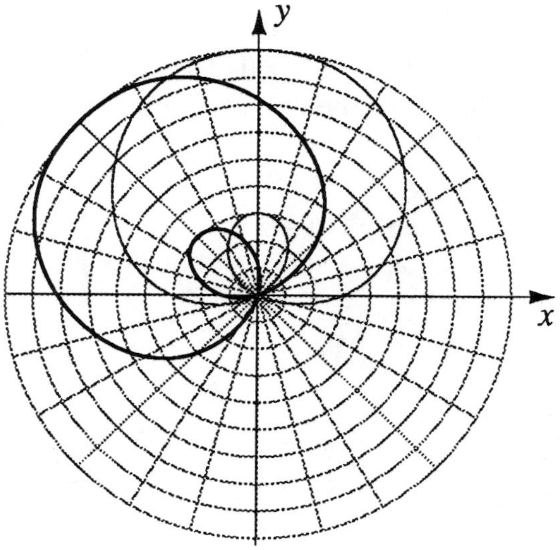

37. We plot points for values of $t = -4, -2, 0, 2, 4$ and sketch the curve: To find the rectangular equation, we solve the second equation for t and substitute this in the first equation:

$$x = t + 2 \Rightarrow t = x - 2$$
$$y = \frac{1}{2}t - 3 = \frac{1}{2}(x - 2) - 3$$
$$\Rightarrow y = \frac{1}{2}x - 4$$

t	x	y
–4	–2	–5
–2	0	–4
0	2	–3
2	4	–2
4	6	–1

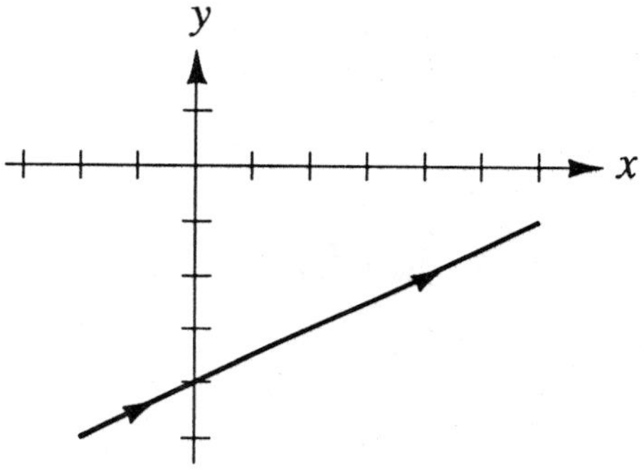

41. We plot points for values of and sketch the curve. To find the rectangular equation, we solve the second equation for t and substitute this in the first equation.

$$y = \sqrt{t} - 2 \Rightarrow t = (y + 2)^2$$
$$x = t^2 = ((y + 2)^2)^2 \Rightarrow x = (y + 2)^4$$

t	x	y
0	0	–2
$\frac{1}{4}$	$\frac{1}{16}$	$-\frac{3}{2}$
1	2	–1
2	4	$\sqrt{2} - 2$

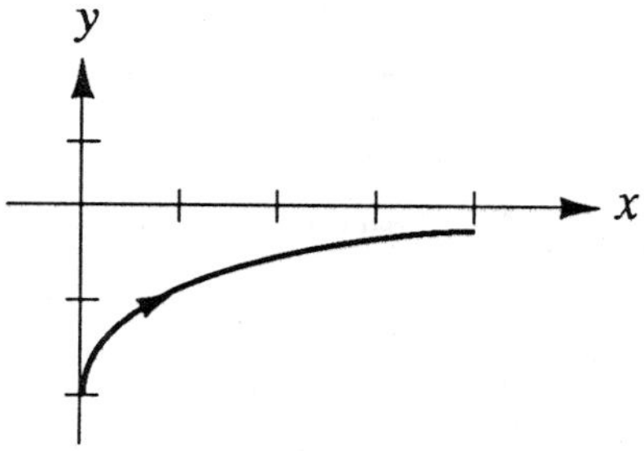

45. We use the Pythagorean identity $\cos^2\theta + \sin^2\theta = 1$ and let $\theta = 2t$:

$$\cos^2 2t + \sin^2 2t = 1$$
$$\Rightarrow x^2 + y^2 = 1$$

The graph is a circle of radius 1. The graph with its orientation is shown.

t	x	y
$-\frac{\pi}{2}$	–1	0
$-\frac{\pi}{4}$	0	–1
0	1	0
$\frac{\pi}{4}$	0	1
$\frac{\pi}{2}$	-1	0

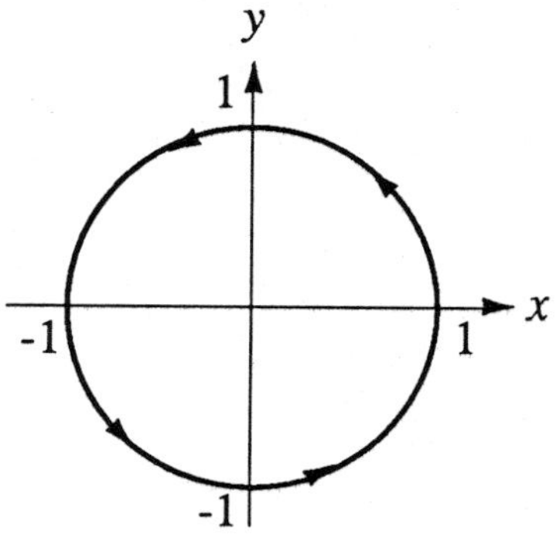

49. We use the identity
$\cos 2\theta = 1 - 2\sin^2\theta$ to find
the corresponding rectangular
equation:
$x = \cos 2t = 1 - 2\sin^2 t$
$\Rightarrow y = 1 - 2y^2.$
The graph and orientation are
shown.

t	x	y
$-\dfrac{\pi}{2}$	-1	-1
$-\dfrac{\pi}{4}$	0	$-\dfrac{\sqrt{2}}{2}$
0	1	0
$\dfrac{\pi}{4}$	0	$\dfrac{\sqrt{2}}{2}$
$\dfrac{\pi}{2}$	-1	1

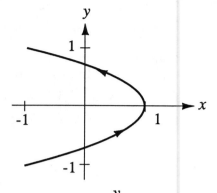

53. We use the identity $\ln(x^2) = 2\ln x$ to find the
corresponding rectangular equation:
$y = \ln(t^2) = 2\ln t \;\Rightarrow\; y = 2x$
Over the given domain, $\ln(t^2)$ varies from 0 to 4 and $\ln t$
varies from 0 to 2 . The graph and orientation are shown.

57. Using the table of parametric representations on page 527,
we get
$$x = t \text{ and } y = 2t^2 - 8$$
The domain of t is $[-2, 2]$.

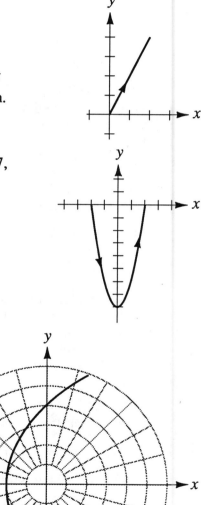

61. First, note that
$$r = \frac{4}{1 - \cos\theta} = \frac{4(1)}{1 - 1\cos\theta}$$
The eccentricity e is 1, so the graph of this
polar equation is a parabola. The value of
p is 4, so by the catalog on page 538, the
directrix is $r = -4\sec\theta$.

65. The equation can be rewritten as
$$r = \frac{16}{1 + 2\cos\theta} = \frac{2(8)}{1 + 2\cos\theta}$$
The eccentricity e is 2, so the graph of this polar equation is a hyperbola. The value of p is 8, so by the catalog on page 538, the directrix is $r = 8\sec\theta$.

69. First, notice that
$$\sin\left(\theta - \frac{3\pi}{2}\right) = \sin\theta\cos\frac{3\pi}{2} - \cos\theta\sin\frac{3\pi}{2}$$
$$= \sin\theta(0) - \cos\theta(-1)$$
$$= \cos\theta$$
Thus,
$$r = \frac{4}{1 + 2\sin\left(\theta - \frac{3}{2}\pi\right)} = \frac{2(2)}{1 + 2\cos\theta}$$
The eccentricity e is 2, so the graph of this polar equation is a hyperbola. The value of p is 2, so by the catalog on page 538, the directrix is $r = 2\sec\theta$.

73. From the graphs in Figure 58, the parameter t ranges from 0 to 10. For a given value of t, we can determine the value of x and y from these graphs. We use these values to sketch the graph.

t	x	y
0	5	0
3	5	1 (approx)
7	3	2 (approx)
10	1	3

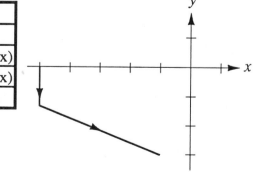

77. For $R = 2$, the parametric representation is $\begin{cases} x = 3\cos\theta - \cos 3\theta \\ y = 3\sin\theta - \sin 3\theta \end{cases}$. The desired result

follows if we can prove the two identities:

$$6\cos\theta - 4\cos^3\theta \overset{?}{=} 3\cos\theta + \cos 3\theta \quad\text{and}\quad 4\sin^3\theta \overset{?}{=} 3\sin\theta + \sin 3\theta$$

$$6\cos\theta - 4\cos^3\theta = 6\cos\theta - 4\cos\theta\left[\cos^2\theta\right] \qquad 4\sin^3\theta = 4\sin\theta\left[\sin^2\theta\right]$$

$$= 6\cos\theta - 4\cos\theta\left[\tfrac{1}{2}(1 + \cos 2\theta)\right] \qquad\qquad = 4\sin\theta\left[\tfrac{1}{2}(1 - \cos 2\theta)\right]$$

(square of function identity) $\qquad\qquad$ (square of function identity)

$$= 4\cos\theta - 2\cos\theta\cos 2\theta \qquad\qquad\qquad = 2\sin\theta + 2\sin\theta\cos 2\theta$$

$$= 4\cos\theta - 2\left(\tfrac{1}{2}\cos 3\theta + \tfrac{1}{2}\cos\theta\right) \qquad\qquad = 2\sin\theta + 2\left(\tfrac{1}{2}\sin 3\theta + \tfrac{1}{2}\sin\theta\right)$$

(product to sum identity) $\qquad\qquad\qquad$ (product to sum identity)

$$= 3\cos\theta - \cos 3\theta \qquad\qquad\qquad\qquad = 3\sin\theta + \sin 3\theta$$

This proves the result.

81. For $R = 4$, the parametric representation is $\begin{cases} x = 3\cos\theta + \cos 3\theta \\ y = 3\sin\theta + \sin 3\theta \end{cases}$. The desired result

follows if we can prove the two identities:

$$4\cos^3\theta \overset{?}{=} 3\cos\theta + \cos 3\theta \quad\text{and}\quad 4\sin^3\theta \overset{?}{=} 3\sin\theta + \sin 3\theta$$

$$4\cos^3\theta = 4\cos\theta\left[\cos^2\theta\right] \qquad\qquad 4\sin^3\theta = 4\sin\theta\left[\sin^2\theta\right]$$

$$= 4\cos\theta\left[\tfrac{1}{2}(1 + \cos 2\theta)\right] \qquad\qquad = 4\sin\theta\left[\tfrac{1}{2}(1 - \cos 2\theta)\right]$$

(square of function identity) $\qquad\qquad$ (square of function identity)

$$= 2\cos\theta + 2\cos\theta\cos 2\theta \qquad\qquad = 2\sin\theta + 2\sin\theta\cos 2\theta$$

$$= 2\cos\theta + 2\left(\tfrac{1}{2}\cos 3\theta + \tfrac{1}{2}\cos\theta\right) \qquad = 2\sin\theta + 2\left(\tfrac{1}{2}\sin 3\theta + \tfrac{1}{2}\sin\theta\right)$$

This proves the result.

Chapter 10

Section 10.1

1. In the form $y = mx + b$, the first equation is $y = -x + 2$; the graph has a y-intercept of $(0, 2)$ and a slope of $\frac{-1}{1}$. In slope-intercept form, the second equation is $y = x - 4$; the graph has a y-intercept of $(0, -4)$ and a slope of $\frac{1}{1}$. The graphs appear to cross at $(3, -1)$, and this can be verified by substituting $x = 3$, $y = -1$ into the system of equations. Of course, this means the system is consistent.

5. The graph of the first equation has a y-intercept of $(0, -5)$ and a slope of $\frac{-2}{1}$. The graph of the second equation has a y-intercept of $(0, 5)$ and a slope of $\frac{1}{2}$. The graphs appear to cross at $(-4, 3)$, and this can be verified by substituting $x = -4$, $y = 3$ into the system of equations. Of course, this means the system is consistent.

9. In slope-intercept form, the first equation is $y = -\frac{3}{2}x + \frac{5}{8}$; the graph has a y-intercept of $(0, \frac{5}{8})$ and a slope of $\frac{-3}{2}$. In slope-intercept form, the second equation is $y = -\frac{3}{2}x - \frac{5}{6}$; the graph has a y-intercept of $(0, -\frac{5}{6})$ and a slope of $\frac{-3}{2}$. The graphs have the same slope, but different y-intercepts. Therefore the lines are parallel and distinct; that is, the system is inconsistent.

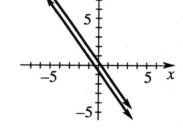

13. There are several ways to proceed; one way is to first solve for x in the second equation: $x - y = 2 \Rightarrow x = y + 2$. If we substitute $y + 2$ for x into the first equation, we get $2x + y = 10 \Rightarrow 2(y + 2) + y = 10 \Rightarrow 3y + 4 = 10 \Rightarrow y = 2$. Since $x = y + 2$, $x = (2) + 2 = 4$. Substituting $y = 2$ into the third equation we get $y + z = 3 \Rightarrow 2 + z = 3 \Rightarrow z = 1$. The solution is $(x, y, z) = (4, 2, 1)$.

17. There are several ways to proceed; one way is to first solve for z in the first equation:
$2x - y + z = 4 \Rightarrow z = -2x + y + 4$. If we substitute $-2x + y + 4$ for z into the second and third equations, we will get two equations in two unknowns.

Second equation: $2x + 2y + 3z = 3 \Rightarrow 2x + 2y + 3(-2x + y + 4) = 3 \Rightarrow -4x + 5y = -9$

Third equation: $6x - 9y - 2z = 17 \Rightarrow 6x - 9y - 2(-2x + y + 4) = 17 \Rightarrow 10x - 11y = 25$

We have reduced the problem to finding the solution of $\begin{cases} -4x + 5y = -9 \\ 10x - 11y = 25 \end{cases}$.

If we solve for y in the first of the two equations, we get $5y = 4x - 9 \Rightarrow y = \dfrac{4x - 9}{5}$.

Substituting $y = \dfrac{4x - 9}{5}$ into the second of the two equations, we get

$10x - 11y = 25 \Rightarrow 10x - 11(\dfrac{4x - 9}{5}) = 25$. Multiplying both sides by 5 and simplifying,

$50x - 11(4x - 9) = 125 \Rightarrow 6x = 26 \Rightarrow x = \dfrac{13}{3}$. If we substitute this for x in $y = \dfrac{4x - 9}{5}$,

we get $y = \dfrac{4(\frac{13}{3}) - 9}{5} = \dfrac{5}{3}$. Finally, substituting $x = \dfrac{13}{3}$ and $y = \dfrac{5}{3}$ into $z = -2x + y + 4$

we get $z = -2(\dfrac{13}{3}) + (\dfrac{5}{3}) + 4 = -3$. The solution is $(x, y, z) = (\dfrac{13}{3}, \dfrac{5}{3}, -3)$.

21. Solving the second equation, we have $3y = 4 \Rightarrow y = \dfrac{4}{3}$. Substituting $y = \dfrac{4}{3}$ into the first

equation gives $7x - 2y = 2 \Rightarrow 7x - 2(\dfrac{4}{3}) = 2 \Rightarrow 7x = \dfrac{14}{3} \Rightarrow x = \dfrac{2}{3}$. The solution is

$(x, y) = (\dfrac{2}{3}, \dfrac{4}{3})$.

25. We can eliminate the $-4x$ term in the second equation by multiplying the first equation by 4

and then adding the equations: $\begin{cases} x + 3y = 5 \\ -4x + y = -7 \end{cases} \Rightarrow \begin{array}{r} \begin{cases} 4x + 12y = 20 \\ +\ \underline{-4x + y = -7} \end{cases} \\ 13y = 13 \end{array}$

Replacing the second equation of the original system with this new equation, we have

$\begin{cases} x + 3y = 5 \\ 13y = 13 \end{cases}$. Now that we have an equivalent system that is upper-triangular, we

can solve for y in the second equation and then back-substitute: $13y = 13 \Rightarrow y = 1$;

$x + 3y = 5 \Rightarrow x + 3(1) = 5 \Rightarrow x = 2$. The solution is $(x, y) = (2, 1)$.

29. We can eliminate the $3x$ term in the second equation by multiplying the first equation by 3, and multiplying the second equation by -2, and then adding the equations:

$$\begin{cases} 2x - 4y = -6 \\ 3x - 5y = 3 \end{cases} \Rightarrow + \begin{cases} 6x - 12y = -18 \\ \underline{-6x + 10y = -6} \\ -2y = -24 \end{cases}$$

Replacing the second equation of the original system with this new equation, we have

$$\begin{cases} 2x - 4y = -6 \\ -2y = -24 \end{cases}.$$ Now that we have an equivalent system that is upper-triangular,

we can solve for y in the second equation and then back-substitute: $-2y = -24 \Rightarrow y = 12$; $2x - 4y = -6 \Rightarrow 2x - 4(12) = -6 \Rightarrow x = 21$. The solution is $(x, y) = (21, 12)$.

33. $\begin{cases} x - 2y + 4z = -2 \\ -y + z = -3 \\ 3x - 3y + z = 3 \end{cases} \Rightarrow \underset{-3E_1 + E_3 \rightarrow E_3}{} \begin{cases} x - 2y + 4z = -2 \\ -y + z = -3 \\ 3y - 11z = 9 \end{cases}$

$\Rightarrow \underset{3E_2 + E_3 \rightarrow E_3}{} \begin{cases} x - 2y + 4z = -2 \\ -y + z = -3 \\ -8z = 0 \end{cases}$

By back-substituting, we obtain

$$-8z = 0 \Rightarrow z = 0$$
$$-y + 0 = -3 \Rightarrow y = 3$$
$$x - 2(3) + 4(0) = -2 \Rightarrow x - 6 = -2 \Rightarrow x = 4. \text{ The solution is } (x, y, z) = (4, 3, 0).$$

37. $\begin{cases} x + 2y - z = -3 \\ 2x - 4y + z = -7 \\ -2x + 2y - 3z = 4 \end{cases} \Rightarrow \underset{\substack{-2E_1 + E_2 \rightarrow E_2 \\ 2E_1 + E_3 \rightarrow E_3}}{} \begin{cases} x + 2y - z = -3 \\ -8y + 3z = -1 \\ 6y - 5z = -2 \end{cases}$

$\Rightarrow \underset{\frac{3}{4}E_2 + E_3 \rightarrow E_3}{} \begin{cases} x + 2y - z = -3 \\ -8y + 3z = -1 \\ -1\frac{1}{4}z = -1\frac{1}{4} \end{cases}$

By back-substituting, we obtain

$$-\frac{11}{4}z = -\frac{11}{4} \Rightarrow z = 1$$
$$-8y + 3(1) = -1 \Rightarrow -8y = -4 \Rightarrow y = \frac{1}{2}$$
$$x + 2\left(\frac{1}{2}\right) - 1 = -3 \Rightarrow x + 1 - 1 = -3 \Rightarrow x = -3. \text{ The solution is } (x, y, z) = \left(-3, \frac{1}{2}, 1\right).$$

41. First we will multiply both sides of each equation by the least common denominator to avoid fractional coefficients.

$$\begin{matrix} 4E_1 \to E_1 \\ 2E_2 \to E_2 \\ 2E_3 \to E_3 \end{matrix} \begin{cases} 2x - 3y - z = 11 \\ x - 2y - 4z = -4 \\ 3x - y - 2z = 8 \end{cases} \Rightarrow \begin{matrix} E_2 \to E_1 \\ E_1 \to E_2 \end{matrix} \begin{cases} x - 2y - 4z = -4 \\ 2x - 3y - z = 11 \\ 3x - y - 2z = 8 \end{cases}$$

$$\Rightarrow \begin{matrix} -2E_1 + E_2 \to E_2 \\ -3E_1 + E_3 \to E_3 \end{matrix} \begin{cases} x - 2y - 4z = -4 \\ y + 7z = 19 \\ 5y + 10z = 20 \end{cases} \Rightarrow \begin{matrix} \\ -5E_2 + E_3 \to E_3 \end{matrix} \begin{cases} x - 2y - 4z = -4 \\ y + 7z = 19 \\ -25z = -75 \end{cases}$$

By back-substitution, we obtain

$$-25z = -75 \Rightarrow z = 3$$
$$y + 7(3) = 19 \Rightarrow y = -2$$
$$x - 2(-2) - 4(3) = -4 \Rightarrow x - 8 = -4 \Rightarrow x = 4.$$ The solution is $(x, y, z) = (4, -2, 3)$.

45. Following the hint we have $\begin{cases} u + v = 169 \\ u - v = 119 \end{cases} \Rightarrow \begin{matrix} \\ 1E_1 + E_2 \to E_2 \end{matrix} \begin{cases} u + v = 169 \\ -2v = -50 \end{cases}$.

Back-substitution yields

$$-2v = -50 \Rightarrow v = 25$$
$$u + v = 169 \Rightarrow u + 25 = 169 \Rightarrow u = 144$$

Since $u = x^2$ and $v = y^2$, we have

$x = \pm \sqrt{u} = \pm \sqrt{144} = \pm 12$ and $y = \pm \sqrt{v} = \pm \sqrt{25} = \pm 5$. There are four solutions (x, y): $(12, 5)$, $(12, -5)$, $(-12, 5)$, and $(-12, -5)$.

49. The given points on the circle represent solutions to the equation $x^2 + y^2 + ax + by + c = 0$. Substituting $(1, 0)$ gives

$(1)^2 + (0)^2 + a(1) + b(0) + c = 0 \Rightarrow a + c = -1$;

substituting $(0, 1)$ gives

$(0)^2 + (1)^2 + a(0) + b(1) + c = 0 \Rightarrow b + c = -1$;

and substituting $(1, -2)$ gives

$(1)^2 + (-2)^2 + a(1) + b(-2) + c = 0 \Rightarrow a - 2b + c = -5$. We need to solve the system:

$$\begin{cases} a \quad\quad + c = -1 \\ \quad b + c = -1 \\ a - 2b + c = -5 \end{cases} \Rightarrow \begin{matrix} \\ \\ -E_1 + E_3 \to E_3 \end{matrix} \begin{cases} a \quad\quad + c = -1 \\ \quad b + c = -1 \\ \quad -2b \quad = -4 \end{cases}$$

Although this is not upper-triangular, we can solve for b in the third equation and back-substitute.

$$-2b = -4 \Rightarrow b = 2$$
$$b + c = -1 \Rightarrow 2 + c = -1 \Rightarrow c = -3$$
$$a + c = -1 \Rightarrow a + (-3) = -1 \Rightarrow a = 2.$$ Thus the equation of the circle is
$$x^2 + y^2 + ax + by + c = 0 \Rightarrow x^2 + y^2 + 2x + 2y - 3 = 0.$$

53. Since the weight of the mixture must be 100 lb, we have

$$\left(\begin{array}{c}\text{pounds of}\\\text{peanuts}\end{array}\right) + \left(\begin{array}{c}\text{pounds of}\\\text{filberts}\end{array}\right) + \left(\begin{array}{c}\text{pounds of}\\\text{cashews}\end{array}\right) = 100$$

Letting P, F, and C represent the pounds of peanuts, filberts, and cashews, respectively, we have

$$P + F + C = 100.$$

Now consider the cost of the mixture:

$$\left(\begin{array}{c}\text{cost of}\\\text{peanuts}\end{array}\right) + \left(\begin{array}{c}\text{cost of}\\\text{filberts}\end{array}\right) + \left(\begin{array}{c}\text{cost of}\\\text{cashews}\end{array}\right) = \left(\begin{array}{c}\text{cost of}\\\text{mix}\end{array}\right)$$

$$\Rightarrow \quad 0.80\left(\begin{array}{c}\text{pounds of}\\\text{peanuts}\end{array}\right) + 1.60\left(\begin{array}{c}\text{pounds of}\\\text{filberts}\end{array}\right) + 2.40\left(\begin{array}{c}\text{pounds of}\\\text{cashews}\end{array}\right) = 1.32(100)$$

$$\Rightarrow \quad 0.80P + 1.60F + 2.40C = 100$$

We are also given that

$$\left(\begin{array}{c}\text{pounds of}\\\text{filberts}\end{array}\right) + \left(\begin{array}{c}\text{pounds of}\\\text{cashews}\end{array}\right) = \left(\begin{array}{c}\text{pounds of}\\\text{peanuts}\end{array}\right)$$

$$\Rightarrow \quad F + C = P \Rightarrow -P + F + C = 0$$

We need to solve the following system:

$$\begin{cases} P + F + C = 100 \\ 0.80P + 1.60F + 2.40C = 132 \\ -P + F + C = 0 \end{cases}$$

$$\Rightarrow 10E_2 + E_2 \to E_2 \begin{cases} P + F + C = 100 \\ 8P + 16F + 24C = 1320 \\ -P + F + C = 0 \end{cases}$$

$$\Rightarrow \begin{array}{c} -8E_1 + E_2 \to E_2 \\ E_1 + E_3 \to E_3 \end{array} \begin{cases} P + F + C = 100 \\ 8F + 16C = 520 \\ 2F + 2C = 100 \end{cases}$$

$$\Rightarrow \begin{array}{c} \frac{1}{8}E_2 \to E_2 \\ \frac{1}{2}E_3 \to E_3 \end{array} \begin{cases} P + F + C = 100 \\ F + 2C = 65 \\ F + C = 50 \end{cases}$$

$$\Rightarrow \begin{array}{c} \\ -E_2 + E_3 \to E_3 \end{array} \begin{cases} P + F + C = 100 \\ F + 2C = 65 \\ -C = -15 \end{cases}$$

Back-substituting, we get

$$-C = -15 \Rightarrow C = 15$$

$$F + 2(15) = 65 \Rightarrow F + 30 = 65 \Rightarrow F = 35$$

$$P + 35 + 15 = 100 \Rightarrow P + 35 + 15 = 100 \Rightarrow P + 50 = 100 \Rightarrow P = 50$$

The answer to the question is 50 lbs of peanuts, 35 lbs of filberts, and 15 lbs of cashews.

57. a) Substituting $x = 0$, $y = 0$, and $z = 0$ into each equation gives a true statement.

b) We are given that
$$\begin{cases} a_{11}x_0 + a_{12}y_0 + a_{13}z_0 = 0 \\ a_{21}x_0 + a_{22}y_0 + a_{23}z_0 = 0 \\ a_{31}x_0 + a_{32}y_0 + a_{33}z_0 = 0 \end{cases}$$

Multiplying both sides of each equation by 2 gives
$$\begin{cases} a_{11}(2x_0) + a_{12}(2y_0) + a_{13}(2z_0) = 0 \\ a_{21}(2x_0) + a_{22}(2y_0) + a_{23}(2z_0) = 0 \\ a_{31}(2x_0) + a_{32}(2y_0) + a_{33}(2z_0) = 0 \end{cases}$$. Therefore $(2x_0, 2y_0, 2z_0)$ is also a solution.

c) Using the same reasoning as in part b), if (x_0, y_0, z_0) is a non-trivial solution to the system, then (kx_0, ky_0, kz_0) is also a solution to the system (k is any constant). Therefore there are infinitely many solutions to the system.

Section 10.2

1. Refer to page 562 in the text. $\begin{cases} x + 2y = 7 \\ 3x - 4y = 12 \end{cases} \Rightarrow \begin{bmatrix} 1 & 2 & | & 7 \\ 3 & -4 & | & 12 \end{bmatrix}$

5. $\begin{cases} 2x + 3y - 5z + 4w = 34 \\ 6x - y + 10z = 21 \\ 3y - 5z = 3 \\ 9w = 12 \end{cases} \Rightarrow \begin{bmatrix} 2 & 3 & -5 & 4 & | & 34 \\ 6 & -1 & 10 & 0 & | & 21 \\ 0 & 3 & -5 & 0 & | & 3 \\ 0 & 0 & 0 & 9 & | & 12 \end{bmatrix}$

9. $\begin{cases} 4x + 2y = 1 \\ 3x + 2y = -9 \end{cases} \Rightarrow \begin{bmatrix} 4 & 2 & | & 1 \\ 3 & 2 & | & -9 \end{bmatrix} \quad -(\tfrac{3}{4})R_1 + R_2 \rightarrow R_2 \begin{bmatrix} 4 & 2 & | & 1 \\ 0 & \tfrac{1}{2} & | & \tfrac{39}{4} \end{bmatrix}$

$2R_2 \rightarrow R_2 \begin{bmatrix} 4 & 2 & | & 1 \\ 0 & 1 & | & \tfrac{39}{2} \end{bmatrix} \qquad -2R_2 + R_1 \rightarrow R_1 \begin{bmatrix} 4 & 0 & | & 40 \\ 0 & 1 & | & \tfrac{39}{2} \end{bmatrix}$

$(\tfrac{1}{4})R_1 \rightarrow R_1 \begin{bmatrix} 1 & 0 & | & 10 \\ 0 & 1 & | & \tfrac{39}{2} \end{bmatrix}$ Thus, the solution is $(x, y) = (10, -\tfrac{39}{2})$.

13. Refer to the box on page 566 (in the text). The given matrix satisfies all four conditions; it is in row reduced form.

17. Refer to the box on page 566 (in the text). The given matrix satisfies all four conditions; it is in row reduced form.

21. $\begin{cases} x - 2y + 4z = -2 \\ - y + z = -3 \\ 3x - 3y + z = 3 \end{cases}$ $\Rightarrow$ $\begin{bmatrix} 1 & -2 & 4 & | & -2 \\ 0 & -1 & 1 & | & -3 \\ 3 & -3 & 1 & | & 3 \end{bmatrix}$ $-3R_1 + R_3 \rightarrow R_3$ $\begin{bmatrix} 1 & -2 & 4 & | & -2 \\ 0 & -1 & 1 & | & -3 \\ 0 & 3 & -11 & | & 9 \end{bmatrix}$

$-R_2 \rightarrow R_2$ $\begin{bmatrix} 1 & -2 & 4 & | & -2 \\ 0 & 1 & -1 & | & 3 \\ 0 & 3 & -11 & | & 9 \end{bmatrix}$ $\begin{matrix} 2R_2 + R_1 \rightarrow R_1 \\ \\ -3R_2 + R_3 \rightarrow R_3 \end{matrix}$ $\begin{bmatrix} 1 & 0 & 2 & | & 4 \\ 0 & 1 & -1 & | & 3 \\ 0 & 0 & -8 & | & 0 \end{bmatrix}$

$\Rightarrow -(\tfrac{1}{8})R_3 \rightarrow R_3$ $\begin{bmatrix} 1 & 0 & 2 & | & 4 \\ 0 & 1 & -1 & | & 3 \\ 0 & 0 & 1 & | & 0 \end{bmatrix}$ $\begin{matrix} -2R_3 + R_1 \rightarrow R_1 \\ R_3 + R_2 \rightarrow R_2 \end{matrix}$ $\begin{bmatrix} 1 & 0 & 0 & | & 4 \\ 0 & 1 & 0 & | & 3 \\ 0 & 0 & 1 & | & 0 \end{bmatrix}$

The solution is $(x, y, z) = (4, 3, 0)$.

25. $\begin{cases} x + 2y - z = -3 \\ 2x - 4y + z = -7 \\ -2x + 2y - 3z = 4 \end{cases}$ $\Rightarrow$ $\begin{bmatrix} 1 & 2 & -1 & | & -3 \\ 2 & -4 & 1 & | & -7 \\ -2 & 2 & -3 & | & 4 \end{bmatrix}$ $\begin{matrix} -2R_1 + R_2 \rightarrow R_2 \\ 2R_1 + R_3 \rightarrow R_3 \end{matrix}$ $\begin{bmatrix} 1 & 2 & -1 & | & -3 \\ 0 & -8 & 3 & | & -1 \\ 0 & 6 & -5 & | & -2 \end{bmatrix}$

$-(\tfrac{1}{8})R_2 \rightarrow R_2$ $\begin{bmatrix} 1 & 2 & -1 & | & -3 \\ 0 & 1 & -\tfrac{3}{8} & | & \tfrac{1}{8} \\ 0 & 6 & -5 & | & -2 \end{bmatrix}$ $\begin{matrix} -2R_2 + R_1 \rightarrow R_1 \\ \\ -6R_2 + R_3 \rightarrow R_3 \end{matrix}$ $\begin{bmatrix} 1 & 0 & -\tfrac{1}{4} & | & -\tfrac{13}{4} \\ 0 & 1 & -\tfrac{3}{8} & | & \tfrac{1}{8} \\ 0 & 0 & -\tfrac{11}{4} & | & -\tfrac{11}{4} \end{bmatrix}$

$-(\tfrac{4}{11})R_3 \rightarrow R_3$ $\begin{bmatrix} 1 & 0 & -\tfrac{1}{4} & | & -\tfrac{13}{4} \\ 0 & 1 & -\tfrac{3}{8} & | & \tfrac{1}{8} \\ 0 & 0 & 1 & | & 1 \end{bmatrix}$ $\begin{matrix} (\tfrac{1}{4})R_3 + R_1 \rightarrow R_1 \\ (\tfrac{3}{8})R_3 + R_2 \rightarrow R_2 \end{matrix}$ $\begin{bmatrix} 1 & 0 & 0 & | & -3 \\ 0 & 1 & 0 & | & \tfrac{1}{2} \\ 0 & 0 & 1 & | & 1 \end{bmatrix}$

The solution is $(x, y, z) = (-3, \tfrac{1}{2}, 1)$.

29. First we will multiply each equation by the least common denominator to clear fractional coefficients and then use row operations to reduce to row reduced form.

$$\begin{cases} \frac{1}{2}x - \frac{3}{4}y - \frac{1}{4}z = \frac{11}{4} \\ \frac{1}{2}x - y - 2z = -2 \\ \frac{3}{2}x - \frac{1}{2}y - z = 4 \end{cases} \Rightarrow \begin{cases} 2x - 3y - z = 11 \\ x - 2y - 4z = -4 \\ 3x - y - 2z = 8 \end{cases} \Rightarrow \begin{bmatrix} 2 & -3 & -1 & | & 11 \\ 1 & -2 & -4 & | & -4 \\ 3 & -1 & -2 & | & 8 \end{bmatrix}$$

$$\begin{matrix} R_2 \to R_1 \\ R_1 \to R_2 \end{matrix} \begin{bmatrix} 1 & -2 & -4 & | & -4 \\ 2 & -3 & -1 & | & 11 \\ 3 & -1 & -2 & | & 8 \end{bmatrix} \qquad \begin{matrix} -2R_1 + R_2 \to R_2 \\ -3R_1 + R_3 \to R_3 \end{matrix} \begin{bmatrix} 1 & -2 & -4 & | & -4 \\ 0 & 1 & 7 & | & 19 \\ 0 & 5 & 10 & | & 20 \end{bmatrix}$$

$$\begin{matrix} 2R_2 + R_1 \to R_1 \\ \\ -5R_2 + R_3 \to R_3 \end{matrix} \begin{bmatrix} 1 & 0 & 10 & | & 34 \\ 0 & 1 & 7 & | & 19 \\ 0 & 0 & -25 & | & -75 \end{bmatrix} \qquad -(\tfrac{1}{25})R_3 \to R_3 \begin{bmatrix} 1 & 0 & 10 & | & 34 \\ 0 & 1 & 7 & | & 19 \\ 0 & 0 & 1 & | & 3 \end{bmatrix}$$

$$\begin{matrix} -10R_3 + R_1 \to R_1 \\ -7R_3 + R_2 \to R_2 \end{matrix} \begin{bmatrix} 1 & 0 & 0 & | & 4 \\ 0 & 1 & 0 & | & -2 \\ 0 & 0 & 1 & | & 3 \end{bmatrix} \qquad \text{The solution is } (x, y, z) = (4, -2, 3).$$

33. The augmented matrix is $\begin{bmatrix} 6 & 12 & | & -18 \\ -7 & -14 & | & 21 \end{bmatrix} \Rightarrow -(\tfrac{1}{6})R_1 \to R_1 \begin{bmatrix} 1 & 2 & | & -3 \\ -7 & -14 & | & 21 \end{bmatrix}$

$$7R_3 + R_2 \to R_2 \begin{bmatrix} 1 & 2 & | & -3 \\ 0 & 0 & | & 0 \end{bmatrix}$$

This represents the system $\begin{cases} x + 2y = -3 \\ 0y = 0 \end{cases}$. Since the last equation is true for any values we might choose for y, it is a dependent system. If we represent y with the parameter t, we have $x + 2t = -3 \Rightarrow x = -3 - 2t$. The solution is $(x, y) = (-3 - 2t, t)$.

37. The augmented matrix is $\begin{bmatrix} 7 & -1 & 3 & | & 14 \\ 2 & -1 & 1 & | & 2 \\ 3 & 1 & 1 & | & 10 \end{bmatrix}$ $\begin{matrix} (\frac{1}{2})R_2 \to R_1 \\ R_1 \to R_2 \end{matrix}$ $\begin{bmatrix} 1 & -\frac{1}{2} & \frac{1}{2} & | & 1 \\ 7 & -1 & 3 & | & 14 \\ 3 & 1 & 1 & | & 10 \end{bmatrix}$

$\begin{matrix} -7R_1 + R_2 \to R_2 \\ -3R_1 + R_3 \to R_3 \end{matrix}$ $\begin{bmatrix} 1 & -\frac{1}{2} & \frac{1}{2} & | & 1 \\ 0 & \frac{5}{2} & -\frac{1}{2} & | & 7 \\ 0 & \frac{5}{2} & -\frac{1}{2} & | & 7 \end{bmatrix}$ $\begin{matrix} (\frac{2}{5})R_2 \to R_2 \\ 2R_3 \to R_3 \end{matrix}$ $\begin{bmatrix} 1 & -\frac{1}{2} & \frac{1}{2} & | & 1 \\ 0 & 1 & -\frac{1}{5} & | & \frac{14}{5} \\ 0 & 5 & -1 & | & 14 \end{bmatrix}$

$\begin{matrix} (\frac{1}{2})R_2 + R_1 \to R_1 \\ \\ -5R_2 + R_3 \to R_3 \end{matrix}$ $\begin{bmatrix} 1 & 0 & \frac{2}{5} & | & \frac{12}{5} \\ 0 & 1 & -\frac{1}{5} & | & \frac{14}{5} \\ 0 & 0 & 0 & | & 0 \end{bmatrix}$ This last matrix is in row reduced form.

This represents the system $\begin{cases} x + \frac{2}{5}z = \frac{12}{5} \\ y - \frac{1}{5}z = \frac{14}{5} \\ 0z = 0 \end{cases}$. Since the last equation is true for any values

we might choose for z, it is a dependent system. If we represent z with the parameter t, we

have $x + \frac{2}{5}t = \frac{12}{5} \Rightarrow x = -\frac{2}{5}t + \frac{12}{5}$

and $y - \frac{1}{5}t = \frac{14}{5} \Rightarrow y = \frac{1}{5}t + \frac{14}{5}$. The solution is $(x, y, z) = (-\frac{2}{5}t + \frac{12}{5}, \frac{1}{5}t + \frac{14}{5}, t)$.

41. The augmented matrix is $\begin{bmatrix} 5 & 1 & -1 & | & 0 \\ -6 & 1 & 1 & | & 0 \\ 6 & -1 & -1 & | & 0 \end{bmatrix}$ $(\frac{1}{5})R_1 \to R_1$ $\begin{bmatrix} 1 & \frac{1}{5} & -\frac{1}{5} & | & 0 \\ -6 & 1 & 1 & | & 0 \\ 6 & -1 & -1 & | & 0 \end{bmatrix}$

$\begin{matrix} 6R_1 + R_2 \to R_2 \\ -6R_1 + R_3 \to R_3 \end{matrix}$ $\begin{bmatrix} 1 & \frac{1}{5} & -\frac{1}{5} & | & 0 \\ 0 & \frac{11}{5} & -\frac{1}{5} & | & 0 \\ 0 & -\frac{11}{5} & \frac{1}{5} & | & 0 \end{bmatrix}$ $\begin{matrix} (\frac{5}{11})R_2 \to R_2 \\ 5R_3 \to R_3 \end{matrix}$ $\begin{bmatrix} 1 & \frac{1}{5} & -\frac{1}{5} & | & 0 \\ 0 & 1 & -\frac{1}{11} & | & 0 \\ 0 & -11 & 1 & | & 0 \end{bmatrix}$

$\begin{matrix} -(\frac{1}{5})R_2 + R_1 \to R_1 \\ \\ 11R_2 + R_3 \to R_3 \end{matrix}$ $\begin{bmatrix} 1 & 0 & -\frac{2}{11} & | & 0 \\ 0 & 1 & -\frac{1}{11} & | & 0 \\ 0 & 0 & 0 & | & 0 \end{bmatrix}$ This last matrix is in row reduced form.

This represents the system $\begin{cases} x - \dfrac{2}{11}z = 0 \\ y - \dfrac{1}{11}z = 0 \\ 0z = 0 \end{cases}$. Since the last equation is true for any values

we might choose for z, it is a dependent system. If we represent z with the parameter t, we

have $\quad x - \dfrac{2}{11}t = 0 \Rightarrow x = \dfrac{2}{11}t$

and $\quad y - \dfrac{1}{11}t = 0 \Rightarrow y = \dfrac{1}{11}t$ The solution is $(x, y, z) = (\dfrac{2}{11}t, \dfrac{1}{11}t, t)$.

45. The system is underdetermined. The augmented matrix is $\begin{bmatrix} 1 & 1 & -2 & | & 2 \\ -3 & 1 & 6 & | & 7 \end{bmatrix}$

$3R_1 + R_2 \rightarrow R_2 \begin{bmatrix} 1 & 1 & -2 & | & 2 \\ 0 & 4 & 0 & | & 13 \end{bmatrix}$ $\qquad (\frac{1}{4})R_2 \rightarrow R_2 \begin{bmatrix} 1 & 1 & -2 & | & 2 \\ 0 & 1 & 0 & | & \frac{13}{4} \end{bmatrix}$

$-R_2 + R_1 \rightarrow R_1 \begin{bmatrix} 1 & 0 & -2 & | & \frac{5}{4} \\ 0 & 1 & 0 & | & \frac{13}{4} \end{bmatrix}$

This last matrix is in row reduced form. The second row represents the equation

$y = \dfrac{13}{4}$. The first row represents the equation $x - 2z = \dfrac{5}{4} \Rightarrow x = \dfrac{5}{4} + 2z$. Letting $z = t$,

we have the solution $(x, y, z) = (\dfrac{5}{4} + 2t, \dfrac{13}{4}, t)$.

49. The augmented matrix is $\begin{bmatrix} 1 & 1 & 1 & 1 & | & 9 \\ 1 & 1 & 2 & 0 & | & 11 \\ 1 & 0 & 2 & -1 & | & 8 \\ 0 & 1 & 3 & -1 & | & 11 \end{bmatrix}$ $\begin{array}{c} -R_1 + R_2 \rightarrow R_2 \\ -R_1 + R_3 \rightarrow R_3 \end{array} \begin{bmatrix} 1 & 1 & 1 & 1 & | & 9 \\ 0 & 0 & 1 & -1 & | & 2 \\ 0 & -1 & 1 & -2 & | & -1 \\ 0 & 1 & 3 & -1 & | & 11 \end{bmatrix}$

$\begin{array}{c} R_4 \rightarrow R_2 \\ \\ R_2 \rightarrow R_4 \end{array} \begin{bmatrix} 1 & 1 & 1 & 1 & | & 9 \\ 0 & 1 & 3 & -1 & | & 11 \\ 0 & -1 & 1 & -2 & | & -1 \\ 0 & 0 & 1 & -1 & | & 2 \end{bmatrix}$ $\begin{array}{c} -R_2 + R_1 \rightarrow R_1 \\ \\ R_2 + R_3 \rightarrow R_3 \end{array} \begin{bmatrix} 1 & 0 & -2 & 2 & | & -2 \\ 0 & 1 & 3 & -1 & | & 11 \\ 0 & 0 & 4 & -3 & | & 10 \\ 0 & 0 & 1 & -1 & | & 2 \end{bmatrix}$

$\begin{array}{c} R_4 \rightarrow R_3 \\ R_3 \rightarrow R_4 \end{array} \begin{bmatrix} 1 & 0 & -2 & 2 & | & -2 \\ 0 & 1 & 3 & -1 & | & 11 \\ 0 & 0 & 1 & -1 & | & 2 \\ 0 & 0 & 4 & -3 & | & 10 \end{bmatrix}$ $\begin{array}{c} 2R_3 + R_1 \rightarrow R_1 \\ -3R_3 + R_2 \rightarrow R_2 \\ \\ -4R_3 + R_4 \rightarrow R_4 \end{array} \begin{bmatrix} 1 & 0 & 0 & 0 & | & 2 \\ 0 & 1 & 0 & 2 & | & 5 \\ 0 & 0 & 1 & -1 & | & 2 \\ 0 & 0 & 0 & 1 & | & 2 \end{bmatrix}$

$\begin{array}{c} -2R_4 + R_2 \rightarrow R_2 \\ R_4 + R_3 \rightarrow R_3 \end{array} \begin{bmatrix} 1 & 0 & 0 & 0 & | & 2 \\ 0 & 1 & 0 & 0 & | & 1 \\ 0 & 0 & 1 & 0 & | & 4 \\ 0 & 0 & 0 & 1 & | & 2 \end{bmatrix}$

The solution is $(x, y, z, w) = (2, 1, 4, 2)$.

53. The augmented matrix is $\begin{bmatrix} 1 & 0 & 1 & 0 & 0 & | & 2 \\ 0 & 1 & 0 & 1 & 0 & | & 4 \\ 0 & 0 & 1 & 0 & 1 & | & 6 \\ 1 & 0 & 0 & 1 & 0 & | & 8 \\ 0 & 1 & 0 & 0 & 1 & | & 0 \end{bmatrix}$ $\begin{array}{l} -R_1 + R_4 \to R_4 \\ -R_2 + R_5 \to R_5 \end{array}$ $\begin{bmatrix} 1 & 0 & 1 & 0 & 0 & | & 2 \\ 0 & 1 & 0 & 1 & 0 & | & 4 \\ 0 & 0 & 1 & 0 & 1 & | & 6 \\ 0 & 0 & -1 & 1 & 0 & | & 6 \\ 0 & 0 & 0 & -1 & 1 & | & -4 \end{bmatrix}$

$\begin{array}{l} -R_3 + R_1 \to R_1 \\ \\ \\ R_3 + R_4 \to R_4 \end{array}$ $\begin{bmatrix} 1 & 0 & 0 & 0 & -1 & | & -4 \\ 0 & 1 & 0 & 1 & 0 & | & 4 \\ 0 & 0 & 1 & 0 & 1 & | & 6 \\ 0 & 0 & 0 & 1 & 1 & | & 12 \\ 0 & 0 & 0 & -1 & 1 & | & -4 \end{bmatrix}$ $\begin{array}{l} -R_4 + R_2 \to R_2 \\ \\ R_4 + R_5 \to R_5 \\ (\frac{1}{2})R_5 \to R_5 \end{array}$ $\begin{bmatrix} 1 & 0 & 0 & 0 & -1 & | & -4 \\ 0 & 1 & 0 & 0 & -1 & | & -8 \\ 0 & 0 & 1 & 0 & 1 & | & 6 \\ 0 & 0 & 0 & 1 & 1 & | & 12 \\ 0 & 0 & 0 & 0 & 1 & | & 4 \end{bmatrix}$

$\begin{array}{l} R_5 + R_1 \to R_1 \\ R_5 + R_2 \to R_2 \\ -R_5 + R_3 \to R_3 \\ -R_5 + R_4 \to R_4 \end{array}$ $\begin{bmatrix} 1 & 0 & 0 & 0 & 0 & | & 0 \\ 0 & 1 & 0 & 0 & 0 & | & -4 \\ 0 & 0 & 1 & 0 & 0 & | & 2 \\ 0 & 0 & 0 & 1 & 0 & | & 8 \\ 0 & 0 & 0 & 0 & 1 & | & 4 \end{bmatrix}$ The solution is $(x, y, z, w, v) = (0, -4, 2, 8, 4)$.

57. The augmented matrix is $\begin{bmatrix} 1 & 2 & 1 & -3 & | & 0 \\ 3 & 4 & -1 & 1 & | & 10 \end{bmatrix}$ $-3R_1 + R_2 \to R_2$ $\begin{bmatrix} 1 & 2 & 1 & -3 & | & 0 \\ 0 & -2 & -4 & 10 & | & 10 \end{bmatrix}$

$-(\frac{1}{2})R_2 \to R_2$ $\begin{bmatrix} 1 & 2 & 1 & -3 & | & 0 \\ 0 & 1 & 2 & -5 & | & -5 \end{bmatrix}$ $-2R_2 + R_1 \to R_1$ $\begin{bmatrix} 1 & 0 & -3 & 7 & | & 10 \\ 0 & 1 & 2 & -5 & | & -5 \end{bmatrix}$

This last matrix is in row reduced form. The second row represents the equation
$y + 2z - 5w = -5 \Rightarrow y = -2z + 5w - 5$. The first row represents the equation
$x - 3z + 7w = 10 \Rightarrow x = 3z - 7w + 10$. Letting $w = t$ and $z = s$, we have the solution
$(x, y, z, w) = (3s - 7t + 10, -2s + 5t - 5, s, t)$.

Section 10.3

1. a) $\begin{vmatrix} 2 & 5 \\ 1 & 7 \end{vmatrix} = (2)(7) - (1)(5) = 9$ b) $\begin{vmatrix} 3 & 1 \\ 4 & 2 \end{vmatrix} = (3)(2) - (4)(1) = 2$

5. a) $\begin{vmatrix} \sqrt{3} + 1 & -\frac{1}{2}\sqrt{2} \\ \sqrt{6} & 4 \end{vmatrix} = (\sqrt{3} + 1)(4) - (\sqrt{6})(-\frac{1}{2}\sqrt{2}) = 4\sqrt{3} + 4 + \frac{1}{2}\sqrt{12} = 5\sqrt{3} + 4$

b) $\begin{vmatrix} \cos x & \sin x \\ \sin y & \cos y \end{vmatrix} = (\cos x)(\cos y) - (\sin x)(\sin y) = \cos(x + y)$

9. Since the entry -7 is in row 1 and column 3, the minor of -7 is the determinant of the matrix that results when we eliminate row 1 and column 3. The cofactor of -7 is $(-1)^{i+j}$(minor of -7), where i and j are the row and column numbers of -7, respectively. Thus, the cofactor is

$$(-1)^{1+3}(\text{minor of} -7) = (+1)\begin{vmatrix} -2 & -3 \\ 3 & 6 \end{vmatrix} = (-2)(6) - (3)(-3) = -3$$

13. We will evaluate the determinant with the entries of the first row:

$$\begin{vmatrix} 4 & 1 & 2 \\ -1 & -3 & -5 \\ 3 & 2 & -1 \end{vmatrix} = 4(\text{cofactor of } 4) + 1(\text{cofactor of } 1) + 2(\text{cofactor of } 2)$$

$$= 4(-1)^{1+1}\begin{vmatrix} -3 & -5 \\ 2 & -1 \end{vmatrix} + 1(-1)^{1+2}\begin{vmatrix} -1 & -5 \\ 3 & -1 \end{vmatrix} + 2(-1)^{1+3}\begin{vmatrix} -1 & -3 \\ 3 & 2 \end{vmatrix}$$

$$= 4(1)(13) + 1(-1)(16) + 2(1)(7) = 50$$

17. We will evaluate the determinant with the entries of the first column (since one entry is 0):

$$\det\begin{bmatrix} 5 & 9 & -3 \\ 0 & 4 & -6 \\ 1 & 0 & 2 \end{bmatrix} = \begin{vmatrix} 5 & 9 & -3 \\ 0 & 4 & -6 \\ 1 & 0 & 2 \end{vmatrix} = 5(\text{cofactor of } 5) + 0(\text{cofactor of } 0) + 1(\text{cofactor of } 1)$$

$$= 5(\text{cofactor of } 5) + 1(\text{cofactor of } 1)$$

$$= 5(-1)^{1+1}\begin{vmatrix} 4 & -6 \\ 0 & 2 \end{vmatrix} + 1(-1)^{3+1}\begin{vmatrix} 9 & -3 \\ 4 & -6 \end{vmatrix}$$

$$= 5(1)(8) + 0 + 1(1)(-42) = -2$$

21. We will evaluate the determinant with the entries of the first row (since one entry is 0):

$$\det\begin{bmatrix} 0 & -8 & 1 & 8 \\ 3 & 5 & -3 & 1 \\ 1 & 2 & -6 & 0 \\ -1 & 6 & 5 & 4 \end{bmatrix} = \begin{vmatrix} 0 & -8 & 1 & 8 \\ 3 & 5 & -3 & 1 \\ 1 & 2 & -6 & 0 \\ -1 & 6 & 5 & 4 \end{vmatrix}$$

$$= 0(\text{cofactor of } 0) + (-8)(\text{cofactor of} -8) + 1(\text{cofactor of } 1) + 8(\text{cofactor of } 8)$$

$$= (-8)(\text{cofactor of} -8) + 1(\text{cofactor of } 1) + 8(\text{cofactor of } 8)$$

$$= (-8)(-1)^{1+2}\begin{vmatrix} 3 & -3 & 1 \\ 1 & -6 & 0 \\ -1 & 5 & 4 \end{vmatrix} + (1)(-1)^{1+3}\begin{vmatrix} 3 & 5 & 1 \\ 1 & 2 & 0 \\ -1 & 6 & 4 \end{vmatrix} + (8)(-1)^{1+4}\begin{vmatrix} 3 & 5 & -3 \\ 1 & 2 & -6 \\ -1 & 6 & 5 \end{vmatrix}$$

We can evaluate each of the three resulting determinants with the second row:

$$\begin{vmatrix} 3 & -3 & 1 \\ 1 & -6 & 0 \\ -1 & 5 & 4 \end{vmatrix} = 1(-1)^{2+1}\begin{vmatrix} -3 & 1 \\ 5 & 4 \end{vmatrix} + (-6)(-1)^{2+2}\begin{vmatrix} 3 & 1 \\ -1 & 4 \end{vmatrix} + 0$$

$$= 1(-1)(-17) + (-6)(1)(13) = -61$$

$$\begin{vmatrix} 3 & 5 & 1 \\ 1 & 2 & 0 \\ -1 & 6 & 4 \end{vmatrix} = 1(-1)^{2+1}\begin{vmatrix} 5 & 1 \\ 6 & 4 \end{vmatrix} + 2(-1)^{2+2}\begin{vmatrix} 3 & 1 \\ -1 & 4 \end{vmatrix} + 0 = 1(-1)(14) + 2(1)(13) = 12$$

$$
\begin{vmatrix} 3 & 5 & -3 \\ 1 & 2 & -6 \\ -1 & 6 & 5 \end{vmatrix} = 1(-1)^{2+1}\begin{vmatrix} 5 & -3 \\ 6 & 5 \end{vmatrix} + 2(-1)^{2+2}\begin{vmatrix} 3 & -3 \\ -1 & 5 \end{vmatrix} + (-6)(-1)^{2+3}\begin{vmatrix} 3 & 5 \\ -1 & 6 \end{vmatrix}
$$

$$
= 1(-1)(43) + 2(1)(12) + (-6)(-1)(23) = 119
$$

Thus, $(-8)(-1)^{1+2}\begin{vmatrix} 3 & -3 & 1 \\ 1 & -6 & 0 \\ -1 & 5 & 4 \end{vmatrix} + (1)(-1)^{1+3}\begin{vmatrix} 3 & 5 & 1 \\ 1 & 2 & 0 \\ -1 & 6 & 4 \end{vmatrix} + (8)(-1)^{1+4}\begin{vmatrix} 3 & 5 & -3 \\ 1 & 2 & -6 \\ -1 & 6 & 5 \end{vmatrix}$

$$
= (-8)(-1)(-61) + (1)(1)(12) + (8)(-1)(119) = -1428
$$

25. The determinant on the left side of the equation can be evaluated using the first column:

$$
\begin{vmatrix} 2-r & 4 & 2 \\ 0 & -3-r & -1 \\ 0 & 0 & -r \end{vmatrix} = (2-r)(-1)^{1+1}\begin{vmatrix} -3-r & -1 \\ 0 & -r \end{vmatrix} + 0 + 0 = (2-r)(-3-r)(-r).
$$

Thus,

$$
\begin{vmatrix} 2-r & 4 & 2 \\ 0 & -3-r & -1 \\ 0 & 0 & -r \end{vmatrix} = 0 \quad\Rightarrow\quad (2-r)(-3-r)(-r) = 0
$$

$$
\Rightarrow 2 - r = 0, -3 - r = 0, \text{ or } -r = 0 \quad\Rightarrow\quad r = 2, -3, \text{ or } 0.
$$

29. Let the coordinates of Q be $Q(x, y)$. Then $x + 6 = 8 \Rightarrow x = 2$, and $y + 2 = 7 \Rightarrow y = 5$. So the coordinates of Q is $Q(2, 5)$. The area of the parallelogram is $\begin{vmatrix} 6 & 2 \\ 2 & 5 \end{vmatrix} = 26$ square units.

33.

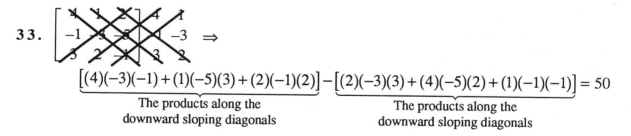

$$
\underbrace{[(4)(-3)(-1) + (1)(-5)(3) + (2)(-1)(2)]}_{\substack{\text{The products along the} \\ \text{downward sloping diagonals}}} - \underbrace{[(2)(-3)(3) + (4)(-5)(2) + (1)(-1)(-1)]}_{\substack{\text{The products along the} \\ \text{downward sloping diagonals}}} = 50
$$

37. We will use the first row to evaluate both determinants:

$$
\det A = 5(-1)^{1+1}\begin{vmatrix} 5 & 6 \\ -7 & 1 \end{vmatrix} + 4(-1)^{1+2}\begin{vmatrix} 1 & 6 \\ 3 & -1 \end{vmatrix} + (-2)(-1)^{1+3}\begin{vmatrix} 1 & 5 \\ 3 & -7 \end{vmatrix}
$$

$$
= 5(1)(37) + 4(-1)(-19) + (-2)(1)(-22) = 305
$$

$$
\det B = 5(-1)^{1+1}\begin{vmatrix} 5 & 6 \\ 3 & 11 \end{vmatrix} + 4(-1)^{1+2}\begin{vmatrix} 1 & 6 \\ 5 & 11 \end{vmatrix} + (-2)(-1)^{1+3}\begin{vmatrix} 1 & 5 \\ 5 & 3 \end{vmatrix}
$$

$$
= 5(1)(37) + 4(-1)(-19) + (-2)(1)(-22) = 305. \text{ Thus, } \det A = \det B.
$$

41. The area is the absolute value of $\dfrac{1}{2}\begin{vmatrix} 3 & 0 & 1 \\ 1 & 1 & 1 \\ -2 & 4 & 1 \end{vmatrix}$ Evaluating the determinant with the first

row, we get $\dfrac{1}{2}\left[3(-1)^{1+1}\begin{vmatrix} 1 & 1 \\ 4 & 1 \end{vmatrix} + 0 + 1(-1)^{1+3}\begin{vmatrix} 1 & 1 \\ -2 & 4 \end{vmatrix}\right] = \dfrac{1}{2}[-3] = -\dfrac{3}{2}$. Taking

the absolute value, we get an area of $\dfrac{3}{2}$ square units.

45. The left side of the equation is $\begin{vmatrix} x^2 + y^2 & x & y & 1 \\ (-3)^2 + 0^2 & -3 & 0 & 1 \\ (-1)^2 + 4^2 & -1 & 4 & 1 \\ 2^2 + (-5)^2 & 2 & -5 & 1 \end{vmatrix} = \begin{vmatrix} x^2 + y^2 & x & y & 1 \\ 9 & -3 & 0 & 1 \\ 17 & -1 & 4 & 1 \\ 29 & 2 & -5 & 1 \end{vmatrix}$.

Evaluating along the first row, we get $(x^2 + y^2)(-1)^{1+1}\begin{vmatrix} -3 & 0 & 1 \\ -1 & 4 & 1 \\ 2 & -5 & 1 \end{vmatrix}$

$+ x(-1)^{1+2}\begin{vmatrix} 9 & 0 & 1 \\ 17 & 4 & 1 \\ 29 & -5 & 1 \end{vmatrix} + y(-1)^{1+3}\begin{vmatrix} 9 & -3 & 1 \\ 17 & -1 & 1 \\ 29 & 2 & 1 \end{vmatrix} + (1)(-1)^{1+4}\begin{vmatrix} 9 & -3 & 0 \\ 17 & -1 & 4 \\ 29 & 2 & -5 \end{vmatrix}$

$(x^2 + y^2)(1)(-30) + x(-1)(-120) + y(1)(0) + (1)(-1)(-630)$

$= -30x^2 - 30y^2 + 120x + 630$. The equation of the circle is $-30x^2 - 30y^2 + 120x + 630$

$= 0 \Rightarrow x^2 + y^2 - 4x - 21 = 0$.

Section 10.4

1. $x = \dfrac{D_x}{D} = \dfrac{\begin{vmatrix} 10 & 2 \\ 22 & -4 \end{vmatrix}}{\begin{vmatrix} 3 & 2 \\ 1 & -4 \end{vmatrix}} = \dfrac{-84}{-14} = 6;$ $y = \dfrac{D_y}{D} = \dfrac{\begin{vmatrix} 3 & 10 \\ 1 & 22 \end{vmatrix}}{\begin{vmatrix} 3 & 2 \\ 1 & -4 \end{vmatrix}} = \dfrac{56}{-14} = -4$

5. First we add 9 to both sides of the second equation so that it is in the form
$a_{21}x + a_{22}y = k$:

$\begin{cases} 3x + y = 4 \\ x + 5y - 9 = 0 \end{cases} \Rightarrow \begin{cases} 3x + y = 4 \\ x + 5y = 9 \end{cases}$. According to Cramer's Rule, the solution is

$x = \dfrac{D_x}{D} = \dfrac{\begin{vmatrix} 4 & 1 \\ 9 & 5 \end{vmatrix}}{\begin{vmatrix} 3 & 1 \\ 1 & 5 \end{vmatrix}} = \dfrac{11}{14};$ $y = \dfrac{D_y}{D} = \dfrac{\begin{vmatrix} 3 & 4 \\ 1 & 9 \end{vmatrix}}{\begin{vmatrix} 3 & 1 \\ 1 & 5 \end{vmatrix}} = \dfrac{23}{14}$

9. $x = \dfrac{D_x}{D} = \dfrac{\begin{vmatrix} 1 & -10 \\ -4 & -15 \end{vmatrix}}{\begin{vmatrix} 6 & -10 \\ 9 & -15 \end{vmatrix}} = \dfrac{-55}{0}$. Since the numerator is nonzero and the denominator is 0,

the system is inconsistent.

13. $D = \begin{vmatrix} 4 & 1 & -4 \\ 1 & -1 & 2 \\ 2 & -1 & 8 \end{vmatrix}, D_x = \begin{vmatrix} 4 & 1 & -4 \\ -2 & -1 & 2 \\ 0 & -1 & 8 \end{vmatrix}, D_y = \begin{vmatrix} 4 & 4 & -4 \\ 1 & -2 & 2 \\ 2 & 0 & 8 \end{vmatrix}, D_z = \begin{vmatrix} 4 & 1 & 4 \\ 1 & -1 & -2 \\ 2 & -1 & 0 \end{vmatrix}.$

To simplify the calculation of D_x and D, add the third column to the first column and

replace the first column with the sum: $D_x = \begin{vmatrix} 4 & 1 & -4 \\ -2 & -1 & 2 \\ 0 & -1 & 8 \end{vmatrix} = \begin{vmatrix} 0 & 1 & -4 \\ 0 & -1 & 2 \\ 8 & -1 & 8 \end{vmatrix}$

$$= 8 \begin{vmatrix} 1 & -4 \\ -1 & 2 \end{vmatrix} = 8(-2) = -16;$$

$$D = \begin{vmatrix} 4 & 1 & -4 \\ 1 & -1 & 2 \\ 2 & -1 & 8 \end{vmatrix} = \begin{vmatrix} 0 & 1 & -4 \\ 3 & -1 & 2 \\ 10 & -1 & 8 \end{vmatrix}$$

$$= 0 + 1(-1)^{1+2} \begin{vmatrix} 3 & 2 \\ 10 & 8 \end{vmatrix} + (-4)(-1)^{1+3} \begin{vmatrix} 3 & -1 \\ 10 & -1 \end{vmatrix} = 1(-1)(4) + (-4)(1)(7) = -32.$$

To simplify the calculation of D_y, replace the first row with the sum of row 1 and twice row 2:

$$D_y = \begin{vmatrix} 4 & 4 & -4 \\ 1 & -2 & 2 \\ 2 & 0 & 8 \end{vmatrix} = \begin{vmatrix} 6 & 0 & 0 \\ 1 & -2 & 2 \\ 2 & 0 & 8 \end{vmatrix} = 6(-1)^{1+1} \begin{vmatrix} -2 & 2 \\ 0 & 8 \end{vmatrix} + 0 + 0 = 6(1)(-16) = -96.$$

Similarly $D_z = \begin{vmatrix} 4 & 1 & 4 \\ 1 & -1 & -2 \\ 2 & -1 & 0 \end{vmatrix} = \begin{vmatrix} 6 & -1 & 0 \\ 1 & -1 & -2 \\ 2 & -1 & 0 \end{vmatrix} = 0 + 0 + (-2)(-1)^{2+3} \begin{vmatrix} 6 & -1 \\ 2 & -1 \end{vmatrix}$

$$= (-2)(-1)(-4) = -8.$$

The solutions are $x = \dfrac{D_x}{D} = \dfrac{-16}{-32} = \dfrac{1}{2}$; $y = \dfrac{D_y}{D} = \dfrac{-96}{-32} = 3$; and $z = \dfrac{D_z}{D} = \dfrac{-8}{-32} = \dfrac{1}{4}$.

17. In order to get a system with three equations, we can add the equation $0x + 0y + 0z = 0$

(this is true for any value of x, y, or z.) $\begin{cases} 3x + y - 4z = 4 \\ x - y + 2z = -2 \end{cases} \Rightarrow \begin{cases} 3x + y - 4z = 4 \\ x - y + 2z = -2 \\ 0x + 0y + 0z = 0 \end{cases}$

According to Cramer's Rule, $D = \begin{vmatrix} 4 & 1 & -4 \\ 1 & -1 & 2 \\ 0 & 0 & 0 \end{vmatrix} = 0$, $D_x = \begin{vmatrix} 4 & 1 & -4 \\ 1 & -2 & 2 \\ 0 & 0 & 0 \end{vmatrix} = 0$,

$D_y = \begin{vmatrix} 4 & 4 & -4 \\ 1 & -2 & 2 \\ 0 & 0 & 0 \end{vmatrix} = 0$, and $D_z = \begin{vmatrix} 4 & 1 & 4 \\ 1 & -1 & -2 \\ 0 & 0 & 0 \end{vmatrix} = 0$. Thus $x = \dfrac{D_x}{D} = \dfrac{0}{0}$;

$y = \dfrac{D_y}{D} = \dfrac{0}{0}$; and $z = \dfrac{D_z}{D} = \dfrac{0}{0}$. This means that the system is dependent (see page 593).

21. Refer to the box on page 588 (row operation 1). $\begin{vmatrix} a_1 & a_2 & a_3 \\ c_1 & c_2 & c_3 \\ b_1 & b_2 & b_3 \end{vmatrix} = -\begin{vmatrix} a_1 & a_2 & a_3 \\ b_1 & b_2 & b_3 \\ c_1 & c_2 & c_3 \end{vmatrix} = -18$

25. The third column has been replaced with 4 times the first column added to the third column. The resulting determinant is equal to the original; that is, the determinant is 18.

29. By Cramer's Rule, $u = \dfrac{\begin{vmatrix} 1 & e^{2x} \\ 0 & 2e^{2x} \end{vmatrix}}{\begin{vmatrix} e^x & e^{2x} \\ e^x & 2e^{2x} \end{vmatrix}} = \dfrac{2e^{2x}}{e^{3x}} = \dfrac{2}{e^x}$; $v = \dfrac{\begin{vmatrix} e^x & 1 \\ e^x & 0 \end{vmatrix}}{\begin{vmatrix} e^x & e^{2x} \\ e^x & 2e^{2x} \end{vmatrix}} = \dfrac{-e^x}{e^{3x}} = \dfrac{-1}{e^{2x}}$

33. $\begin{vmatrix} a_{11} & a_{12} \\ a_{21} & a_{22} \end{vmatrix} = a_{11}a_{22} - a_{21}a_{12}$; $-\begin{vmatrix} a_{12} & a_{11} \\ a_{22} & a_{21} \end{vmatrix} = -(a_{12}a_{21} - a_{22}a_{11})$

$= a_{11}a_{22} - a_{21}a_{12}$.

37. $D = \begin{vmatrix} 1 & -1 & 3 & -2 \\ 3 & 0 & 1 & 1 \\ 6 & -4 & 4 & 0 \\ 5 & 0 & -2 & -1 \end{vmatrix}$, $D_x = \begin{vmatrix} 1 & -1 & 3 & -2 \\ 6 & 0 & 1 & 1 \\ 0 & -4 & 4 & 0 \\ 2 & 0 & -2 & -1 \end{vmatrix}$, $D_y = \begin{vmatrix} 1 & 1 & 3 & -2 \\ 3 & 6 & 1 & 1 \\ 6 & 0 & 4 & 0 \\ 5 & 2 & -2 & -1 \end{vmatrix}$,

$D_z = \begin{vmatrix} 1 & -1 & 1 & -2 \\ 3 & 0 & 6 & 1 \\ 6 & -4 & 0 & 0 \\ 5 & 0 & 2 & -1 \end{vmatrix}$, and $D_w = \begin{vmatrix} 1 & -1 & 3 & -2 \\ 3 & 0 & 1 & 1 \\ 6 & -4 & 4 & 0 \\ 5 & 0 & -2 & -1 \end{vmatrix}$. To simplify the calculation of

D, D_x, D_z, and D_w, replace the third row with the sum of 4 times the row 1 and row 3:

$$D = \begin{vmatrix} 1 & -1 & 3 & -2 \\ 3 & 0 & 1 & 1 \\ 2 & 0 & -8 & 8 \\ 5 & 0 & -2 & -1 \end{vmatrix} = (-1)(-1)^{1+2} \begin{vmatrix} 3 & 1 & 1 \\ 2 & -8 & 8 \\ 5 & -2 & -1 \end{vmatrix} = 150;$$

$$D_x = \begin{vmatrix} 1 & -1 & 3 & -2 \\ 6 & 0 & 1 & 1 \\ -4 & 0 & -8 & 8 \\ 2 & 0 & -2 & -1 \end{vmatrix} = (-1)(-1)^{1+2} \begin{vmatrix} 6 & 1 & 1 \\ -4 & -8 & 8 \\ 2 & -2 & -1 \end{vmatrix} = 180;$$

$$D_z = \begin{vmatrix} 1 & -1 & 1 & -2 \\ 3 & 0 & 6 & 1 \\ 2 & 0 & -4 & 8 \\ 5 & 0 & 2 & -1 \end{vmatrix} = (-1)(-1)^{1+2} \begin{vmatrix} 3 & 6 & 1 \\ 2 & -4 & 8 \\ 5 & 2 & -1 \end{vmatrix} = 240;$$

$$D_w = \begin{vmatrix} 1 & -1 & 3 & 1 \\ 3 & 0 & 1 & 6 \\ 2 & 0 & -8 & -4 \\ 5 & 0 & -2 & 2 \end{vmatrix} = (-1)(-1)^{1+2} \begin{vmatrix} 3 & 1 & 6 \\ 2 & -8 & -4 \\ 5 & -2 & 2 \end{vmatrix} = 120.$$

To evaluate D_y we can use the third row:

$$D_y = 6(-1)^{3+1} \begin{vmatrix} 1 & 3 & -2 \\ 6 & 1 & 1 \\ 2 & -2 & -1 \end{vmatrix} + 4(-1)^{3+3} \begin{vmatrix} 1 & 1 & -2 \\ 3 & 6 & 1 \\ 5 & 2 & -1 \end{vmatrix} = 6(1)(53) + 4(1)(48) = 510. \text{ Thus}$$

$$x = \frac{D_x}{D} = \frac{180}{150} = \frac{6}{5}; \; y = \frac{D_y}{D} = \frac{510}{150} = \frac{17}{5} \; ; \; z = \frac{D_z}{D} = \frac{240}{150} = \frac{8}{5}; \text{ and } w = \frac{D_w}{D} = \frac{120}{150} = \frac{4}{5}.$$

41. According to Cramer's Rule, $D = \begin{vmatrix} a & b \\ ka & kb \end{vmatrix} = 0.$ Therefore the system is either dependent

or inconsistent. To be dependent, D_x must be $0 \Leftrightarrow \begin{vmatrix} c & b \\ d & kb \end{vmatrix} = 0 \Leftrightarrow b(kc - d) = 0$

$\Leftrightarrow d = kc.$

Section 10.5

1. $\dfrac{12}{(x + 5)(x - 1)} = \dfrac{A}{x + 5} + \dfrac{B}{x - 1} \Rightarrow 12 = A(x - 1) + B(x + 5).$ If we let $x = -5$ in

this last equation, we get $12 = A(-6) + B(0) \Rightarrow 12 = -6A \Rightarrow A = -2.$ Letting $x = 1$ in the

equation as before, we get $12 = A(0) + B(6) \Rightarrow 12 = 6B \Rightarrow B = 2.$ The decomposition is

$\dfrac{A}{x + 5} + \dfrac{B}{x - 1} = \dfrac{-2}{x + 5} + \dfrac{2}{x - 1} \; .$

5. $\dfrac{4}{x^2 - 2x} = \dfrac{4}{x(x - 2)} = \dfrac{A}{x} + \dfrac{B}{x - 2} \Rightarrow 4 = A(x - 2) + Bx$. If we let $x = 2$ in this last

equation, we get $4 = A(0) + B(2) \Rightarrow 4 = 2B \Rightarrow B = 2$. Letting $x = 0$ in the equation as

before, we get $4 = A(-2) + B(0) \Rightarrow 4 = -2A \Rightarrow A = -2$. The decomposition is

$\dfrac{A}{x} + \dfrac{B}{x - 2} = \dfrac{-2}{x} + \dfrac{2}{x - 2}$.

9. $\dfrac{1}{x^3 - x} = \dfrac{1}{x(x + 1)(x - 1)} = \dfrac{A}{x} + \dfrac{B}{x + 1} + \dfrac{C}{x - 1}$

$\Rightarrow 1 = A(x + 1)(x - 1) + Bx(x - 1) + Cx(x + 1)$. If we let $x = -1, 0$, and 1 in this last
equation, we get

$x = -1 \Rightarrow 1 = A(0)(-2) + B(-1)(-2) + C(-1)(0) \Rightarrow 1 = 2B \Rightarrow B = \dfrac{1}{2}$;

$x = 0 \Rightarrow 1 = A(1)(-1) + B(0)(-1) + C(0)(1) \Rightarrow 1 = -A \Rightarrow A = -1$; and

$x = 1 \Rightarrow 1 = A(2)(0) + B(1)(0) + C(1)(2) \Rightarrow 1 = 2C \Rightarrow C = \dfrac{1}{2}$. The decomposition is

$\dfrac{A}{x} + \dfrac{B}{x + 1} + \dfrac{C}{x - 1} = \dfrac{-1}{x} + \dfrac{\frac{1}{2}}{x + 1} + \dfrac{\frac{1}{2}}{x - 1}$.

13. $\dfrac{4}{x(x^2 + 1)} = \dfrac{A}{x} + \dfrac{Bx + C}{x^2 + 1} \Rightarrow 4 = A(x^2 + 1) + (Bx + C)x$. Collecting like terms on the

right side of this equation, we have $4 = (A + B)x^2 + Cx + A$. If we compare coefficients of

like terms on each side of this equation, we have $\begin{cases} A + B = 0 \\ \quad\;\; C = 0 \\ \quad\;\; A = 4 \end{cases}$. Back-substituting gives

$B = -4$. The decomposition is $\dfrac{A}{x} + \dfrac{Bx + C}{x^2 + 1} = \dfrac{4}{x} + \dfrac{-4x}{x^2 + 1}$.

17. $\dfrac{x^3}{(x^2 + 1)^2} = \dfrac{Ax + B}{x^2 + 1} + \dfrac{Cx + D}{(x^2 + 1)^2} \Rightarrow x^3 = (Ax + B)(x^2 + 1) + (Cx + D)$. Collecting

like terms on the right side of this equation, we have
$x^3 = Ax^3 + Bx^2 + (A + C)x + (B + D)$. If we compare coefficients of like terms on each

side of this equation, we have $\begin{cases} \quad A = 1 \\ \quad B = 0 \\ A + C = 0 \\ B + D = 0 \end{cases}$. Back-substituting gives $C = -1, D = 0$.

The decomposition is $\dfrac{Ax + B}{x^2 + 1} + \dfrac{Cx + D}{x(x^2 + 1)^2} = \dfrac{x}{x^2 + 1} + \dfrac{-x}{(x^2 + 1)^2}$.

21. $\dfrac{5x^2 - 21x + 13}{(x + 2)(x^2 - 6x + 9)} = \dfrac{5x^2 - 21x + 13}{(x + 2)(x - 3)^2} = \dfrac{A}{x + 2} + \dfrac{B}{x - 3} + \dfrac{C}{(x - 3)^2}$

$\Rightarrow 5x^2 - 21x + 13 = A(x - 3)^2 + B(x + 2)(x - 3) + C(x + 2)$.

Substituting $x = 3$ into this last equation gives

$5(3)^2 - 21(3) + 13 = 0 + 0 + 5C \Rightarrow -5 = 5C \Rightarrow C = -1$

Subtituting $x = -2$, we get

$5(-2)^2 - 21(-2) + 13 = A(-5)^2 + 0 + 0 \Rightarrow 75 = 25A \Rightarrow A = 3$.

To find B, we collect like terms on the right side of the equation

$5x^2 - 21x + 13 = A(x - 3)^2 + B(x + 2)(x - 3) + C(x + 2)$

$\Rightarrow 5x^2 - 21x + 13 = (A + B)x^2 + (-6A - B + C)x + (9A - 6B + 2C)$

$\Rightarrow A + B = 5 \Rightarrow 3 + B = 5 \Rightarrow B = 2$.

The solution to this system is $A = 3$, $B = 2$, and $C = -1$. The decomposition is

$\dfrac{A}{x + 2} + \dfrac{B}{x - 3} + \dfrac{C}{(x - 3)^2} = \dfrac{3}{x + 2} + \dfrac{2}{x - 3} + \dfrac{-1}{(x - 3)^2}$.

25. $\dfrac{13x + 1}{(x^3 - 8)} = \dfrac{13x + 1}{(x - 2)(x^2 + 2x + 4)} = \dfrac{A}{x - 2} + \dfrac{Bx + C}{x^2 + 2x + 4}$

$\Rightarrow 13x + 1 = A(x^2 + 2x + 4) + (Bx + C)(x - 2)$.

Substituting $x = 2$ into this last equation gives

$13(2) + 1 = 12A + 0 \Rightarrow 27 = 12A \Rightarrow A = \dfrac{9}{4}$.

To find B and C, we collect like terms on the right side of this equation:

$13x + 1 = (A + B)x^2 + (2A - 2B + C)x + (4A - 2C)$. Comparing coefficients of like terms

on each side of this equation, we have $\begin{cases} A + B = 0 \\ 2A - 2B + C = 13 \\ 4A - 2C = 1 \end{cases}$.

$A + B = 0 \Rightarrow \dfrac{9}{4} + B = 0 \Rightarrow B = -\dfrac{9}{4}$

$4A - 2C = 1 \Rightarrow 9 - 2C = 1 \Rightarrow C = 4$

The decomposition is

$\dfrac{A}{x - 2} + \dfrac{Bx + C}{x^2 + 2x + 4} = \dfrac{\frac{9}{4}}{x - 2} + \dfrac{-\frac{9}{4}x + 4}{x^2 + 2x + 4}$.

29. Since the degree of the numerator is greater than that of the denominator, we first need to

divide $x^2 - 9$ into x^4 : $\dfrac{x^4}{x^2 - 9} = x^2 + 9 + \dfrac{81}{x^2 - 9}$. Now we determine the partial fraction

decomposition of $\dfrac{81}{x^2 - 9}$: $\dfrac{81}{x^2 - 9} = \dfrac{81}{(x+3)(x-3)} = \dfrac{A}{x+3} + \dfrac{B}{x-3}$

$\Rightarrow 81 = A(x-3) + B(x+3)$. If we let $x = 3$ in this last equation, we get

$81 = 0 + B(6) \Rightarrow B = \dfrac{27}{2}$.

Letting $x = -3$ in the equation as before, we get

$81 = A(-6) + 0 \Rightarrow A = -\dfrac{27}{2}$. The decomposition is

$x^2 + 9 + \dfrac{81}{x^2 - 9} = x^2 + 9 + \dfrac{A}{x+3} + \dfrac{B}{x-3} = x^2 + 9 + \dfrac{-\dfrac{27}{2}}{x+3} + \dfrac{\dfrac{27}{2}}{x-3}$.

33. We first need to factor the denominator. According to the Rational Root Theorem, possible rational roots are $\pm 1, 2, 3, 6,$ or 12. Trying these by synthetic division, leads to $x^4 - 2x^3 - 7x^2 + 8x + 12 = (x+2)(x-3)(x-2)(x+1)$. Thus,

$\dfrac{12}{x^4 - 2x^3 - 7x^2 + 8x + 12} = \dfrac{12}{(x+2)(x-3)(x-2)(x+1)}$

$= \dfrac{A}{x+2} + \dfrac{B}{x-3} + \dfrac{C}{x-2} + \dfrac{D}{x+1}$

$\Rightarrow 12 = A(x-3)(x-2)(x+1) + B(x+2)(x-2)(x+1) + C(x+2)(x-3)(x+1) +$

$D(x+2)(x-3)(x-2)$. If we let $x = -2, -1, 2,$ and 3 in this last equation, we get

$x = -2 \Rightarrow 12 = A(-5)(-4)(-1) + 0 + 0 + 0 \Rightarrow 12 = -20A \Rightarrow A = -\dfrac{3}{5}$;

$x = -1 \Rightarrow 12 = 0 + 0 + 0 + D(1)(-4)(-3) \Rightarrow 12 = 12D \Rightarrow D = 1$;

$x = 2 \Rightarrow 12 = 0 + 0 + C(4)(-1)(3) + 0 \Rightarrow C = -1$; and

$x = 3 \Rightarrow 12 = 0 + B(5)(1)(4) + 0 + 0 \Rightarrow B = \dfrac{3}{5}$. The decomposition is

$\dfrac{A}{x+2} + \dfrac{B}{x-3} + \dfrac{C}{x-2} + \dfrac{D}{x+1} = \dfrac{-\dfrac{3}{5}}{x+2} + \dfrac{\dfrac{3}{5}}{x-3} + \dfrac{-1}{x-2} + \dfrac{1}{x+1}$.

37. We can apply a property of logarithms to rewrite the numerator as $\ln(x^2) = 2\ln x$. Then if

we let $u = \ln x$, we have $\dfrac{2\ln x}{(\ln x + 2)(\ln x - 2)} = \dfrac{2u}{(u+2)(u-2)} = \dfrac{A}{u+2} + \dfrac{B}{u-2}$

$\Rightarrow 2u = A(u-2) + B(u+2)$. Substituting $u = 2$ in this last equation gives

$4 = 4B \Rightarrow B = 1$,

and substituting $u = -2$ in the equation gives

$-4 = -4A \Rightarrow A = 1$. Thus,

$\dfrac{2\ln x}{(\ln x + 2)(\ln x - 2)} = \dfrac{A}{u+2} + \dfrac{B}{u-2} = \dfrac{1}{u+2} + \dfrac{1}{u-2} = \dfrac{1}{\ln x + 2} + \dfrac{1}{\ln x - 2}$.

41. $\dfrac{2}{(x-1)(x^2+4)} = \dfrac{A}{x-1} + \dfrac{Bx+C}{x^2+4} \Rightarrow 2 = A(x^2+4) + (Bx+C)(x-1)$. Substituting

$x = 1$ in this last equation, we get $2 = A(5) + (B+C)(0) \Rightarrow A = \dfrac{2}{5}$. Substituting $x = 2i$ (the

imaginary number, $2\sqrt{-1}$) in this last equation, we get

$2 = A(0) + (2Bi + C)(2i - 1) \Rightarrow 2 + 0i = (-4B - C) + (2C - 2B)i$

$\Rightarrow -4B - C = 2$ and $-2B + 2C = 0$. Solving this system of two equations in B and C

gives $B = -\dfrac{2}{5}$ and $C = -\dfrac{2}{5}$. Therefore the decomposition is

$\dfrac{A}{x-1} + \dfrac{Bx+C}{x^2+4} = \dfrac{\frac{2}{5}}{x-1} + \dfrac{-\frac{2}{5}x - \frac{2}{5}}{x^2+4}$.

45. $\dfrac{3x^5 + 6x^3 + 3x + 4}{(x^2+1)^3} = \dfrac{Ax+B}{x^2+1} + \dfrac{Cx+D}{(x^2+1)^2} + \dfrac{Ex+F}{(x^2+1)^3}$

$\Rightarrow 3x^5 + 6x^3 + 3x + 4 = (Ax+B)(x^2+1)^2 + (Cx+D)(x^2+1) + Ex + F$. Collecting like

terms on the right side of this equation, we have $3x^5 + 6x^3 + 3x + 4$

$= Ax^5 + Bx^4 + (2A+C)x^3 + (2B+D)x^2 + (A+C+E)x + (B+D+F)$. Comparing

coefficients of like terms on each side of this equation, we have

$$\begin{cases} A = 3 \\ B = 0 \\ 2A + C = 6 \\ 2B + D = 0 \\ A + C + E = 3 \\ B + D + F = 4 \end{cases}$$. The solution to this system is $A = 3, B = 0, C = 0, D = 0,$

$E = 0$, and $F = 4$. The decomposition is

$\dfrac{Ax+B}{x^2+1} + \dfrac{Cx+D}{(x^2+1)^2} + \dfrac{Ex+F}{(x^2+1)^3} = \dfrac{3x}{x^2+1} + \dfrac{4}{(x^2+1)^3}$.

Section 10.6

1. Since $y = x - 1$, we can substitute $x - 1$ for y in the
first equation: $x^2 + y^2 = 25 \Rightarrow x^2 + (x-1)^2 = 25$
$\Rightarrow 2x^2 - 2x + 1 = 25 \Rightarrow x^2 - x - 12 = 0$
$\Rightarrow (x-4)(x+3) = 0 \Rightarrow x = 4$ or $x = -3$. Since
$y = x - 1$, the points of intersection are $(4, 3)$ and
$(-3, -4)$.

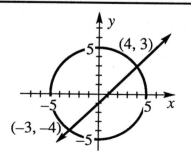

5. Since $y = x^2$, we can substitute y for x^2 in the first equation:
$x^2 - y^2 = 1 \Rightarrow y - y^2 = 1 \Rightarrow 2x^2 - 2x + 1 = 25$
$\Rightarrow y^2 - y + 1 = 0$. By the quadratic formula, we obtain
$y = \dfrac{1 \pm \sqrt{1^2 - 4(1)(1)}}{2} = \dfrac{1 \pm \sqrt{-3}}{2}$. Since these two solutions to the equation are not real numbers, there are no points of intersection .

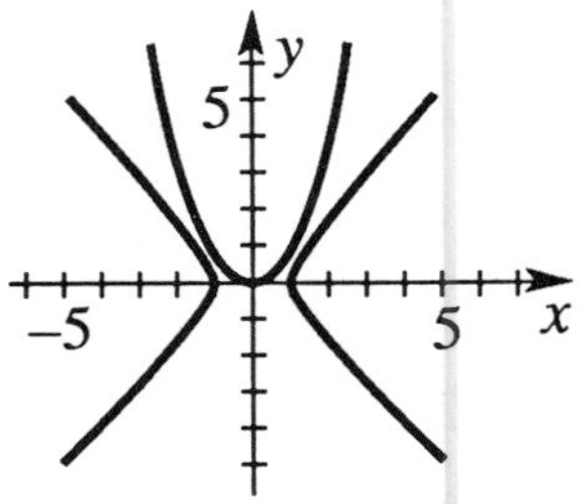

9. If we isolate y in the second equation, we get $y = 11 - 2x$. Substituting $11 - 2x$ for y in the first equation, we get
$(x - 2)^2 + (12 - 2x)^2 = 9$
$\Rightarrow 5x^2 - 52x + 139 = 0$. By the quadratic formula, we obtain
$y = \dfrac{52 \pm \sqrt{52^2 - 4(5)(139)}}{10} = \dfrac{52 \pm \sqrt{-76}}{10}$. Since these two solutions to the equation are not real numbers, there are no points of intersection .

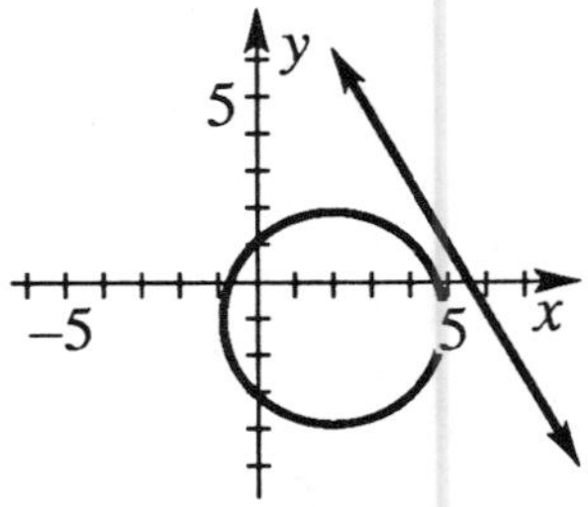

13. Since $y = \log_2 x$, we can substitute $\log_2 x$ for y in the second equation: $\log_2 x = 4 - \log_2 x \Rightarrow 2\log_2 x = 4$
$\Rightarrow \log_2 x^2 = 4 \Rightarrow x^2 = 2^4 \Rightarrow x = \pm 4$. However, the domain of $\log_2 x$ is $x > 0 \Rightarrow x = 4$. Since $y = \log_2 x$, $y = \log_2 4 = 2$. The solution is (4, 2).

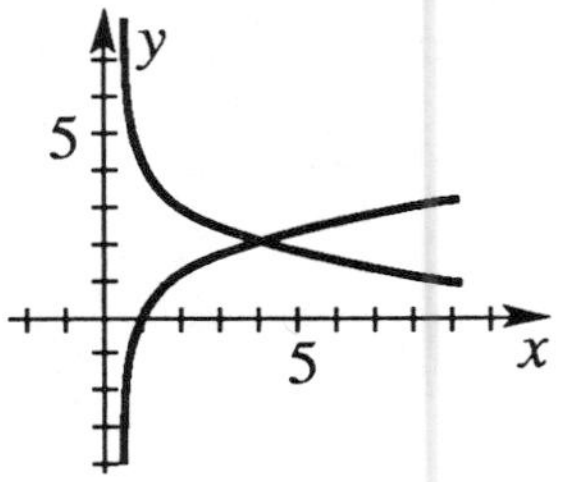

17. Substituting x for y in the first equation, we get
$x = \dfrac{x}{x^2 - 4} \Rightarrow x(x^2 - 4) = x \Rightarrow x^3 - 4x - x = 0$
$\Rightarrow x(x^2 - 5) = 0 \Rightarrow x = 0, \sqrt{5}$, or $-\sqrt{5}$. Therefore the solutions are $(0, 0)$, $(\sqrt{5}, \sqrt{5})$, and $(-\sqrt{5}, -\sqrt{5})$.

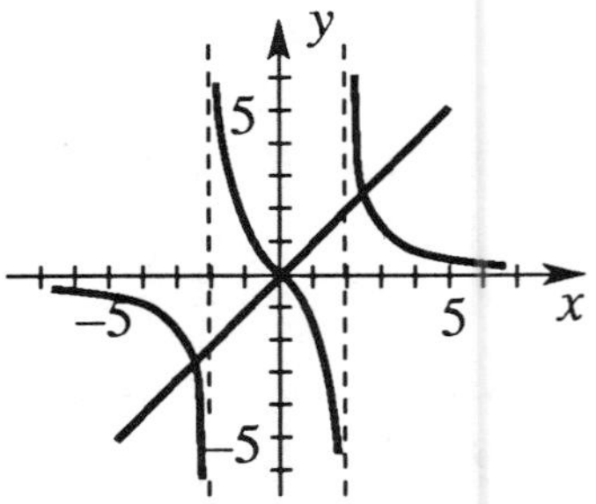

21. We can write both equations in exponential form:
$\log_2(2x - y) = 3 \Rightarrow 2x - y = 2^3 = 8$; and $\log_3(x + y) = 2 \Rightarrow x + y = 3^2 = 9$. An equivalent system, then, is $\begin{cases} 2x - y = 8 \\ x + y = 9 \end{cases}$.

Adding these two equations, we get $3x = 17 \Rightarrow x = \dfrac{17}{3}$.

Substituting back into either equation gives $y = \dfrac{10}{3}$.

25. Substituting 2 for r in the first equation, we get

$2 = 4\cos\theta \Rightarrow \dfrac{1}{2} = \cos\theta \Rightarrow \theta = 60°$ or $300°$. If we graph the

two equations, we see that their graphs intersect exactly twice. (See page 510 in the text for the graph of $r = 4\cos\theta$.) The points of intersection are $[2, 60°]$ and $[2, 300°]$.

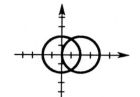

29. Substituting $2\sqrt{2}$ for r in the first equation, we get

$(2\sqrt{2})^2 = 16\cos2\theta \Rightarrow \dfrac{1}{2} = \cos2\theta \Rightarrow 2\theta = 60°,\ 300°,$

$420°$, or $660° \Rightarrow \theta = 30°,\ 150°,\ 210°,$ or $330°$.   If we graph the two equations, we see that their graphs intersect exactly four times (See Example 9, Section 9.2 in the text for the graph of $r^2 = 16\cos2\theta$.). The points of intersection are $[2\sqrt{2}, 30°]$, $[2\sqrt{2}, 150°]$, $[2\sqrt{2}, 210°]$, and $[2\sqrt{2}, 330°]$.

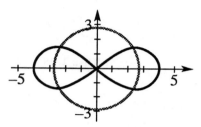

33. Let x be the width as well as the length of the square bottom, and let y be the height of the box. The volume is

$$\text{(width)(length)(height)} = (x)(x)(y) \Rightarrow 20 = x^2 y$$

The surface area is (area of the bottom square) + (area of the four rectangular sides) $= x^2 + 4xy \Rightarrow 44 = x^2 + 4xy$. We need to solve

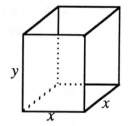

the system $\begin{cases} 20 = x^2 y \\ 44 = x^2 + 4xy \end{cases}$. Solving the first equation for y, we get $y = \dfrac{20}{x^2}$. Substituting

$\dfrac{20}{x^2}$ for y in the second equation gives $44 = x^2 + 4x(\dfrac{20}{x^2}) = x^2 + \dfrac{80}{x}$.

Multiplying both sides by x, we get $44x = x^3 + 4 \Rightarrow x^3 - 44x + 80 = 0$. This is a third degree polynomial equation, and you might want to refer to Section 3.3 for techniques of finding the zeros of polynomials. If we try synthetic division to test if $x = 2$ is a zero of $x^3 - 44x + 80$, we get the factorization $x^3 - 44x + 80 = (x - 2)(x^2 + 2x - 40)$. Thus, $x = 2$ is a zero. Applying the quadratic formula to the second factor, we get $x = -1 \pm \sqrt{41}$. Since x represents a positive length, $-1 - \sqrt{41}$ is not a possibility.

For $x = 2$, we have $y = \dfrac{20}{x^2} = \dfrac{20}{2^2} = 5$. For $x = -1 + \sqrt{41}$, we have

$$y = \frac{20}{x^2} = \frac{20}{(-1 + \sqrt{41})^2} = \frac{20}{21 - \sqrt{41}} = \frac{20(21 + \sqrt{41})}{(21 - \sqrt{41})(21 + \sqrt{41})} = \frac{21 + \sqrt{41}}{40} .$$

Therefore, there are two possible boxes: one that is 2 by 2 by 5 feet, and another that is $-1 + \sqrt{41}$ by $-1 + \sqrt{41}$ by $\dfrac{21 + \sqrt{41}}{40}$ feet.

37. If we solve for y in the second equation, we get $y = \dfrac{3^x - 3}{2}$. Substituting this into the

first equation, we get $9^x + \dfrac{(3^x - 3)^2}{4} = 90 \Rightarrow 4(9^x) + (3^x - 3)^2 = 360$

$\Rightarrow 4(3^x)^2 + [(3^x)^2 - 6(3^x) + 9] = 360 \Rightarrow 5(3^x)^2 - 6(3^x) - 351 = 0$. This equation is

quadratic in 3^x. If we let $u = 3^x$, we have $5u^2 - 6u - 351 = 0 \Rightarrow (5u + 39)(u - 9) = 0$

$\Rightarrow u = -\dfrac{39}{5}$ or $u = 9$. Since $u = 3^x > 0$, the only solution is $u = 9 \Rightarrow 3^x = 9 \Rightarrow x = 2$.

Substituting this in $y = \dfrac{3^x - 3}{2}$ gives $y = 3$. The solution is $(2, 3)$.

41. The graph of $y = \sqrt{3}\cos x$ has amplitude $\sqrt{3}$
and period 2π; the graph of $y = \sin 2x$ has
amplitude 1 and period π.(Refer to Section 5.5
for the general graphs of
$y = A\cos Bx$ and $y = A\cos Bx$.) To solve the
system, we substitute $\sqrt{3}\cos x$ for y in the
second equation: $\sqrt{3}\cos x = \sin 2x$

$\Rightarrow \sqrt{3}\cos x = 2\sin x\cos x$

$\Rightarrow 0 = 2\sin x\cos x - \sqrt{3}\cos x$

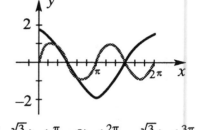

$\left(\dfrac{\pi}{3}, \dfrac{\sqrt{3}}{2}\right), \left(\dfrac{\pi}{2}, 0\right), \left(\dfrac{2\pi}{3}, -\dfrac{\sqrt{3}}{2}\right), \left(\dfrac{3\pi}{2}, 0\right)$

$\Rightarrow 0 = \cos x(2\sin x - \sqrt{3}) \Rightarrow \cos x = 0$ or $\sin x = \dfrac{\sqrt{3}}{2} \Rightarrow x = \dfrac{\pi}{2}$ or $\dfrac{3\pi}{2}$, or $x = \dfrac{\pi}{3}$ or $\dfrac{2\pi}{3}$.

Substituting these four values for x in either equation of the system gives the corresponding
y values.

45. The graph of $y = \sin x$ is on page 306 in the text; the
graph of $y = \dfrac{1}{2}x$ is a line with slope $\dfrac{1}{2}$ and y-intercept
$(0, 0)$. The number of solutions to the system
corresponds to the number of points of intersection of
the two graphs. Clearly there are three.

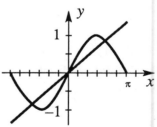

49. a) Substituting mx for y in the two equations gives
$$\begin{cases} x^2 + 4x(mx) - (mx)^2 = 16 \\ 2x^2 - 7x(mx) + 2(mx)^2 = -4 \end{cases} \Rightarrow \begin{cases} x^2 + 4mx^2 - m^2x^2 = 16 \\ 2x^2 - 7mx^2 + 2m^2x^2 = -4 \end{cases}$$

b) In both equations, we can factor out x^2 to solve for x^2 : $\begin{cases} x^2(1 + 4m - m^2) = 16 \\ x^2(2 - 7m + 2m^2) = -4 \end{cases}$

$\Rightarrow \begin{cases} x^2 = \dfrac{16}{1 + 4m - m^2} \\ x^2 = \dfrac{-4}{2 - 7m + 2m^2} \end{cases}$. Therefore, $\dfrac{16}{1 + 4m - m^2} = \dfrac{-4}{2 - 7m + 2m^2}$, and clearing

fractions, we get $16(2 - 7m + 2m^2) = -4(1 + 4m - m^2)$. This leads to $7m^2 - 24m + 9 = 0$

$\Rightarrow (7m - 3)(m - 3) = 0 \Rightarrow m = 3$ or $\frac{3}{7}$. Substituting $m = 3$ back into $x^2 = \dfrac{16}{1 + 4m - m^2}$,

we get $x^2 = 4 \Rightarrow x = \pm 2$. Substituting $m = \frac{3}{7}$ back into $x^2 = \dfrac{16}{1 + 4m - m^2}$, we get x^2

$= \pm \dfrac{196}{31} \Rightarrow x = \pm \sqrt{\dfrac{196}{31}} = \pm \dfrac{14\sqrt{31}}{31}$.

c) The solutions (x, y) are $(2, 6)$, $(-2, -6)$, $\left(\dfrac{14\sqrt{31}}{31}, \dfrac{6\sqrt{31}}{31} \right)$, and $\left(-\dfrac{14\sqrt{31}}{31}, -\dfrac{6\sqrt{31}}{31} \right)$.

Section 10.7

---

1. We start by graphing the associated equation $y = \frac{1}{2}x - 4$; this is

a line with y-intercept $(0, -4)$ and slope $\frac{1}{2}$. Since the

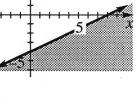

inequality specifies "less than or equal to," the sketch of the inequality is the set of points on or below the line. As a check, we choose a test point not on the line, say $(0, 0)$. Substituting 0 for x and y in the inequality, we get a false statement:

$0 \le \frac{1}{2}(0) - 4$, and this verifies that we are correct in shading the

side of the line that does not contain the point $(0, 0)$. The line is solid because the inequality specifies "less than <u>or equal to</u>."

5. We start by graphing the associated equation

$1 + x^2 = y$; this is a parabola that opens upward with vertex $(0, 1)$. (Refer to Section 2.4 to review sketching parabolas.) If

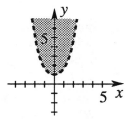

we write the inequality as $y > 1 + x^2$, we see the inequality specifies "greater than."; therefore the sketch of the inequality is the set of points above the parabola. As a check, we choose a test point not on the parabola, say $(0, 0)$. Substituting 0 for x and y

in the inequality, we get a false statement: $0 > 1 + 0^2$, and this verifies that we are correct in shading the side of the parabola that does not contain the point $(0, 0)$. The parabola is dashed because the inequality specifies strictly "less than."

9. We start by graphing the associated equation

$16(x - 4)^2 + 9(y + 3)^2 = 144 \Rightarrow \dfrac{(x - 4)^2}{9} + \dfrac{(y + 3)^2}{16} = 1$

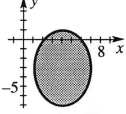

this is an ellipse centered at $(4, -3)$. (Refer to Section 8.3 to review sketching ellipses.) The graph of the ellipse is solid because the inequality specifies "less than <u>or equal to</u>." The solution to the inequality is either the region inside or the region

outside the ellipse. To determine the correct region, choose a test point not on the ellipse, say $(4, -3)$. Substituting 4 for x and -3 for y in the inequality, we get a true statement: $16(4 - 4)^2 + 9(-3 + 3)^2 \le 144$. Thus, we shade the region inside the ellipse.

13. The graph of the continued inequality is the same as the
graph of the system $\begin{cases} y \geq x^2 \\ y \leq 4 \end{cases}$. The graph of this system is

the set of points that are above or on the parabola $y = x^2$
and on or below the line $y = 4$.

17. The graph of the continued inequality is the same as the
graph of the system $\begin{cases} y \geq \log_2 x \\ y \leq 3 \end{cases}$. The graph of this system

is the set of points that are above or on the graph of
$y = \log_2 x$ (see Section 4.3) and on or below the line $y = 3$.

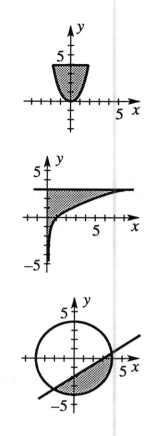

21. The first inequality has an associated equation
$2x - 3y = 6 \Rightarrow y = \frac{2}{3}x - 2$; this is a line with y-intercept

$(0, -2)$ and slope $\frac{2}{3}$. Using a point not on the line, say

$(0,0)$, as a test point, we have $2(0) - 3(0) \geq 6$, which is
false. Thus the graph of the first inequality is the set of all

points on or below the line $y = \frac{2}{3}x - 2$. The second

inequality has an associated equation $x^2 + y^2 = 16$; this is a circle centered at $(0,0)$ with
radius 4. The solution to the second inequality is either the region inside or the region
outside the circle. Choosing a test point not on the circle, say $(0, 0)$ and substituting into
the inequality, we get a true statement: $0^2 + 0^2 \leq 16$. Thus, the solution to the second
inequality is the region inside circle. The solution to the system is the set of all points that
are on or below the line and on or inside the circle.

25. The graph of the first inequality is the set of all points on or
inside the circle $x^2 + y^2 = 16$ (see solution to Problem 21).
The associated equation of the second inequality is

$9x^2 + 16y^2 = 144 \Rightarrow \frac{x^2}{16} + \frac{y^2}{9} = 1$. This is an ellipse

centered at $(0,0)$ with intercepts $(\pm 4,0)$ and $(0, \pm 3)$.
Testing the point $(0,0)$ in the inequality, we get a true

statement: $9(0)^2 + 16(0)^2 \leq 144$. Therefore, the solution to
the system is the set of all points that are on or inside the
circle and on or inside the ellipse.

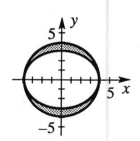

29. The graph of this system is the set of points that are on or below the graph of $y = \log_2 x$ and on or above the graph of $y = \log_4 x$.

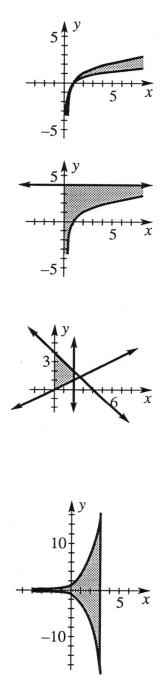

33. The graph of the first inequality is the set of points that are on or to the right of the graph of $x = 0$ (the y-axis). The graph of the second inequality is the set of points that are on or below the graph of $y = 4$. The graph of the third inequality is the set of points that are above or on the graph of $y = \log_2 x$. The graph of the system is the set of all points that are common to all three of these regions. The y-axis ($x = 0$) is not included in the final graph, since the domain of $\log_2 x$ is $x > 0$.

37. The first compound inequality represents the set of all points (x,y) such that $y \geq \dfrac{1}{2}x$ and $y \leq 4 - x$. This is the set of all points that are on or above the line $y = \dfrac{1}{2}x$ and on or below the line $y = 4 - x$. The second compound inequality represents the set of all points that are both on or to the right of the vertical line $x = 0$ and on or to the left of the vertical line $x = 2$. The solution to the system is the set of all points that are common to both of these regions.

41. The first compound inequality represents the set of all points (x,y) such that $y \geq -2^x$ and $y \leq 2^x$. This is the region on or above the graph of $y = -2^x$ and on or below the graph of $y = 2^x$ (see Section 4.1). The second compound inequality represents the set of all points that are both on or to the right of the vertical line $x = -4$ and on or to the left of the vertical line $x = 4$. The solution to the system is the set of all points that are common to both of these regions.

45. a) The associated equation is
$y = \log_2 x^2$; which is an even function (See Problem 53 in Section 4.4 or Problem 25 Miscellaneous Exercises in Chapter 4). For $x > 0$, $y = \log_2 x^2 = 2\log_2 x$, so we can graph $y = 2\log_2 x$, and reflect about the y-axis. The graph of $y \le \log_2 x^2$ is the set of all points on or below the graph.

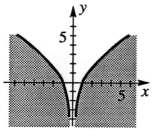

b) The graph of the associated equation
$y = 2\log_2 x$ is <u>not</u> even as in part a), since the domain of $y = 2\log_2 x$ is $x > 0$. The graph of $y = 2\log_2 x$ is the same as the right "half" of the graph in part a).

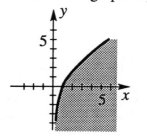

49. The expression within a square root must not be negative, so $x^2 - y \ge 0 \Rightarrow y \le x^2$. The set of all points (x,y) that satisfy this inequality is precisely the region on or below the parabola $y = x^2$.

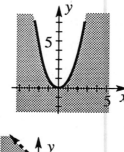

53. The expression within a square root must be nonnegative and the denominator cannot be 0, so
$6 - x - y > 0 \Rightarrow y < 6 - x$. The set of all points (x,y) that satisfy this inequality is the region below the line $y = 6 - x$.

57. The graph of the continued inequality is the same as the graph of the system $\begin{cases} r \ge 3 \\ r \le 4 + 2\sin\theta \end{cases}$. The graph of this system is the set of points that are on or outside the circle $r = 3$ and on or inside the limaçon $r = 4 + 2\sin\theta$ (see Figure 24, page 519 in the text).

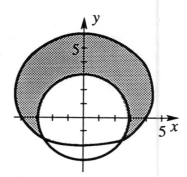

61. The first compound inequality represents the set of all points $[r,\theta]$ such that $r \geq 0$ and $r \leq 4\cos\theta$. This is the region on or inside the circle of $r = 4\cos\theta$. The second compound inequality represents the set of all points that are between the lines $\theta = 0$ and $\theta = \dfrac{\pi}{3}$. The solution to the system is the set of all points that are common to both of these regions.

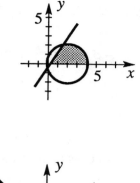

65. The first compound inequality represents the set of all points $[r,\theta]$ such that $r \geq 0$ and $r \leq \theta$. This is the region on or inside the spiral $r = \theta$. The second compound inequality represents the region between the lines $\theta = \dfrac{\pi}{3}$ and $\theta = \dfrac{2\pi}{3}$. The solution to the system is the set of all points that are common to both of these regions.

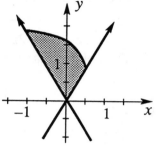

Miscellaneous Exercises for Chapter 10

1. In the form $y = mx + b$, the first equation is $y = -2x$; the graph has a y-intercept of $(0, 0)$ and a slope of $\dfrac{-2}{1}$. In the form $y = mx + b$, the second equation is $y = 3x - \dfrac{5}{2}$; the graph has a y-intercept of $(0, -\dfrac{5}{2})$ and a slope of $\dfrac{3}{1}$. The point of intersection can be determined by replacing y with $-2x$ in the second equation: $6x - 2y = 5$

$\Rightarrow 6x - 2(-2x) = 5 \Rightarrow x = \dfrac{1}{2}$. Since $y = -2x$, $y = -1$.

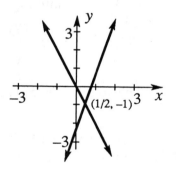

5. In the form $y = mx + b$, the first equation is $y = -\dfrac{5}{4}x + 2$; the graph has a y-intercept of $(0,2)$ and a slope of $\dfrac{-5}{4}$. In the form $y = mx + b$, the second equation is $y = -\dfrac{5}{4}x + 2$; the graph is identical to that of the first equation. Therefore, the system is dependent. If we isolate x, we obtain : $5x + 4y = 8 \Rightarrow x = \dfrac{8}{5} - \dfrac{4}{5}y \Rightarrow$ a parametric representation of the solutions is $(\dfrac{8}{5} - \dfrac{4}{5}t, t)$.

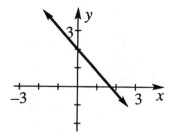

9. We eliminate the $3y$ term in the third equation by multiplying the second equation by -3 and then adding the equations: $\begin{cases} x + 3y + 2z = -6 \\ y + z = 3 \\ 3y + 4z = 20 \end{cases}$

$-3E_2 + E_3 \rightarrow E_3 \quad \begin{cases} x + 3y + 2z = -6 \\ y + z = 3 \\ z = 11 \end{cases}$

Back-substitution gives:

$y + z = 3 \Rightarrow y + 11 = 3 \Rightarrow y = -8$

$x + 3y + 2z = -6 \Rightarrow x + 3(-8) + 2(11) = -6 \Rightarrow x - 2 = -6 \Rightarrow x = -4$

The solution is $(x, y, z) = (-4, -8, 11)$.

13. $\begin{cases} x + y - z = 1 \\ 2x - 2z = 1 \\ 8x + 2y - 8z = 5 \end{cases} \Rightarrow \quad \begin{matrix} -2E_1 + E_2 \rightarrow E_2 \\ -8E_1 + E_3 \rightarrow E_3 \end{matrix} \quad \begin{cases} x + y - z = 1 \\ -2y = -1 \\ -6y = -3 \end{cases}$

$-3E_2 + E_3 \rightarrow E_3 \quad \begin{cases} x + y - z = 1 \\ -2y = -1 \\ 0z = 0 \end{cases}$. This is a dependent system.

The last equation is true for any value we choose for z, so we let the parameter t represent z. Back-substituting gives

$$-2y = -1 \Rightarrow y = \frac{1}{2}$$

$$x + y - z = 1 \Rightarrow x + \left(\frac{1}{2}\right) - t = 1 \Rightarrow x = \frac{1}{2} + t$$

The solution is $(x, y, z) = \left(\frac{1}{2} + t, \frac{1}{2}, t\right)$.

17. $\begin{cases} I_1 - I_2 + I_3 = 0 \\ I_1 - 2I_3 = 6 \\ I_2 + 2I_3 = 11 \end{cases} \Rightarrow \quad -1E_1 + E_2 \rightarrow E_2 \quad \begin{cases} I_1 - I_2 + I_3 = 0 \\ I_2 - 3I_3 = 6 \\ I_2 + 2I_3 = 11 \end{cases}$

$-1E_2 + E_3 \rightarrow E_3 \quad \begin{cases} I_1 - I_2 + I_3 = 0 \\ I_2 - 3I_3 = 6 \\ 5I_3 = 5 \end{cases}$

Back-substituting, we have

$$5I_3 = 1 \Rightarrow I_3 = 1$$

$$I_2 - 3I_3 = 6 \Rightarrow I_2 - 3(1) = 6 \Rightarrow I_2 - 3 = 6 \Rightarrow I_2 = 9$$

$$I_1 - I_2 + I_3 = 0 \Rightarrow I_1 - (9) + (1) = 0 \Rightarrow I_1 - 8 = 0 \Rightarrow I_1 = 8.$$

21. The augmented matrix is $\begin{bmatrix} -4 & 6 & | & -8 \\ 6 & -9 & | & 12 \end{bmatrix} \Rightarrow -(\frac{1}{4})R_1 \to R_1 \begin{bmatrix} 1 & -\frac{3}{2} & | & 2 \\ 6 & -9 & | & 12 \end{bmatrix}$

$-6R_1 + R_2 \to R_2 \begin{bmatrix} 1 & -\frac{3}{2} & | & 2 \\ 0 & 0 & | & 0 \end{bmatrix}$ The system is dependent.

The last equation represents $0y = 0$. Since this equation is true for any value we choose for y, we let the parameter t represent y. Back-substituting, we get

$$x - \frac{3}{2}t = 2 \Rightarrow x = 2 + \frac{3}{2}t. \text{ The solution is } (2 + \frac{3}{2}t, t).$$

25. First, "swap" the first and third equations:

$\begin{cases} x - 2y - 2z = -2 \\ 9x + 2y + z = 3 \\ 4x - 2y = -4 \end{cases} \Rightarrow \begin{bmatrix} 1 & -2 & -2 & | & -2 \\ 9 & 2 & 1 & | & 3 \\ 4 & -2 & 0 & | & -4 \end{bmatrix} \begin{array}{c} -9R_1 + R_2 \to R_2 \\ -4R_1 + R_3 \to R_3 \end{array} \begin{bmatrix} 1 & -2 & -2 & | & -2 \\ 0 & 20 & 19 & | & 21 \\ 0 & 6 & 8 & | & 4 \end{bmatrix}$

$\begin{array}{c} R_3 \to R_2 \\ R_2 \to R_3 \end{array} \begin{bmatrix} 1 & -2 & -2 & | & -2 \\ 0 & 6 & 8 & | & 4 \\ 0 & 20 & 19 & | & 21 \end{bmatrix}$ $(\frac{1}{6})R_2 \to R_2 \begin{bmatrix} 1 & -2 & -2 & | & -2 \\ 0 & 1 & \frac{4}{3} & | & \frac{2}{3} \\ 0 & 20 & 19 & | & 21 \end{bmatrix}$

$\begin{array}{c} 2R_2 + R_1 \to R_1 \\ \\ -20R_2 + R_3 \to R_3 \end{array} \begin{bmatrix} 1 & 0 & \frac{2}{3} & | & -\frac{2}{3} \\ 0 & 1 & \frac{4}{3} & | & \frac{2}{3} \\ 0 & 0 & -\frac{23}{3} & | & \frac{23}{3} \end{bmatrix}$ $-(\frac{3}{23})R_3 \to R_3 \begin{bmatrix} 1 & 0 & \frac{2}{3} & | & -\frac{2}{3} \\ 0 & 1 & \frac{4}{3} & | & \frac{2}{3} \\ 0 & 0 & 1 & | & -1 \end{bmatrix}$

$\begin{array}{c} -(\frac{2}{3})R_3 + R_1 \to R_1 \\ -(\frac{4}{3})R_3 + R_2 \to R_2 \end{array} \begin{bmatrix} 1 & 0 & 0 & | & 0 \\ 0 & 1 & 0 & | & 2 \\ 0 & 0 & 1 & | & -1 \end{bmatrix}$

The solution is $(x, y, z) = (0, 2, -1)$.

29. The system is underdetermined. The augmented matrix is

$\begin{bmatrix} 1 & 2 & -1 & | & 0 \\ 5 & 1 & -3 & | & -6 \end{bmatrix}$ $-5R_1 + R_2 \to R_2 \begin{bmatrix} 1 & 2 & -1 & | & 0 \\ 0 & -9 & 2 & | & -6 \end{bmatrix}$

$-(\frac{1}{9})R_2 \to R_2 \begin{bmatrix} 1 & 2 & -1 & | & 0 \\ 0 & 1 & -\frac{2}{9} & | & \frac{2}{3} \end{bmatrix}$ $-2R_2 + R_1 \to R_1 \begin{bmatrix} 1 & 0 & -\frac{5}{9} & | & -\frac{4}{3} \\ 0 & 1 & -\frac{2}{9} & | & \frac{2}{3} \end{bmatrix}$

The first row represents $x - \frac{5}{9}z = -\frac{4}{3} \Rightarrow x = -\frac{4}{3} + \frac{5}{9}z$. The second row represents

$y - \frac{2}{9}z = \frac{2}{3} \Rightarrow y = \frac{2}{3} + \frac{2}{9}z$. Letting t represent z, we have the solution

$(x, y, z) = (-\frac{4}{3} + \frac{5}{9}t, \frac{2}{3} + \frac{2}{9}t, t)$.

33. Let a, b, and c represent $\dfrac{1}{x}$, $\dfrac{1}{y}$, and $\dfrac{1}{z}$ respectively. Substituting into the system, we have

$$\begin{cases} a + 2b - c = 3 \\ 2a + 4b - 3c = 5 \\ a \quad\quad - 6c = -1 \end{cases} \Rightarrow \left[\begin{array}{ccc|c} 1 & 2 & -1 & 3 \\ 2 & 4 & -3 & 5 \\ 1 & 0 & -6 & -1 \end{array}\right]$$

$$\begin{array}{c} -2R_1 + R_2 \to R_2 \\ -1R_1 + R_3 \to R_3 \end{array} \left[\begin{array}{ccc|c} 1 & 2 & -1 & 3 \\ 0 & 0 & -1 & -1 \\ 0 & -2 & -5 & -4 \end{array}\right] \qquad \begin{array}{c} R_3 \to R_2 \\ R_2 \to R_3 \end{array} \left[\begin{array}{ccc|c} 1 & 2 & -1 & 3 \\ 0 & -2 & -5 & -4 \\ 0 & 0 & -1 & -1 \end{array}\right]$$

$$\begin{array}{c} R_2 + R_1 \to R_1 \\ -(\frac{1}{2})R_2 \to R_2 \\ -1R_3 \to R_3 \end{array} \left[\begin{array}{ccc|c} 1 & 0 & -6 & -1 \\ 0 & 1 & \frac{5}{2} & 2 \\ 0 & 0 & 1 & 1 \end{array}\right] \qquad \begin{array}{c} 6R_3 + R_1 \to R_1 \\ -(\frac{5}{2})R_3 + R_2 \to R_2 \end{array} \left[\begin{array}{ccc|c} 1 & 0 & 0 & 5 \\ 0 & 1 & 0 & -\frac{1}{2} \\ 0 & 0 & 1 & 1 \end{array}\right]$$

Thus, $(a, b, c) = (5, -\dfrac{1}{2}, 1) \Rightarrow (x, y, z) = (\dfrac{1}{5}, -2, 1)$.

37. $\begin{vmatrix} \tan\theta & \sec\theta \\ \sec\theta & \tan\theta \end{vmatrix} = (\tan\theta)(\tan\theta) - (\sec\theta)(\sec\theta) = \tan^2\theta - \sec^2\theta = -1$

41. Multiplying row 2 by 2 and adding to row 1 ($2R_2 + R_1 \to R_1$), does not change the value of the determinant:

$$\begin{vmatrix} 4 & 10 & -2 & 0 \\ -2 & -5 & 1 & 1 \\ 9 & 6 & 3 & -2 \\ -2 & -5 & 12 & 7 \end{vmatrix} = \begin{vmatrix} 0 & 0 & 0 & 2 \\ -2 & -5 & 1 & 1 \\ 9 & 6 & 3 & -2 \\ -2 & -5 & 12 & 7 \end{vmatrix} \quad \text{Using the top row, we have}$$

$$= 0 + 0 + 0 + 2(\text{cofactor of 2})$$

$$= 2(-1)^{1+4} \begin{vmatrix} -2 & -5 & 1 \\ 9 & 6 & 3 \\ -2 & -5 & 12 \end{vmatrix} = -2 \begin{vmatrix} -2 & -5 & 1 \\ 9 & 6 & 3 \\ -2 & -5 & 12 \end{vmatrix}$$

We can multiply row 1 by -1 and add to row 3 ($-R_1 + R_3 \to R_3$):

$$-2 \begin{vmatrix} -2 & -5 & 1 \\ 9 & 6 & 3 \\ -2 & -5 & 12 \end{vmatrix} = -2 \begin{vmatrix} -2 & -5 & 1 \\ 9 & 6 & 3 \\ 0 & 0 & 11 \end{vmatrix}$$

Evaluating this using the third row, we have

$$-2 \begin{vmatrix} -2 & -5 & 1 \\ 9 & 6 & 3 \\ 0 & 0 & 11 \end{vmatrix} = -2 \left[0 + 0 + (11)(-1)^{1+3} \begin{vmatrix} -2 & -5 \\ 9 & 6 \end{vmatrix} \right] = -2[11(33)] = -726$$

45. The area is the absolute value of $\frac{1}{2}\begin{vmatrix} 1 & -1 & 1 \\ 3 & 2 & 1 \\ 0 & 1 & 1 \end{vmatrix}$. Multiplying row 1 by -3 and adding to

row 2 ($-3R_1 + R_2 \rightarrow R_2$), does not change the value of the determinant: $\frac{1}{2}\begin{vmatrix} 1 & -1 & 1 \\ 3 & 2 & 1 \\ 0 & 1 & 1 \end{vmatrix} =$

$\frac{1}{2}\begin{vmatrix} 1 & -1 & 1 \\ 0 & 5 & -2 \\ 0 & 1 & 1 \end{vmatrix}$. Evaluating using the first column, this is $\frac{1}{2}\begin{vmatrix} 5 & -2 \\ 1 & 1 \end{vmatrix} = \frac{7}{2}$.

49. $x = \dfrac{D_x}{D} = \dfrac{\begin{vmatrix} 3 & -3 \\ -2 & 7 \end{vmatrix}}{\begin{vmatrix} 1 & -3 \\ -2 & 7 \end{vmatrix}} = \dfrac{15}{1} = 15;$ $y = \dfrac{D_y}{D} = \dfrac{\begin{vmatrix} 1 & 3 \\ -2 & -2 \end{vmatrix}}{\begin{vmatrix} 1 & -3 \\ -2 & 7 \end{vmatrix}} = \dfrac{4}{1} = 4$

53. $x = \dfrac{D_x}{D} = \dfrac{\begin{vmatrix} 0 & 4 & -1 \\ -31 & 1 & 0 \\ 13 & 6 & 5 \end{vmatrix}}{\begin{vmatrix} 2 & 4 & -1 \\ -4 & 1 & 0 \\ 3 & 6 & 5 \end{vmatrix}} = \dfrac{819}{117} = 7;$ $y = \dfrac{D_y}{D} = \dfrac{\begin{vmatrix} 2 & 0 & -1 \\ -4 & -31 & 0 \\ 3 & 13 & 5 \end{vmatrix}}{D} = \dfrac{-351}{117} = -3;$

$z = \dfrac{D_z}{D} = \dfrac{\begin{vmatrix} 2 & 4 & 0 \\ -4 & 1 & -31 \\ 3 & 6 & 13 \end{vmatrix}}{D} = \dfrac{234}{117} = 2$

57. $x = \dfrac{D_x}{D} = \dfrac{\begin{vmatrix} 4 & 3 & -2 \\ 6 & -1 & 1 \\ 10 & 0 & 7 \end{vmatrix}}{\begin{vmatrix} 2 & 3 & -2 \\ 1 & -1 & 1 \\ 3 & 0 & 7 \end{vmatrix}} = \dfrac{-144}{-32} = \dfrac{9}{2};$ $y = \dfrac{D_y}{D} = \dfrac{\begin{vmatrix} 2 & 4 & -2 \\ 1 & 6 & 1 \\ 3 & 10 & 7 \end{vmatrix}}{D} = \dfrac{64}{-32} = -2;$

$z = \dfrac{D_z}{D} = \dfrac{\begin{vmatrix} 2 & 3 & 4 \\ 1 & -1 & 6 \\ 3 & 0 & 10 \end{vmatrix}}{D} = \dfrac{16}{-32} = -\dfrac{1}{2}$

61. $\dfrac{1}{2}\left[\begin{vmatrix} 6 & 3 \\ 1 & 5 \end{vmatrix} + \begin{vmatrix} 3 & -4 \\ 5 & 2 \end{vmatrix} + \begin{vmatrix} -4 & -2 \\ 2 & 1 \end{vmatrix} + \begin{vmatrix} -2 & 1 \\ 1 & 1 \end{vmatrix} + \begin{vmatrix} 1 & 2 \\ 1 & -1 \end{vmatrix} + \begin{vmatrix} 2 & 6 \\ -1 & 1 \end{vmatrix} \right]$

$= \dfrac{1}{2}[27 + 26 + 0 + (-3) + (-3) + 8] = \dfrac{55}{2}$ square units.

65. $\dfrac{-x-9}{(x-1)(x+4)} = \dfrac{A}{x-1} + \dfrac{B}{x+4} \Rightarrow -x-9 = A(x+4) + B(x-1).$

$x = -4 \Rightarrow -(-4) - 9 = A(0) + B(-5) \Rightarrow -5 = -5B \Rightarrow B = 1.$

$x = 1 \Rightarrow -(1) - 9 = A(5) + B(0) \Rightarrow -10 = 5A \Rightarrow A = -2.$ The decomposition is

$\dfrac{A}{x-1} + \dfrac{B}{x+4} = \dfrac{-2}{x-1} + \dfrac{1}{x+4}.$

69. $\dfrac{4x^2 - x + 18}{(x+2)(x^2 + x + 2)} = \dfrac{A}{x+2} + \dfrac{Bx + C}{x^2 + x + 2}$

$\Rightarrow 4x^2 - x + 18 = A(x^2 + x + 2) + (Bx + C)(x + 2).$

$x = -2 \Rightarrow 4(-2)^2 - (-2) + 18 = A(4) + 0 \Rightarrow 36 = 4A \Rightarrow A = 9$

To find B and C, we collect like terms on the right side of this equation:

$4x^2 - x + 18 = (A + B)x^2 + (A + 2B + C)x + (2A + 2C).$ Comparing coefficients of like terms on each side of this equation, we have

$$\begin{cases} A + B = 4 \\ A + 2B + C = -1 \\ 2A + 2C = 18 \end{cases}$$

$A + B = 4 \Rightarrow 9 + B = 4 \Rightarrow B = -5$

$2A + 2C = 18 \Rightarrow 18 + 2C = 18 \Rightarrow 2C = 0 \Rightarrow C = 0.$ The decomposition is

$\dfrac{A}{x+2} + \dfrac{Bx + C}{x^2 + x + 2} = \dfrac{9}{x+2} + \dfrac{-5x}{x^2 + x + 2}.$

73. Since the degree of the numerator is greater than that of the denominator, we first need to

divide $4x^2 + 3x$ into $8x^3 - x^2 - 10x + 3$: $\dfrac{8x^3 - x^2 - 10x + 3}{4x^2 + 3x} = 2x - 3 + \dfrac{-x + 3}{4x^2 + 3x}.$

Now we determine the partial fraction decomposition of $\dfrac{-x + 3}{4x^2 + 3x}$:

$\dfrac{-x + 3}{4x^2 + 3x} = \dfrac{-x + 3}{x(4x + 3)} = \dfrac{A}{x} + \dfrac{B}{4x + 3} \Rightarrow -x + 3 = A(4x + 3) + Bx$

$x = 0 \Rightarrow -0 + 3 = A(3) + 0 \Rightarrow 3 = 3A \Rightarrow A = 1$

$x = -\dfrac{3}{4} \Rightarrow -(-\dfrac{3}{4}) + 3 = 0 + B(-\dfrac{3}{4}) \Rightarrow \dfrac{15}{4} = -\dfrac{3}{4}B \Rightarrow B = -5.$

Thus, the decomposition is

$2x - 3 + \dfrac{-x + 3}{4x^2 + 3x} = 2x - 3 + \dfrac{A}{x} + \dfrac{B}{4x + 3} = 2x - 3 + \dfrac{1}{x} + \dfrac{-5}{4x + 3}.$

77. From the second equation, we get $y = x + 3$, and we can substitute $x + 3$ for y in the first equation:
$x^2 + 2x = y - 1 \Rightarrow x^2 + 2x = (x + 3) - 1$
$\Rightarrow x^2 + x - 2 = 0 \Rightarrow (x - 1)(x + 2) = 0 \Rightarrow x = 1$ or $x = -2$. Since $y = x + 3$, the points of intersection are $(1, 4)$ and $(-2, 1)$.

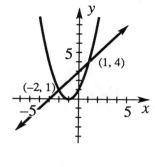

81. From the second equation, we get $x = 14 - 4y$, and we can substitute $14 - 4y$ for x in the first equation:
$xy = 6 \Rightarrow (14 - 4y)y = 6 \Rightarrow 4y^2 - 14y + 6 = 0$
$\Rightarrow 2y^2 - 7y + 3 = 0 \Rightarrow (2y - 1)(y - 3) = 0 \Rightarrow y = \frac{1}{2}$ or $y = 3$. Since $x = 14 - 4y$, the points of intersection are $(12, \frac{1}{2})$ and $(2, 3)$.

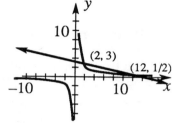

85. From the second equation, we get $x = 16y$, and we can substitute $16y$ for x in the first equation: $\sqrt{x} - \sqrt{y} = 6 \Rightarrow \sqrt{16y} - \sqrt{y} = 6 \Rightarrow 4\sqrt{y} - \sqrt{y} = 6 \Rightarrow 3\sqrt{y} = 6 \Rightarrow \sqrt{y} = 2$
$\Rightarrow y = 4$. Since $x = 16y$, $x = 16(4) = 64$. The solution is $(64, 4)$.

89. $\tan(x + y) = 1 \Rightarrow x + y = \frac{\pi}{4} + s\pi$, where s is an integer. Similarly, $\cos(x - y) = 0$

$\Rightarrow x - y = \frac{\pi}{2} + t\pi$, where t is an integer. If we add the equations $x + y = \frac{\pi}{4} + s\pi$ and

$x - y = \frac{\pi}{2} + t\pi$, we get $2x = \frac{3\pi}{4} + (s + t)\pi \Rightarrow x = \frac{3\pi}{8} + \frac{(s + t)\pi}{2}$. Letting $s + t = k$,

this is equivalent to $x = \frac{3\pi}{8} + \frac{k\pi}{2}$, where k is an integer. Returning to the equations

$x + y = \frac{\pi}{4} + s\pi$ and $x - y = \frac{\pi}{2} + t\pi$, if we subtract them we obtain $2y = -\frac{\pi}{4} + (s - t)\pi$

$\Rightarrow y = -\frac{\pi}{8} + \frac{(s + t)\pi}{2}$. Letting $s + t = n$, this is equivalent to $y = -\frac{\pi}{8} + \frac{n\pi}{2}$, where n is

an integer. This last equation is also equivalent to $y = \frac{3\pi}{8} + \frac{n\pi}{2}$.

93. Substituting 2 for r in the second equation, we get

$$2(3 + 2\sin\theta) = 4 \Rightarrow 3 + 2\sin\theta = 2 \Rightarrow \sin\theta = -\frac{1}{2}$$

$$\Rightarrow \theta = \frac{7\pi}{6} \text{ or } \frac{11\pi}{6}.$$

The graph of the first equation is a circle centered at the pole with radius 2. The graph of the second equation is easier to identify if we isolate r:

$$r(3 + 2\sin\theta) = 4 \Rightarrow r = \frac{\dfrac{4}{3}}{1 + \dfrac{2}{3}\sin\theta}$$

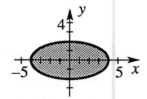

Referring to Section 9.4, this is an ellipse. From the two graphs, we see that their graphs intersect exactly twice . The points of intersection are $[2, \frac{7\pi}{6}]$ and $[2, \frac{11\pi}{6}]$.

97. We start by graphing the associated equation

$$x^2 + 4y^2 = 16 \Rightarrow \frac{x^2}{16} + \frac{y^2}{4} = 1;$$ this is an ellipse

centered at the origin. (Refer to Section 8.3 to review sketching ellipses.) The graph of the ellipse is solid because the inequality specifies "less than <u>or equal to</u>."
The solution to the inequality is either the region inside or the region outside the ellipse. To determine the correct region, choose a test point not on the ellipse, say $(0, 0)$. Substituting 0 for x and 0 for y in the inequality, we get a true statement: $0^2 + 4(0)^2 \leq 16$. Thus , we shade the region inside the ellipse.

101. The first inequality has an associated equation $y - 1 = -(x + 2)^2$; this is a parabola that opens down with vertex at $(-2, 1)$. If we set $y = 0$ and solve for x, we get the x-intercepts $(-3, 0)$ and $(-1, 0)$. Using a point not on the parabola, say $(0,0)$, as a test point , we have $0 - 1 < -(0 + 2)^2$, which is false. Thus the graph of the first inequality is the set of all points below the parabola $y - 1 = -(x + 2)^2$. The second inequality has an associated equation $(x + 2)^2 + (y - 1)^2 = 2$; this is a circle centered at $(-2, 1)$ with radius $\sqrt{2}$.
The solution to the second inequality is either the region inside or the region outside the circle. Choosing a test point not on the circle, say $(0, 0)$ and substituting into the inequality, we get a false statement: $(0 + 2)^2 + (0 - 1)^2 < 2$. Thus, the solution to the second inequality is the region inside circle. The solution to the system is the set of all points that are below the parabola and inside the circle.

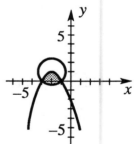

105. The first equation has the associated equation

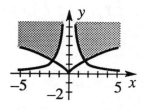

$x^2 y = 4 \Rightarrow y = \dfrac{4}{x^2}$. This is a rational function with vertical

asymptote $x = 0$. This function is also even. (Refer to Example 4, Section 3.4 .) The graph of $x^2 y \geq 4$ is equivalent to the

graph of $y \geq \dfrac{4}{x^2}$, this can be seen by dividing both sides of the

first inequality by the nonnegative expression x^2. The graph of $y = x^{2/3}$ is discussed in Example 2, Section 3.5. Since in both cases we have $y \geq f(x)$, the solution to the system is the region that is above both curves.

109. The expression within a square root must not be negative,

so $x - y \geq 0 \Rightarrow y \leq x$. The set of all points (x, y) that satisfy this inequality is precisely the region on or below the line $y = x$.

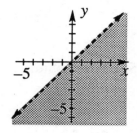

113. The first compound inequality represents the the region on or outside the circle $r = 2$ and on or inside the circle $r = 5$. The second compound inequality represents the set of all points $[r, \theta]$

such that θ is between $\dfrac{\pi}{6}$ and $\dfrac{2\pi}{3}$ (inclusive). This is the region

on or between the lines

$$\theta = \frac{\pi}{6} \text{ and } \theta = \frac{2\pi}{3} .$$

The solution to the system is the set of all points common to both of these regions.

Chapter 11

Section 11.1

1. a) Substituting 1, 2, 3, 4, and 5 for n, we get
$a_1 = 3(1) - 1 = 2$, $a_2 = 3(2) - 1 = 5$, $a_3 = 3(3) - 1 = 8$,
$a_4 = 3(4) - 1 = 11$, and $a_5 = 3(5) - 1 = 14$. Thus, the
first five terms are 2, 5, 8, 11, and 14.
b) The ordered pairs of a sequence are (n, a_n); the first
five are (1, 2), (2, 5), (3, 8), (4, 11), and (5, 14).

5. a) Substituting 1, 2, 3, 4, and 5 for n, we get
$a_1 = 1 - \cos(1)\pi = 2$, $a_2 = 1 - \cos 2\pi = 0$,
$a_3 = 1 - \cos 3\pi = 2$, $a_4 = 1 - \cos 4\pi = 0$, and
$a_5 = 1 - \cos 5\pi = 2$. Thus, the first five terms are 2, 0,
2, 0, and 2.
b) The ordered pairs of a sequence are (n, a_n); the first
five are (1, 2), (2, 0), (3, 2), (4, 0), and (5, 2).

9. a) Substituting 1, 2, 3, 4, and 5 for n, we get
$a_1 = \dfrac{1}{2(1) + 1} = \dfrac{1}{3}$, $a_2 = \dfrac{2}{2(2) + 1} = \dfrac{2}{5}$,
$a_3 = \dfrac{3}{2(3) + 1} = \dfrac{3}{7}$, $a_4 = \dfrac{4}{2(4) + 1} = \dfrac{4}{9}$, and
$a_5 = \dfrac{5}{2(5) + 1} = \dfrac{5}{11}$. Thus, the first five terms are
$\dfrac{1}{3}, \dfrac{2}{5}, \dfrac{3}{7}, \dfrac{4}{9}$, and $\dfrac{5}{11}$.

b) The ordered pairs of a sequence are (n, a_n); the first five are $(1, \frac{1}{3})$, $(2, \frac{2}{5})$, $(3, \frac{3}{7})$,
$(4, \frac{4}{9})$, and $(5, \frac{5}{11})$.

13. a) Substituting 1, 2, 3, 4, and 5 for n, we get
$a_1 = 2^1 = 2$, $a_2 = 2^2 = 4$, $a_3 = 2^3 = 8$,
$a_4 = 2^4 = 16$, and $a_5 = 2^5 = 32$. Thus, the first five
terms are 2, 4, 8, 16, and 32.
b) The ordered pairs of a sequence are (n, a_n); the first
five are (1, 2), (2, 4), (3, 8), (4, 16), and (5, 32).

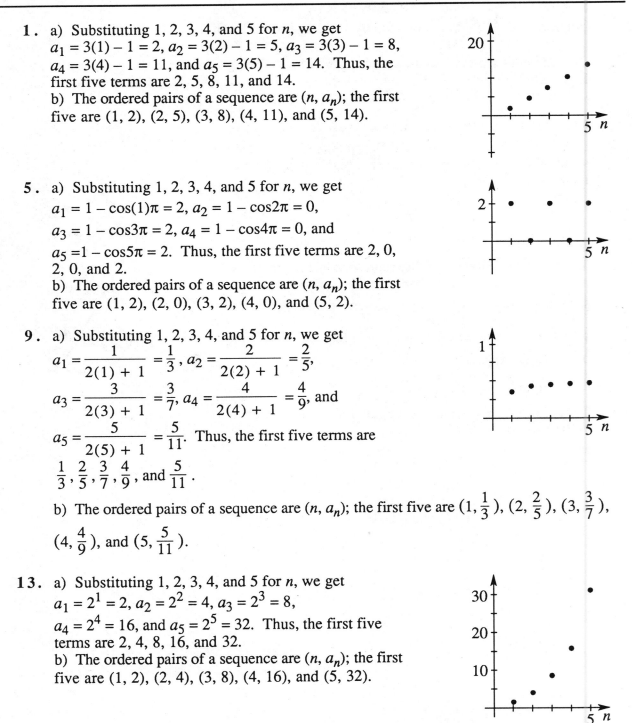

17. a) Substituting 1, 2, 3, 4, and 5 for n, we get

$$a_1 = (-\tfrac{1}{2})^1 = -\tfrac{1}{2}, \ a_2 = (-\tfrac{1}{2})^2 = \tfrac{1}{4}, \ a_3 = (-\tfrac{1}{2})^3$$

$$= -\tfrac{1}{8}, \ a_4 = (-\tfrac{1}{2})^4 = \tfrac{1}{16}, \text{ and } a_5 = (-\tfrac{1}{2})^5 = -\tfrac{1}{32}.$$

The first five terms are $-\tfrac{1}{2}, \tfrac{1}{4}, -\tfrac{1}{8}, \tfrac{1}{16},$ and $-\tfrac{1}{32}$

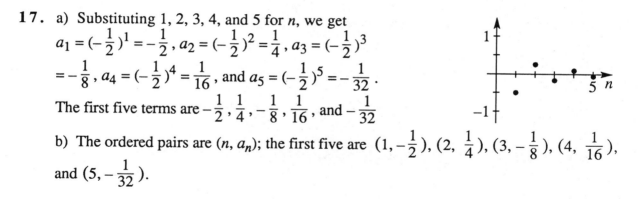

b) The ordered pairs are (n, a_n); the first five are $(1, -\tfrac{1}{2}), (2, \tfrac{1}{4}), (3, -\tfrac{1}{8}), (4, \tfrac{1}{16}),$ and $(5, -\tfrac{1}{32})$.

21. a) $f(x) = \dfrac{2x}{x+1}$ is a rational function, and we can use

the techniques discussed in Section 3.4 of the text to sketch its graph. The graph crosses the axes at $(0, 0)$ only. There is one vertical asymptote : $x = -1$. Dividing the denominator into the numerator, we get

$$\dfrac{2x}{x+1} = 2 + \dfrac{-2}{x+1} \Rightarrow \text{ the graph has a horizontal}$$

asymptote at $y = 2$.

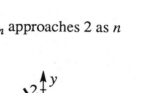

b) Since the graph in part a) has a horizontal asymptote at $y = 2$, a_n approaches 2 as n grows without bound.

25. a) $f(x) = (\tfrac{1}{2})^x$ is an exponential function, so we know

the characteristic shape (See Section 4.1, Example 4.). The graph crosses the y-axis at $(0, 1)$. A few ordered

pairs on the graph are $(-2, 4), (-1, 2), (1, \tfrac{1}{2}), (2, \tfrac{1}{4})$.

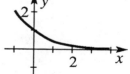

The x-axis is a horizontal asymptote for large positive values of x.

b) Since the graph in part a) has a horizontal asymptote at $y = 0$, a_n approaches 0 as n grows without bound.

29. We are given the first two terms, so we first substitute $n = 3$ to find the third term: $a_3 = 3a_{3-1} + 2a_{3-2} = 3a_2 + 2a_1 = 3(1) + 2(2) = 7$. Now we substitute $n = 4$ to find the fourth term: $a_4 = 3a_{4-1} + 2a_{4-2} = 3a_3 + 2a_2 = 3(7) + 2(1) = 23$. Substituting $n = 5$, we get the fifth term: $a_5 = 3a_{5-1} + 2a_{5-2} = 3a_4 + 2a_3 = 3(23) + 2(7) = 83$.

33. Substituting $n = 1, 2, 3, 4, 5$ into the formula, we get

$$a_1 = \sum_{k=1}^{1} (3k - 5) = 3(1) - 5 = -2$$

$$a_2 = \sum_{k=1}^{2} (3k - 5) = [3(1) - 5] + [3(2) - 5] = (-2) + 1 = -1$$

$$a_3 = \sum_{k=1}^{3} (3k - 5) = [3(1) - 5] + [3(2) - 5] + [3(3) - 5] = (-2) + 1 + 4 = 3$$

$$a_4 = \sum_{k=1}^{4} (3k - 5) = [3(1) - 5] + [3(2) - 5] + [3(3) - 5] + [3(4) - 5]$$

$$= (-2) + 1 + 4 + 7 = 10$$

$$a_5 = \sum_{k=1}^{5} (3k - 5) = [3(1) - 5] + [3(2) - 5] + [3(3) - 5] + [3(4) - 5] + [3(5) - 5]$$

$$= (-2) + 1 + 4 + 7 + 10 = 20$$

37. Substituting $n = 1, 2, 3, 4, 5$ into the formula, we get

$$a_1 = \sum_{k=1}^{1} \frac{x^{k-1}}{k} = \frac{x^{1-1}}{1} = 1$$

$$a_2 = \sum_{k=1}^{2} \frac{x^{k-1}}{k} = \frac{x^{1-1}}{1} + \frac{x^{2-1}}{2} = 1 + \frac{x}{2}$$

$$a_3 = \sum_{k=1}^{3} \frac{x^{k-1}}{k} = \frac{x^{1-1}}{1} + \frac{x^{2-1}}{2} + \frac{x^{3-1}}{3} = 1 + \frac{x}{2} + \frac{x^2}{3}$$

$$a_4 = \sum_{k=1}^{4} \frac{x^{k-1}}{k} = \frac{x^{1-1}}{1} + \frac{x^{2-1}}{2} + \frac{x^{3-1}}{3} + \frac{x^{4-1}}{4} = 1 + \frac{x}{2} + \frac{x^2}{3} + \frac{x^3}{4}$$

$$a_5 = \sum_{k=1}^{5} \frac{x^{k-1}}{k} = \frac{x^{1-1}}{1} + \frac{x^{2-1}}{2} + \frac{x^{3-1}}{3} + \frac{x^{4-1}}{4} + \frac{x^{5-1}}{5} = 1 + \frac{x}{2} + \frac{x^2}{3} + \frac{x^3}{4} + \frac{x^4}{5}.$$

41. $$\sum_{k=1}^{30} (k^2 - 5k + 2) = \sum_{k=1}^{30} k^2 - \sum_{k=1}^{30} 5k + \sum_{k=1}^{30} 2 = \sum_{k=1}^{30} k^2 - 5\sum_{k=1}^{30} k + \sum_{k=1}^{30} 2$$

$$= \frac{30(30+1)(2(30)+1)}{6} - 5\frac{30(30+1)}{2} + 30(2) = 9455 - 2325 + 60 = 7190$$

45. $\displaystyle\sum_{k=1}^{100}(k+1)(k^2-3) = \sum_{k=1}^{100}(k^3+k^2-3k-3)$

$\displaystyle = \sum_{k=1}^{100}k^3 + \sum_{k=1}^{100}k^2 - 3\sum_{k=1}^{100}k - \sum_{k=1}^{100}3$

$\displaystyle = \left[\frac{100(100+1)}{2}\right]^2 + \frac{100(100+1)(2(100)+1)}{6} - (3)\frac{100(100+1)}{2} - 100(3)$

$= 25502500 + 338350 - 15150 - 300 = 25,825,400$

49. The first term is $3 = 3(1)$, the second term is $6 = 3(2)$, the third is $9 = 3(3)$, etc. Thus, since

there are 6 terms, $3 + 6 + 9 + 12 + 15 + 18 = \displaystyle\sum_{k=1}^{6}3k$ (This answer is not unique.)

53. Notice that the numerators are consecutive even integers; the first numerator is $2 = 2(1)$, the second numerator is $4 = 2(2)$, the third numerator is $6 = 2(3)$, and the fourth is $8 = 2(4)$. Similarly, the denominators are one more than their respective numerators; so the first denominator is $3 = 2(1) + 1$, the second denominator is $5 = 2(2) + 1$, the third denominator is $7 = 2(3) + 1$, the fourth is $9 = 2(4) + 1$. Since there are 4 terms,

$\dfrac{2}{3} + \dfrac{4}{5} + \dfrac{6}{7} + \dfrac{8}{9} = \displaystyle\sum_{k=1}^{4}\frac{2k}{2k+1}$ (This answer is not unique.)

57. Taking our cue from the last term, $\dfrac{x^n}{(2n+2)(2n+4)}$, we will write the sum as

$\displaystyle\sum \frac{x^k}{(2k+2)(2k+4)}$. We must be careful that the first and last terms are correctly

accounted for; since the first term is $\dfrac{1}{2(4)} = \dfrac{x^0}{(2(0)+2)(2(0)+4)}$, we need to start the

summation at $k = 0$ and end at $k = n$. Thus, a correct representation is

$\displaystyle\sum_{k=0}^{n}\frac{x^k}{(2k+2)(2k+4)}$.

61. According to the given information, $a_1 = \dfrac{P}{2} = \dfrac{5}{2}$. The next term is $a_2 = \dfrac{1}{2}a_{2-1} + \dfrac{P}{2a_{2-1}}$

$= \dfrac{1}{2}a_1 + \dfrac{5}{2a_1} = \dfrac{1}{2}\cdot\dfrac{5}{2} + \dfrac{5}{2(5/2)} = \dfrac{9}{4}$. The third term is $a_3 = \dfrac{1}{2}a_{3-1} + \dfrac{P}{2a_{3-1}}$

$= \dfrac{1}{2}a_2 + \dfrac{5}{2a_2} = \dfrac{1}{2}\cdot\dfrac{9}{4} + \dfrac{5}{2(9/4)} = \dfrac{161}{72}$. The decimal for this is $2.236\overline{1}$.

65. The first part of the figure shows three pieces, each with $1 + 4 + 9$ blocks. The idea is to imagine each piece having $1 + 4 + 9 + \ldots + n$ blocks. Putting these three pieces together as shown, we end up with a block that is n by $n + 1$ by $n + \frac{1}{2}$ (3 by $3 + 1$ by $3 + \frac{1}{2}$ is shown). Therefore $3(1^2 + 2^2 + 3^2 + \ldots + n^2) = n(n + 1)(n + \frac{1}{2})$, and dividing both sides of this last equation by 3 gives the final result.

Section 11.2
---

1. The first term is $2 = a_1$. The common difference is $a_2 - a_1 = 9 - 2 = 7 = d$. To determine the fifth term, add the common difference to the fourth term: $a_5 = a_4 + d = 23 + 7 = 30$. The general term is $a_n = a_1 + (n - 1)d = 2 + (n - 1)7 = 7n - 5$.

5. The first term is $\sqrt{3} - 2 = a_1$. The common difference is $a_2 - a_1 = 2\sqrt{3} - (\sqrt{3} - 2)$ $= \sqrt{3} + 2 = d$. To determine the fifth term, add the common difference to the fourth term: $a_5 = a_4 + d = (4\sqrt{3} + 4) + (\sqrt{3} + 2) = 5\sqrt{3} + 6$. The general term is $a_n = a_1 + (n - 1)d$ $= (\sqrt{3} - 2) + (n - 1)(\sqrt{3} + 2) = n(\sqrt{3} + 2) - 4$.

9. The first term is $\dfrac{x^2 - 1}{x} = a_1$. The common difference is $a_2 - a_1 = x - \dfrac{x^2 - 1}{x} = \dfrac{1}{x}$ $= d$. To determine the fifth term, add the common difference to the fourth term: $a_5 = a_4 + d = \dfrac{x^2 + 2}{x} + \dfrac{1}{x} = \dfrac{x^2 + 3}{x}$. The general term is $a_n = a_1 + (n - 1)d$ $= \dfrac{x^2 - 1}{x} + (n - 1)\dfrac{1}{x} = \dfrac{n + x^2 - 2}{x}$.

13. The first term is $125 = a_1$. The common ratio is $\dfrac{a_2}{a_1} = \dfrac{-100}{125} = -\dfrac{4}{5} = r$. To determine the fifth term, multiply the the fourth term by the common ratio: $a_5 = a_4 r = (-64)(-\dfrac{4}{5})$ $= \dfrac{256}{5}$. The general term is $a_n = a_1 r^{n-1} = 125(-\dfrac{4}{5})^{n-1}$.

17. The first term is $1 = a_1$. The common ratio is $\dfrac{a_2}{a_1} = \dfrac{1 + \sqrt{2}}{1} = 1 + \sqrt{2} = r$. The fifth term is the fourth term times the common ratio: $a_5 = a_4 r = (7 + 5\sqrt{2})(1 + \sqrt{2}) = 17 + 12\sqrt{2}$. The general term is $a_n = a_1 r^{n-1} = 1(1 + \sqrt{2})^{n-1} = (1 + \sqrt{2})^{n-1}$.

21. To see if the sequence is arithmetic, we calculate the differences between successive terms:

$a_2 - a_1 = \dfrac{1}{2} - \dfrac{1}{3} = \dfrac{1}{6}$; $a_3 - a_2 = \dfrac{2}{3} - \dfrac{1}{2} = \dfrac{1}{6}$; $a_4 - a_3 = \dfrac{5}{6} - \dfrac{2}{3} = \dfrac{1}{6}$.

Since each difference is the same, the sequence is arithmetic with common difference $\dfrac{1}{6}$. The general term is

$a_n = a_1 + (n - 1)d \Rightarrow a_{10} = a_1 + (10 - 1)d = \dfrac{1}{3} + (9)\dfrac{1}{6} = \dfrac{11}{6}$. To see if the sequence is geometric, we calculate the ratios of successive terms: $\dfrac{a_2}{a_1} = \dfrac{1/2}{1/3} = \dfrac{3}{2}$; $\dfrac{a_3}{a_2} = \dfrac{2/3}{1/2} = \dfrac{4}{3}$.

Since these two ratios are not equal, the sequence is not geometric.

25. To see if the sequence is arithmetic, we calculate the differences between successive terms:

$a_2 - a_1 = -\dfrac{3}{2} - 3 = -\dfrac{9}{2}$; $a_3 - a_2 = \dfrac{3}{4} - (-\dfrac{3}{2}) = \dfrac{9}{4}$. Since these two differences are not

equal, the sequence is not arithmetic. To see if the sequence is geometric, we calculate the

ratios of successive terms: $\dfrac{a_2}{a_1} = \dfrac{-3/2}{3} = -\dfrac{1}{2}$; $\dfrac{a_3}{a_2} = \dfrac{3/4}{-3/2} = -\dfrac{1}{2}$; $\dfrac{a_4}{a_3} = \dfrac{-3/8}{3/4} = -\dfrac{1}{2}$.

Since each ratio is the same, the sequence is geometric with common ratio $-\dfrac{1}{2}$. The

general term is $a_n = a_1 r^{n-1} \Rightarrow a_{10} = 3(-\dfrac{1}{2})^{10-1} = 3(-\dfrac{1}{2})^9 = -\dfrac{3}{512}$.

29. First, we recognize that the terms are from an arithmetic sequence with common difference $d = 3$. Next, we determine the number of terms from the relationship $a_n = a_1 + (n - 1)d$ $\Rightarrow 32 = a_n \Rightarrow 32 = a_1 + (n - 1)d = 2 + (n - 1)3$. Solving $32 = 2 + (n - 1)3 \Rightarrow n = 11$. Therefore, since $S_n = \dfrac{n}{2}[a_1 + a_n]$, the sum is $S_{11} = \dfrac{11}{2}[a_1 + a_{11}] = \dfrac{11}{2}[2 + 32] = 187$.

33. The sum in expanded form is $(-15) + (-11) + (-7) + \ldots + 29$. The terms are from an arithmetic sequence with common difference $d = 4$, and there are 12 terms in the sum. Therefore, since $S_n = \dfrac{n}{2}[a_1 + a_n]$, the sum is $S_{12} = \dfrac{12}{2}[a_1 + a_{12}] = 6[(-15) + 29] = 84$.

37. The sum in expanded form is $8 + (-16) + 32 + \ldots + 2048$. The terms are from a geometric sequence with common ratio $r = -2$, and there are 9 terms in the sum. Since $S_n = \dfrac{a_1 - a_1 r^n}{1 - r}$, the sum is $S_9 = \dfrac{8 - 8(-2)^9}{1 - (-2)} = 1368$.

41. The terms are from a geometric sequence with common difference $r = -x$, and there are 12 terms in the sum. Since $S_n = \dfrac{a_1 - a_1 r^n}{1 - r}$, the sum is $S_{12} = \dfrac{1 - 1(x)^{12}}{1 - (x)} = \dfrac{1 - x^{12}}{1 - x}$.

45. $\displaystyle\sum_{k=1}^{7} (\tfrac{1}{2}^k + 12(\tfrac{1}{3})^{k-1}) = \sum_{k=1}^{7} (\tfrac{1}{2}^k) + \sum_{k=1}^{7} 12(\tfrac{1}{3})^{k-1} = \sum_{k=1}^{7} (\tfrac{1}{2})^k + 12\sum_{k=1}^{7} (\tfrac{1}{3})^{k-1}$

$\displaystyle = \frac{(1/2) - (1/2)(1/2)^7}{1 - (1/2)} + 12[\frac{1 - 1(1/3)^7}{1 - (1/3)}] = \frac{127}{128} + 12[\frac{4372}{243}] = \frac{590477}{31104}$.

49. An arithmetic sequence is a linear function whose slope is the common difference d. In this case we are given the ordered pairs $(3, 16)$ and $(12, 13)$. The slope of the line containing these points is $\frac{13 - 16}{12 - 3} = -\frac{1}{3}$. Using the point-slope equation for a line, with the point $(3, 16)$, we have

$y - k = m(x - h) \Rightarrow y - 16 = -\frac{1}{3}(x - 3)$.

For the sequence, a_n plays the role of y and n plays the role of x. Therefore,

$a_n - 16 = -\frac{1}{3}(n - 3)$. Isolating a_n we have

$a_n = -\frac{1}{3}n + 17$. Thus, the 20 term is $a_{20} = -\frac{1}{3}(20) + 17 = \frac{31}{3}$.

53. The multiples of 3 between 10 and 200 are 12, 15, ..., 198. This is an arithmetic

sequence with common difference $d = 3$. We can determine the number of terms from the relationship $a_n = a_1 + (n - 1)d$:

$198 = a_n \Rightarrow 198 = a_1 + (n - 1)d = 12 + (n - 1)3$. Solving $198 = 12 + (n - 1)3 \Rightarrow n = 63$.

There are 63 multiples of 3 between 10 and 200. Their sum is $S_n = \frac{n}{2}[a_1 + a_n] \Rightarrow S_{63} = \frac{63}{2}$

$[a_1 + a_{63}] = \frac{63}{2}[12 + 198] = 6615$.

57. Refer to the formula for compound interest in Section 4.1 of the text. The first deposit will earn interest for 59 months and therefore will grow to $P(1 + \frac{r}{n})^{nt}$

$= 350(1 + \frac{0.10}{12})^{59}$ dollars. The second deposit will earn interest for 58 months and therefore will grow to $P(1 + \frac{r}{n})^{nt} = 350(1 + \frac{0.10}{12})^{58}$ dollars; the third deposit will grow to

$350(1 + \frac{0.10}{12})^{57}$ dollars, and so on. The total amount accumulated after 5 years will be

$350 + 350(1 + \frac{0.10}{12})^1 + 350(1 + \frac{0.10}{12})^2 + ... + 350(1 + \frac{0.10}{12})^{59}$. The terms we are

adding are from the geometric sequence with first term 350 and common ratio $(1 + \frac{0.10}{12})$.

The sum, $S_n = \frac{a_1 - a_1 r^n}{1 - r}$, is

$$S_{60} = \frac{350 - 350(1 + \frac{0.10}{12})^{60}}{1 - (1 + \frac{0.10}{12})} = \frac{350 - 350(\frac{12.1}{12})^{60}}{(\frac{0.10}{12})} \approx 27102.98 \text{ (dollars)}.$$

Keystrokes: 350 $\boxed{-}$ 350 $\boxed{\times}$ $\boxed{(}$ 12.1 $\boxed{\div}$ 12 $\boxed{)}$ $\boxed{y^x}$

60 $\boxed{=}$ $\boxed{\div}$ $\boxed{(}$ 0.10 $\boxed{\div}$ 12 $\boxed{)}$ $\boxed{=}$

61. The multiples of 2 between 1 and 1000 are 2, 4, 6, 8, ...,1000; this is an arithmetic sequence with common difference 2. The number of terms is 500. The multiples of 3 form an arithmetic sequence with common difference 3: 3, 6, 9, ..., 996. There are 332 terms in this sequence. To determine the number of multiples of 2 and 3, we need to consider the multiples of 6 between 1 and 1000: 6, 12, 15,..., 996. There are 166 terms in this sequence. If we add the number of multiples of 2 (500) with the number of multiples of 3 (332), we will be counting the multiples of 6 twice (once too many). Therefore the answer to our problem is $500 + 332 - 166 = 666$.

Section 11.3

1. $r = \frac{30}{60} = \frac{1}{2}$. Since $-1 < r < 1$, the series converges to $\frac{a_1}{1 - r} = \frac{60}{1 - \frac{1}{2}} = 120$.

5. $r = \frac{36}{24} = \frac{3}{2}$. Since $r > 1$, the series diverges.

9. $\displaystyle\sum_{k=1}^{\infty} 10(0.1)^{k-1} = 10 + 10(0.1) + 10(0.1)^2 + ... \Rightarrow r = 0.1$. Since $-1 < r < 1$, the series

converges to $\frac{a_1}{1 - r} = \frac{10}{1 - (0.1)} = \frac{100}{9}$.

13. $\sum_{k=1}^{\infty}(-\frac{3}{5})^k=-\frac{3}{5}+(-\frac{3}{5})^2+\dots \Rightarrow r=-\frac{3}{5}$. Since $-1<r<1$, the series converges to

$$\frac{a_1}{1-r}=\frac{-\frac{3}{5}}{1-(-\frac{3}{5})}=-\frac{3}{8}.$$

17. $0.\overline{5}=0.5+0.05+0.005+\dots$. This is an infinite geometric series with $r=0.1=\frac{1}{10}$.

The sum is $\dfrac{a_1}{1-r}=\dfrac{0.5}{1-(1/10)}=\dfrac{1/2}{1-(1/10)}=\dfrac{5}{9}$.

21. $6.3\overline{42}=6.3+(0.042+0.00042+0.0000042+\dots)$. The sum in the parentheses is an

infinite geometric series with $r=0.01=\dfrac{1}{100}$. The sum (of the series in the parentheses) is

$\dfrac{0.042}{1-(1/100)}=\dfrac{42/1000}{1-(1/100)}=\dfrac{7}{165}$. Thus, $6.3\overline{42}=6.3+\dfrac{7}{165}=\dfrac{63}{10}+\dfrac{7}{165}=\dfrac{2093}{330}$.

25. Comparing $\dfrac{4}{1-3x}$ to $\dfrac{a_1}{1-r}$ suggests letting $a_1=4$ and $r=3x$. We have

$$\frac{a_1}{1-r}=a_1+a_1r+a_1r^2+a_1r^3+\dots \Rightarrow \frac{4}{1-3x}=4+4(3x)+4(3x)^2+4(3x)^3+\dots$$

$=4+12x+36x^2+108x^3+\dots$. Since $-1<r<1, -1<3x<1 \Rightarrow -\dfrac{1}{3}<x<\dfrac{1}{3} \Rightarrow |x|<\dfrac{1}{3}$.

29. You might want to refer to "Finding the Horizontal Asymptotes of a Rational Function" on

page 195 of the text. Since $S_n=\dfrac{2n-1}{3n+5}$ is an improper rational function of n, we can

perform long division and rewrite the function as $S_n=\dfrac{2}{3}+\dfrac{-13/3}{3n+5}$. As n grows

without bound, $\dfrac{-\frac{13}{3}}{3n+5}$ approaches zero. Therefore, $S_n \to \dfrac{2}{3}+0=\dfrac{2}{3}$.

33. As n grows without bound, $\dfrac{1}{n}$ approaches zero. Therefore, $S_n \to 3-0=3$.

37. First we need the partial fraction decomposition of $\dfrac{2}{(2k-1)(2k+1)} = \dfrac{A}{2k-1} + \dfrac{B}{2k+1}$.

Multiplying both sides by $(2k-1)(2k+1)$, we get $2 = A(2k+1) + B(2k-1)$. Letting $k = \dfrac{1}{2}$ gives $2 = A(2) + B(0) \Rightarrow A = 1$. Letting $k = -\dfrac{1}{2}$ gives $2 = A(0) + B(-2)$

$\Rightarrow B = -1$. Thus, $S_n = \displaystyle\sum_{k=1}^{n} \dfrac{2}{(2k-1)(2k+1)} = \dfrac{2}{1(3)} + \dfrac{2}{3(5)} + \ldots + \dfrac{2}{(2n-1)(2n+1)}$

$= \displaystyle\sum_{k=1}^{n} \left(\dfrac{1}{2k-1} + \dfrac{-1}{2k+1} \right) = \left(1 - \dfrac{1}{3}\right) + \left(\dfrac{1}{3} - \dfrac{1}{5}\right) + \left(\dfrac{1}{5} - \dfrac{1}{7}\right) + \ldots + \left(\dfrac{1}{2n-1} - \dfrac{1}{2n+1}\right)$

$= 1 + \left(-\dfrac{1}{3} + \dfrac{1}{3}\right) + \left(-\dfrac{1}{5} + \dfrac{1}{5}\right) + \ldots + \left(-\dfrac{1}{2n-1} + \dfrac{1}{2n-1}\right) - \dfrac{1}{2n+1} = 1 - \dfrac{1}{2n+1}$. As n

grows without bound, $S_n \to 1 - 0 = 1$.

41. $S_n = \displaystyle\sum_{k=1}^{n} (3k+1) = 4 + 7 + 10 + 13 + \ldots + (3n+1)$. This is the sum of the first n terms

of an arithmetic sequence with first term $a_1 = 4$ and common difference $d = 3$. Thus,

$S_n = \dfrac{n}{2}[a_1 + a_n] = \dfrac{n}{2}[4 + (3n+1)] = \dfrac{3}{2}n^2 + \dfrac{5}{2}n$.

As n grows without bound,

$S_n = \dfrac{3}{2}n^2 + \dfrac{5}{2}n$ grows without bound also. The series diverges.

45. The ball travels 16 feet to the ground, bounces $(0.64)16$ feet high, and falls $(0.64)16$ feet to the ground. Then it bounces $(0.64)[(0.64)16] = 16(0.64)^2$ feet high and falls the same amount, etc. The total distance travelled is $16 + [32(0.64) + 32(0.64)^2 + 32(0.64)^3 + \ldots]$. The expression in the brackets is an infinite geometric series with $r = 0.64$. Since $-1 < r < 1$, the series converges. The sum is

$\dfrac{a_1}{1-r} = \dfrac{32(0.64)}{1 - (0.64)} = \dfrac{2048/100}{1 - (64/100)} = \dfrac{512}{9}$.

Therefore the total distance travelled is $16 + \dfrac{512}{9} = 72\dfrac{8}{9}$.

49. $\cos(0.4) \approx 1 - \dfrac{0.4^2}{2!} + \dfrac{0.4^4}{4!} = 1 - 0.08 + \dfrac{0.0256}{24} \approx 0.92107$

53. a) $y = 1$ is a horizontal line.

b) $y = 1 - \frac{1}{2}x^2$ is a parabola with vertex $(0, 1)$ that opens downward. The x-intercepts are $(\pm \sqrt{2}, 0)$, and the endpoints of the parabola are

$(\pm\pi, 1 - \frac{\pi^2}{2}) \approx (\pm 3.1, -3.9)$.

c) $y = 1 - \frac{1}{2}x^2 + \frac{1}{24}x^4$ is a fourth degree polynomial. Some of the ordered pairs are (approx.) $(0, 1)$, $(0.5, 0.88)$, $(1, 0.54)$, $(1.5, 0.09)$, $(2, -0.33)$, $(2.5, -0.50)$, $(3, -0.13)$. Note that the function is even and therefore symmetric with respect to the y-axis.

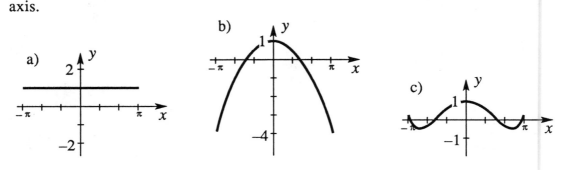

57. The width of the box must be $\frac{1}{2} + \frac{1}{4} + \frac{1}{8} + \ldots$. This is a geometric series with $r = \frac{1}{2}$, and

since $-1 < r < 1$, the sum is $\dfrac{a_1}{1 - r} = \dfrac{\frac{1}{2}}{1 - \frac{1}{2}} = 1$.

61. The sequence of partial sums is $S_1 = \frac{1}{2!} = 1 - \frac{1}{2!}$; $S_2 = \frac{5}{3!} = 1 - \frac{1}{3!}$; $S_3 = \frac{23}{4!} = 1 - \frac{1}{4!}$;

$\ldots$. Following the pattern, we get $S_n = 1 - \dfrac{1}{(n + 1)!}$. Therefore as n grows without

bound, $S_n = 1 - \dfrac{1}{(n + 1)!} \to 1 - 0$. The sum of the series is 1.

Section 11.4

1. $\dbinom{5}{2} = \dfrac{5!}{2!3!} = \dfrac{5 \cdot 4 \cdot 3!}{2!3!} = \dfrac{5 \cdot 4}{2 \cdot 1} = 10$

5. $\dbinom{15}{6} = \dfrac{15!}{6!9!} = \dfrac{15 \cdot 14 \cdot 13 \cdot 12 \cdot 11 \cdot 10 \cdot 9!}{6!9!} = \dfrac{15 \cdot 14 \cdot 13 \cdot 12 \cdot 11 \cdot 10}{6 \cdot 5 \cdot 4 \cdot 3 \cdot 2 \cdot 1} = 5005$

9. $(x + h)^4 = \binom{4}{0}x^4 + \binom{4}{1}x^3h + \binom{4}{2}x^2h^2 + \binom{4}{3}xh^3 + \binom{4}{4}h^4$

$= \frac{4!}{0!4!}x^4 + \frac{4!}{1!3!}x^3h + \frac{4!}{2!2!}x^2h^2 + \frac{4!}{3!1!}xh^3 + \frac{4!}{4!0!}h^4$

$= x^4 + 4x^3h + 6x^2h^2 + 4xh^3 + h^4$

13. $(2x + y)^4 = \binom{4}{0}(2x)^4 + \binom{4}{1}(2x)^3y + \binom{4}{2}(2x)^2y^2 + \binom{4}{3}(2x)y^3 + \binom{4}{4}y^4$

$= \frac{4!}{0!4!}(2x)^4 + \frac{4!}{1!3!}(2x)^3y + \frac{4!}{2!2!}(2x)^2y^2 + \frac{4!}{3!1!}(2x)y^3 + \frac{4!}{4!0!}y^4$

$= (2x)^4 + 4(2x)^3y + 6(2x)^2y^2 + 4(2x)y^3 + y^4 = 16x^4 + 32x^3y + 24x^2y^2 + 8xy^3 + y^4$

17. $(2x + (-y))^8 = \binom{8}{0}(2x)^8 + \binom{8}{1}(2x)^7(-y) + \binom{8}{2}(2x)^6(-y)^2 + \binom{8}{3}(2x)^5(-y)^3$

$+ \binom{8}{4}(2x)^4(-y)^4 + \binom{8}{5}(2x)^3(-y)^5 + \binom{8}{6}(2x)^2(-y)^6 + \binom{8}{7}(2x)(-y)^7 + \binom{8}{8}(-y)^8$

$= (2x)^8 + 8(2x)^7(-y) + 28(2x)^6(-y)^2 + 56(2x)^5(-y)^3 + 70(2x)^4(-y)^4 + 56(2x)^3(-y)^5$

$+ 28(2x)^2(-y)^6 + 8(2x)(-y)^7 + (-y)^8$

$= 256x^8 - 1024x^7y + 1792x^6y^2 - 1792x^5y^3 + 1120x^4y^4 - 448x^3y^5 + 112x^2y^6 - 16xy^7 + y^8$

21. $(\sqrt{p} + q^2)^6 = \binom{6}{0}(\sqrt{p})^6 + \binom{6}{1}(\sqrt{p})^5(q^2) + \binom{6}{2}(\sqrt{p})^4(q^2)^2 + \binom{6}{3}(\sqrt{p})^3(q^2)^3$

$\binom{6}{4}(\sqrt{p})^2(q^2)^4 + \binom{6}{5}(\sqrt{p})(q^2)^5 + \binom{6}{6}(q^2)^6$

$= (\sqrt{p})^6 + 6(\sqrt{p})^5(q^2) + 15(\sqrt{p})^4(q^2)^2 + 20(\sqrt{p})^3(q^2)^3 + 15(\sqrt{p})^2(q^2)^4 + 6(\sqrt{p})(q^2)^5 + (q^2)^6$

$= p^3 + 6p^2q^2\sqrt{p} + 15p^2q^4 + 20pq^6\sqrt{p} + 15pq^8 + 6q^{10}\sqrt{p} + q^{12}$

25. Refer to Pascal's triangle on page 665 (Rows 0 - 6 are shown). The next row (row 7) is 1, 7, 21, 35, 35, 21, 7, 1. These numbers are the respective coefficients in the expansion:

$(3x + 2y)^7 = \mathbf{1}(3x)^7 + \mathbf{7}(3x)^6(2y) + \mathbf{21}(3x)^5(2y)^2 + \mathbf{35}(3x)^4(2y)^3 + \mathbf{35}(3x)^3(2y)^4 +$

$\mathbf{21}(3x)^2(2y)^5 + \mathbf{7}(3x)(2y)^6 + \mathbf{1}(2y)^7 = 2187x^7 + 10206x^6y + 20412x^5y^2 + 22680x^4y^3$

$+ 15120x^3y^4 + 6048x^2y^5 + 1344xy^6 + 128y^7$

29. $\binom{11}{3} = \frac{11!}{3!8!} = \frac{11!}{8!3!} = \binom{11}{8} \Rightarrow k = 8$

33. $11^1 = (10 + 1)^1 = 11;$

$11^2 = (10 + 1)^2 = 10^2 + 2(10)(1) + 1^2 = 121;$

$11^3 = (10 + 1)^3 = 10^3 + 3(10)^2(1) + 3(10)(1)^2 + 1^3 = 1331;$

$11^4 = (10 + 1)^4 = 10^4 + 4(10)^3(1) + 6(10)^2(1)^2 + 4(10)(1)^3 + 1^4 = 14641.$ In other words, these powers of 11 are precisely the respective rows of Pascal's triangle.

$11^5 = (10 + 1)^5 = 10^5 + 5(10)^4(1) + 10(10)^3(1)^2 + 10(10)^2(1)^3 + 5(10)(1)^4 + 1^5 = 161051$

37. $(\sqrt{p} + 2q)^{10} = \binom{10}{0}(\sqrt{p})^{10} + \binom{10}{1}(\sqrt{p})^9(2q) + \binom{10}{2}(\sqrt{p})^8(2q)^2 + \ldots + \binom{10}{10}(2q)^{10}$

The p^3q^4 term is $\binom{10}{6}(\sqrt{p})^6(2q)^4 = 210(\sqrt{p})^6(2q)^4 = 3360\,p^3q^4$.

41. $(x^2 + 2yz)^6 = \binom{6}{0}(x^2)^6 + \binom{6}{1}(x^2)^5(2yz) + \binom{6}{2}(x^2)^4(2yz)^2$

$\qquad + \binom{6}{3}(x^2)^3(2yz)^3 + \binom{6}{4}(x^2)^2(2yz)^4 + \binom{6}{5}(x^2)(2yz)^5 + \binom{6}{6}(2yz)^6$

$= (x^2)^6 + 6(x^2)^5(2yz) + 15(x^2)^4(2yz)^2 + 20(x^2)^3(2yz)^3$

$\qquad\qquad\qquad\qquad + 15(x^2)^2(2yz)^4 + 6(x^2)(2yz)^5 + (2yz)^6$

$= x^{12} + 12x^{10}yz + 60x^8y^2z^2 + 160x^6y^3z^3 + 240x^4y^4z^4 + 192x^2y^5z^5 + 64y^6z^6$

45. $(x^3 + (-\frac{1}{x^2}))^6 = \binom{6}{0}(x^3)^6 + \binom{6}{1}(x^3)^5(-\frac{1}{x^2}) + \binom{6}{2}(x^3)^4(-\frac{1}{x^2})^2$

$\qquad + \binom{6}{3}(x^3)^3(-\frac{1}{x^2})^3 + \binom{6}{4}(x^3)^2(-\frac{1}{x^2})^4 + \binom{6}{5}(x^3)(-\frac{1}{x^2})^5 + \binom{6}{6}(-\frac{1}{x^2})^6$

$= (x^3)^6 + 6(x^3)^5(-\frac{1}{x^2}) + 15(x^3)^4(-\frac{1}{x^2})^2 + 20(x^3)^3(-\frac{1}{x^2})^3$

$\qquad\qquad\qquad + 15(x^3)^2(-\frac{1}{x^2})^4 + 6(x^3)(-\frac{1}{x^2})^5 + (-\frac{1}{x^2})^6$

$= x^{18} - 6x^{13} + 15x^8 - 20x^3 + \dfrac{15}{x^2} - \dfrac{6}{x^7} + \dfrac{1}{x^{12}}$

49. $\dfrac{f(x+h) - f(x)}{h} = \dfrac{[2(x+h)^4 + (x+h)^2 - 4] - [2x^4 + x^2 - 4]}{h} =$

$\dfrac{[2(x^4 + 4x^3h + 6x^2h^2 + 4xh^3 + h^4) + (x^2 + 2xh + h^2) - 4] - [2x^4 + x^2 - 4]}{h}$

$= \dfrac{8x^3h + 12x^2h^2 + 8xh^3 + 2h^4 + 2xh + h^2}{h} = 8x^3 + 12x^2h + 8xh^2 + 2h^3 + 2x + h$

53. $\dfrac{1}{\sqrt{1-x}} = (1 + (-x))^{-1/2} = 1 + (-\frac{1}{2})(-x) + \dfrac{(-\frac{1}{2})(-\frac{1}{2}-1)}{2!}(-x)^2$

$\qquad\qquad + \dfrac{(-\frac{1}{2})(-\frac{1}{2}-1)(-\frac{1}{2}-2)}{3!}(-x)^3 + \ldots$

$= 1 + \frac{1}{2}x + \frac{3}{8}x^2 + \frac{5}{16}x^3 + \ldots$

57. $(1 - x^3)^{1/4} = (1 + (-x^3))^{1/4} = 1 + (\frac{1}{4})(-x^3) + \dfrac{(\frac{1}{4})(\frac{1}{4} - 1)}{2!}(-x^3)^2$

$$+ \dfrac{(\frac{1}{4})(\frac{1}{4} - 1)(\frac{1}{4} - 2)}{3!}(-x^3)^3 + \dots$$

$$= 1 - \frac{1}{4}x^3 - \frac{3}{32}x^6 - \frac{7}{128}x^9 + \dots$$

61. $(1 + x)^{1/3} = 1 + \dfrac{x}{3} - \dfrac{x^2}{9} + \dfrac{5x^3}{81} + \dots \Rightarrow (1 + 0.1)^{1/3} \approx 1 + \dfrac{0.1}{3} - \dfrac{0.1^2}{9} + \dfrac{5(0.1)^3}{81}$

$= 1.032284$

Section 11.5

1. The equation holds for $n = 1$: $1 = \dfrac{1(2)}{2}$.

Now assume the equation holds for $n = k$:

$1 + 2 + 3 + \dots + k = \dfrac{k(k + 1)}{2}$. Adding $k + 1$ to both sides gives

$1 + 2 + 3 + \dots + k + (k + 1) = \dfrac{k(k + 1)}{2} + (k + 1) = \dfrac{k(k + 1)}{2} + \dfrac{2(k + 1)}{2}$

$= \dfrac{(k + 1)(k + 2)}{2} = \dfrac{(k + 1)((k + 1) + 1)}{2}$.

Therefore, if the equation holds for $n = k$, then it also holds for $n = k + 1$. This completes the proof.

5. The equation holds for $n = 1$: $2 = 2(2 - 1)$.

Now assume the equation holds for $n = k$:

$2 + 2^2 + 2^3 + \dots + 2^k = 2(2^k - 1)$.

Adding 2^{k+1} to both sides gives

$2 + 2^2 + 2^3 + \dots + 2^k + 2^{k+1} = 2(2^k - 1) + 2^{k+1} = 2^{k+1} - 2 + 2^{k+1} = 2(2^{k+1} - 1)$.

Therefore, if the equation holds for $n = k$, then it also holds for $n = k + 1$. This completes the proof.

9. The equation holds for $n = 1$: $\dfrac{1}{1(2)} = \dfrac{1}{2}$.

Now assume the equation holds for $n = k$:

$$\frac{1}{1(2)} + \frac{1}{2(3)} + \frac{1}{3(4)} + \ldots + \frac{1}{k(k+1)} = \frac{k}{k+1} \ .$$

Adding $\dfrac{1}{(k+1)(k+2)}$ to both sides gives

$$\frac{1}{1(2)} + \frac{1}{2(3)} + \frac{1}{3(4)} + \ldots + \frac{1}{k(k+1)} + \frac{1}{(k+1)(k+2)} = \frac{k}{k+1} + \frac{1}{(k+1)(k+2)}$$

$$= \frac{k(k+2)}{(k+1)(k+2)} + \frac{1}{(k+1)(k+2)} = \frac{(k^2 + 2k + 1)}{(k+1)(k+2)} = \frac{(k+1)^2}{(k+1)(k+2)} = \frac{(k+1)}{(k+2)} \ .$$

Therefore, if the equation holds for $n = k$, then it also holds for $n = k + 1$. This completes the proof.

13. The equation holds for $n = 1$: $\cos(\phi + 1\pi) = (-1)^1 \cos\phi$.

Assume the equation holds for $n = k$:

$\cos(\phi + k\pi) = (-1)^k \cos\phi$.

Now consider $\cos(\phi + (k+1)\pi) = \cos[(\phi + k\pi) + \pi]$

$= \cos(\phi + k\pi)\cos\pi - \sin(\phi + k\pi)\sin\pi = \cos(\phi + k\pi)(-1) = (-1)^k\cos\phi(-1) = (-1)^{k+1}\cos\phi$.

Therefore, if the equation holds for $n = k$, then it also holds for $n = k + 1$. This completes the proof.

17. The inequality holds for $n = 1$: $3^1 > 1 + 1$.

Assume the inequality holds for $n = k$:

$3^k > k + 1$.

Multiplying each side by 3 gives $3^{k+1} > 3(k + 1)$. If $k \geq 1$, then

$3(k + 1) > k + 2$; so $3^{k+1} > 3(k + 1) > k + 2 \Rightarrow 3^{k+1} > (k + 1) + 1$.

Therefore, if the inequality holds for $n = k$, then it also holds for $n = k + 1$. This completes the proof.

21. The equation holds for $n = 1$: $\frac{1}{2!} = 1 - \frac{1}{(1 + 1)!}$.

Now assume the equation holds for $n = k$:

$$\frac{1}{2!} + \frac{2}{3!} + \frac{3}{4!} + \ldots + \frac{k}{(k + 1)!} = 1 - \frac{1}{(k + 1)!} .$$

Adding $\frac{k + 1}{(k + 2)!}$ to both sides gives

$$\frac{1}{2!} + \frac{2}{3!} + \frac{3}{4!} + \ldots + \frac{k}{(k + 1)!} + \frac{k + 1}{(k + 2)!} = 1 - \frac{1}{(k + 1)!} + \frac{k + 1}{(k + 2)!}$$

$$= 1 - \frac{k + 2}{(k + 2)!} + \frac{k + 1}{(k + 2)!} = 1 - \frac{1}{(k + 2)!} .$$

Therefore, if the equation holds for $n = k$, then it also holds for $n = k + 1$. This completes the proof.

25. The inequality holds for $n = 4$: $4! > 2^4$.

Assume the inequality holds for $n = k$ ($k \geq 4$): $k! > 2^k$.

Multiplying each side by $k + 1$ gives $(k + 1)! > 2^k(k + 1)$. If

$k \geq 4$, then $2^k(k + 1) > 2^k(2) = 2^{k+1}$; so $(k + 1)! > 2^k(k + 1) > 2^{k+1}$. This implies

that $(k + 1)! > 2^{k+1}$. Therefore, if the inequality holds for $n = k$, then it also holds for

$n = k + 1$. This completes the proof.

29. The first few terms are

$$a_1 = \frac{1}{2}, a_2 = (a_1 + \frac{1}{1})(\frac{1}{1 + 1}) = \frac{3}{4}, a_3 = (a_2 + \frac{1}{2})(\frac{2}{2 + 1}) = \frac{5}{6},$$

$a_4 = (a_3 + \frac{1}{3})(\frac{3}{3 + 1}) = \frac{7}{8}$. Continuing, it seems as though $a_n = \frac{2n - 1}{2n}$. To prove

that this is the case by mathematical induction, we have already shown the pattern holds for

$n = 1$. Now we assume that $a_k = \frac{2k - 1}{2k}$. By the given formula for the next term, we

have $a_{k+1} = (a_k + \frac{1}{k})(\frac{k}{k + 1}) = (\frac{2k - 1}{2k} + \frac{1}{k})(\frac{k}{k + 1}) = (\frac{2k + 1}{2k})(\frac{k}{k + 1})$

$= \frac{2k + 1}{2(k + 1)} = \frac{2(k + 1) - 1}{2(k + 1)}$. Therefore, if the equation holds for $n = k$, then it also

holds for $n = k + 1$.

33. Even though the inequality holds for $n = 1, 2, \ldots, 9$; it is <u>not</u> true for $n = 10$: $2^{10} > 10^3 + 2$. This counterexample disproves the statement.

37. The statement holds for $n = 1$, but it is <u>not</u> true for $n = 2$: $\frac{3}{1(2)^2} + \frac{5}{2(3)^2} \neq \frac{2(2 + 2)}{(2 + 1)^2}$.

This counterexample disproves the statement.

41. Starting with $n = 3$, the sum of the interior angles of triangle is $180(3 - 2)°$. Assume thst the sum of the interior angles of a polygon with k sides is $180(k - 2)°$. Now consider a ploygon with $k + 1$ sides. Select any three consecutive vertices, and draw a diagonal from the first to the third vertex, so that a triangle is formed by the diagonal and the two adjacent sides. This diagonal divides the $k + 1$ sided polygon into a triangle and a polygon with k sides. The sum of the interior angles is $180° + 180(k - 2)° = 180[(k + 1) - 2]°$. Since this shows that the assertion is true for $n = k + 1$ whenever it is true for $n = k$, the conjecture is true for all positive integers $n \geq 3$.

Miscellaneous Exercises for Chapter 11

1. a) Substituting 1, 2, 3, 4, and 5 for n, we get
$a_1 = 4(1) - 3 = 1$, $a_2 = 4(2) - 3 = 5$, $a_3 = 4(3) - 3 = 9$,
$a_4 = 4(4) - 3 = 13$, and $a_5 = 4(5) - 3 = 17$. Thus, the first five terms are 1, 5, 9, 13, and 17.
b) The ordered pairs of the sequence are (n, a_n); the first five are $(1, 1)$, $(2, 5)$, $(3, 9)$, $(4, 13)$, and $(5, 17)$.

5. a) Substituting 1, 2, 3, 4, and 5 for n, we get
$a_1 = 48(\frac{1}{2})^1 = 24$, $a_2 = 48(\frac{1}{2})^2 = 12$, $a_3 = 48(\frac{1}{2})^3$
$= 6$, $a_4 = 48(\frac{1}{2})^4 = 3$, and $a_5 = 48(\frac{1}{2})^5 = \frac{3}{2}$. Thus,
the first five terms are 24, 12, 6, 3, and $\frac{3}{2}$.
b) The ordered pairs of a sequence are (n, a_n); the first five are $(1, 24)$, $(2, 12)$, $(3, 6)$, $(4, 3)$, and $(5, \frac{3}{2})$.

9. a) $f(x) = (\frac{3}{4})^x$ is an exponential function, so we know the characteristic shape (See Section 4.1, Example 4.). The graph crosses the y-axis at $(0, 1)$. A few ordered pairs on the graph are $(-2, \frac{16}{9})$, $(-1, \frac{4}{3})$, $(1, \frac{3}{4})$, and $(2, \frac{9}{16})$. The x-axis is a horizontal asymptote for large positive values of x.
b) Since the graph in part a) has a horizontal asymptote at $y = 0$, a_n approaches 0 as n grows without bound.

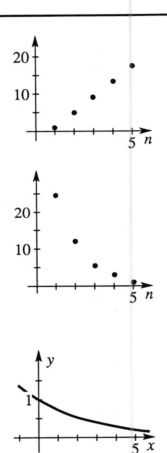

13. a) $f(x) = \sqrt{x}$ is a function in the catalog of graphs on page 102 of the text, so we know the characteristic shape. A few ordered pairs on the graph are $(0,0)$, $(1, 1)$, $(2, 4)$, $(3, 9)$.

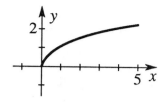

b) Since the graph in part a) grows without bound as x grows without bound, $a_n = \sqrt{n}$ grows without bound as n grows without bound. In other words, a_n does not approach a value.

17. $\displaystyle\sum_{k=1}^{40}(k^3 - 2k + 3) = \sum_{k=1}^{40}k^3 - \sum_{k=1}^{40}2k + \sum_{k=1}^{40}3 = \sum_{k=1}^{40}k^3 - 2\sum_{k=1}^{40}k + \sum_{k=1}^{40}3$

$= \left[\dfrac{40(40+1)}{2}\right]^2 - 2\left[\dfrac{40(40+1)}{2}\right] + 40(3) = 670880$

21. To see if the sequence is arithmetic, we calculate the differences between successive terms: $a_2 - a_1 = 24 - (-18) = 42$; $a_3 - a_2 = -32 - (24) = -56$.

Since these two differences are not equal, the sequence is not arithmetic. To see if the sequence is geometric, we calculate the ratios of successive terms:

$\dfrac{a_2}{a_1} = \dfrac{24}{-18} = -\dfrac{4}{3}$; $\dfrac{a_3}{a_2} = \dfrac{-32}{24} = -\dfrac{4}{3}$; $\dfrac{a_4}{a_3} = \dfrac{\frac{128}{3}}{-32} = -\dfrac{4}{3}$.

Since each ratio is the same, the sequence is geometric with common ratio $-\dfrac{4}{3}$. The first

term is $a_1 = -18$, the general term is $a_n = a_1 r^{n-1} = (-18)(-\frac{4}{3})^{n-1}$

$\Rightarrow a_6 = (-18)(-\frac{4}{3})^{6-1} = (-18)(-\frac{4}{3})^5 = \dfrac{2048}{27}$.

25. To see if the sequence is arithmetic, we calculate the differences between successive terms:

$a_2 - a_1 = (3\sqrt{2} + \sqrt{6}) - (\sqrt{3} + 1) = 3\sqrt{2} + \sqrt{6} - \sqrt{3} - 1$;

$a_3 - a_2 = (6\sqrt{3} + 6) - (3\sqrt{2} + \sqrt{6}) = 6\sqrt{3} + 6 - 3\sqrt{2} - \sqrt{6}$.

Since these two differences are not equal, the sequence is not arithmetic. To see if the sequence is geometric, we calculate the ratios of successive terms:

$\dfrac{a_2}{a_1} = \dfrac{3\sqrt{2} + \sqrt{6}}{\sqrt{3} + 1} = \dfrac{3\sqrt{2} + \sqrt{6}}{\sqrt{3} + 1} \cdot \dfrac{\sqrt{3} - 1}{\sqrt{3} - 1} = \sqrt{6}$;

$\dfrac{a_3}{a_2} = \dfrac{6\sqrt{3} + 6}{3\sqrt{2} + \sqrt{6}} = \dfrac{6\sqrt{3} + 6}{3\sqrt{2} + \sqrt{6}} \cdot \dfrac{3\sqrt{2} - \sqrt{6}}{3\sqrt{2} - \sqrt{6}} = \sqrt{6}$; $\dfrac{a_4}{a_3} = \dfrac{18\sqrt{2} + 6\sqrt{6}}{6\sqrt{3} + 6} = \sqrt{6}$.

Since each ratio is the same, the sequence is geometric with common ratio $r = \sqrt{6}$. The first term is $a_1 = \sqrt{3} + 1$, the general term is $a_n = a_1 r^{n-1} = (\sqrt{3} + 1)(\sqrt{6})^{n-1}$

$\Rightarrow a_6 = (\sqrt{3} + 1)(\sqrt{6})^{6-1} = (\sqrt{3} + 1)(36\sqrt{6}) = 108\sqrt{2} + 36\sqrt{6}$.

29. We are given two ordered pairs of the sequence: $(3, a_3) = (3, 2)$ and $(9, a_9) = (9, 4)$. Since an arithmetic sequence is a linear function, we can find a linear equation that contains the ordered pairs $(3, 2)$ and $(9, 4)$. The slope is $m = \dfrac{4 - 2}{9 - 3} = \dfrac{1}{3}$.

Substituting into the point-slope form of a linear equation with the point $(3, 2)$, we get

$$y - k = m (x - h) \Rightarrow y - 2 = \frac{1}{3} (x - 3) \Rightarrow y = \frac{1}{3} x + 1.$$

In terms of the sequence, this means that the general term is $a_n = \dfrac{1}{3} n + 1$. The first four terms are

$$a_1 = \frac{1}{3} (1) + 1 = \frac{4}{3}, \, a_2 = \frac{1}{3} (2) + 1 = \frac{5}{3}, \, a_3 = \frac{1}{3} (3) + 1 = 2, \, a_4 = \frac{1}{3} (4) + 1 = \frac{7}{3}.$$

33. Since the general term for a geometric sequence is $a_n = a_1 r^{n-1}$, we are given that $3 = a_3 = a_1 r^2$ and $1 = a_5 = a_1 r^4$.

If we multiply both sides of the equation $3 = a_1 r^2$ by r^2, we have

$$3r^2 = a_1 r^4 \Rightarrow 3r^2 = 1 \Rightarrow r = \pm \frac{\sqrt{3}}{3}.$$

This means that there are two possible sequences. Substituting this into $3 = a_1 r^2$ and solving for a_1, we get $a_1 = 9$. Thus, the general term is

$$a_n = a_1 r^{n-1} = 9(\frac{\sqrt{3}}{3})^{n-1}, \text{ or } a_n = a_1 r^{n-1} = 9(-\frac{\sqrt{3}}{3})^{n-1}.$$

In the case $r = \dfrac{\sqrt{3}}{3}$, the first four terms are

$$9, \, 9(\frac{\sqrt{3}}{3}) = 3\sqrt{3}, \, 9(\frac{\sqrt{3}}{3})^2 = 3, \, 9(\frac{\sqrt{3}}{3})^3 = \sqrt{3}.$$

Similarly, for $r = \dfrac{\sqrt{3}}{3}$, the first four terms are $9, \, 9(-\frac{\sqrt{3}}{3}) = -3\sqrt{3}, \, 9(-\frac{\sqrt{3}}{3})^2 = 3,$

$$9(-\frac{\sqrt{3}}{3})^3 = -\sqrt{3}.$$

37. First, we recognize that the terms are from an geometric sequence with common ratio $r = 3$. Next, we determine the number of terms from the relationship

$$a_n = a_1 r^{n-1} \Rightarrow 13122 = a_n \Rightarrow 13122 = a_1 r^{n-1} = 2(3)^{n-1}.$$

Solving $13122 = 2(3)^{n-1} \Rightarrow n = 9$. Therefore, since $S_n = \dfrac{a_1 - a_1 r^n}{1 - r}$, the sum is

$$S_9 \, \frac{2 - 2(3)^9}{1 - 3} = 19{,}682.$$

41. The sum in expanded form is $-\frac{5}{2} + 5 + (-10) + \ldots + 1280$. The terms are from a geometric sequence with common ratio $r = -2$, and there are 10 terms in the sum. Since $S_n = \frac{a_1 - a_1 r^n}{1 - r}$, the sum is

$$S_{10} = \frac{-\frac{5}{2} - (-\frac{5}{2})(-2)^{10}}{1 - (-2)} = \frac{1705}{2}.$$

45. First, we write the problem as two sums:

$$\sum_{k=1}^{10} [(-2)^{k-1} + k^2] = \sum_{k=1}^{10} (-2)^{k-1} + \sum_{k=1}^{10} k^2.$$

The first of these two is the sum of a geometric sequence with $a_1 = 1$ and $r = -2$; the second sum is given on page 635 in the text (property 6). Therefore,

$$\sum_{k=1}^{10} (-2)^{k-1} + \sum_{k=1}^{10} k^2 = \frac{a_1 - a_1 r^n}{1 - r} + \frac{n(n+1)(2n+1)}{6}$$

$$= \frac{1 - 1(-2)^{10}}{1 - (-2)} + \frac{10(10+1)(2(10)+1)}{6} = -341 + 385 = 44$$

49. $r = \frac{20}{40} = \frac{1}{2}$. Since $-1 < r < 1$, the series converges to $\frac{a_1}{1-r} = \frac{40}{1 - \frac{1}{2}} = 80$.

53. $r = \frac{-\frac{4}{3}}{1} = -\frac{4}{3}$. Since $r < -1$, the series does not have a sum; the series diverges.

57. $r = \sin^2 x$, and for all x satisfying $0 < x < \frac{\pi}{2}$, $-1 < r < 1$. Therefore the series converges to $\frac{a_1}{1-r} = \frac{\sin^2 x}{1 - \sin^2 x} = \frac{\sin^2 x}{\cos^2 x} = \tan^2 x$.

61. We know that for any geometric series

$$a_1 + a_1 r + a_1 r^2 + a_1 r^3 + \ldots = \frac{a_1}{1-r}, \text{ provided that } -1 < r < 1.$$

Comparing this to our problem, we see that $a_1 = 2$ and $r = x$. Thus the equation is valid for all x satisfying $-1 < x < 1$.

65. We know that for any geometric series

$$\frac{a_1}{1-r} = a_1 + a_1 r + a_1 r^2 + a_1 r^3 + \ldots, \text{ provided that } -1 < r < 1.$$

Comparing this to our problem, we see that $a_1 = 3$ and $r = -2x$. Thus,

$$\frac{3}{1+2x} = \frac{3}{1-(-2x)} = 3 + 3(-2x) + 3(-2x)^2 + 3(-2x)^3 + \ldots = 3 - 6x + 12x^2 - 24x^3 + \ldots.$$

69. First we need the partial fraction decomposition of $\dfrac{10}{(5k-2)(5k+3)} = \dfrac{A}{5k-2} + \dfrac{B}{5k+3}$.

Multiplying both sides by $(5k-2)(5k+3)$, we get $10 = A(5k+3) + B(5k-2)$.

$k = \dfrac{2}{5} \Rightarrow 10 = A(5) + B(0) \Rightarrow A = 2$.

$k = -\dfrac{3}{5} \Rightarrow 2 = A(0) + B(-5) \Rightarrow B = -2$

Thus, $S_n = \displaystyle\sum_{k=1}^{n} \dfrac{10}{(5k-2)(5k+3)} = \sum_{k=1}^{n} \left(\dfrac{2}{5k-2} + \dfrac{-2}{5k+3}\right)$

$\qquad = \left(\dfrac{2}{3} - \dfrac{2}{8}\right) + \left(\dfrac{2}{8} - \dfrac{2}{13}\right) + \left(\dfrac{2}{13} - \dfrac{2}{18}\right) + \dots + \left(\dfrac{2}{5n-2} - \dfrac{2}{5n+3}\right)$

$\qquad = \dfrac{2}{3} + \left(-\dfrac{2}{8} + \dfrac{2}{8}\right) + \left(-\dfrac{2}{13} + \dfrac{2}{13}\right) + \dots + \left(-\dfrac{2}{5n-2} + \dfrac{2}{5n-2}\right) - \dfrac{2}{5n+3} = \dfrac{2}{3} - \dfrac{2}{5n+3}$.

As n grows without bound, $S_n \to \dfrac{2}{3} - 0 = \dfrac{2}{3}$.

73. $(x^2 + 2y)^5 = \dbinom{5}{0}(x^2)^5 + \dbinom{5}{1}(x^2)^4(2y) + \dbinom{5}{2}(x^2)^3(2y)^2$

$\qquad\qquad\qquad + \dbinom{5}{3}(x^2)^2(2y)^3 + \dbinom{5}{4}(x^2)(2y)^4 + \dbinom{5}{5}(2y)^5$

$\quad = (x^2)^5 + 5(x^2)^4(2y) + 10(x^2)^3(2y)^2 + 10(x^2)^2(2y)^3 + 5(x^2)(2y)^4 + (2y)^5$

$\quad = x^{10} + 10x^8y + 40x^6y^2 + 80x^4y^3 + 80x^2y^4 + 32y^5$

77. $(x^2 + 3y)^6 = \dbinom{6}{0}(x^2)^6 + \dbinom{6}{1}(x^2)^5(3y) + \dbinom{6}{2}(x^2)^4(3y)^2$

$\qquad\qquad\qquad + \dbinom{6}{3}(x^2)^3(3y)^3 + \dbinom{6}{4}(x^2)^2(3y)^4 + \dbinom{6}{5}(x^2)(3y)^5 + \dbinom{6}{6}(3y)^6$

The x^6y^3 term is $\dbinom{6}{3}(x^2)^3(3y)^3 = 20(x^2)^3(3y)^3 = 20x^6(27y^3) = 540x^6y^3$.

81. $\dfrac{1}{\sqrt[3]{1-x}} = [1 + (-x)]^{-1/3}$

$\quad = 1 + \left(-\dfrac{1}{3}\right)(-x) + \dfrac{\left(-\dfrac{1}{3}\right)\left(-\dfrac{1}{3} - 1\right)}{2!}(-x)^2 + \dfrac{\left(-\dfrac{1}{3}\right)\left(-\dfrac{1}{3} - 1\right)\left(-\dfrac{1}{3} - 2\right)}{3!}(-x)^3 + \dots$

$\quad = 1 + \dfrac{1}{3}x + \dfrac{2}{9}x^2 + \dfrac{14}{81}x^3 + \dots$

85. The equation holds for $n = 1$: $1 = (2(1) - 1)^2$.
Now assume the equation holds for $n = k$:
$$1 + 8 + 16 + \ldots + 8(k - 1) = (2k - 1)^2.$$
Adding $8k$ to both sides gives
$$1 + 8 + 16 + \ldots + 8(k - 1) + 8k = (2k - 1)^2 + 8k = 4k^2 - 4k + 1 + 8k$$
$$= 4k^2 + 4k + 1 = (2k + 1)^2 = (2(k + 1) - 1)^2.$$
Therefore, if the equation holds for $n = k$, then it also holds for $n = k + 1$. This completes the proof.

89. The equation holds for $n = 1$: $\dfrac{1}{1 \cdot 2 \cdot 3} = \dfrac{1(1 + 3)}{4(1 + 1)(1 + 2)}$.
Now assume the equation holds for $n = k$:
$$\frac{1}{1 \cdot 2 \cdot 3} + \frac{1}{2 \cdot 3 \cdot 4} + \frac{1}{3 \cdot 4 \cdot 5} + \ldots + \frac{1}{k(k + 1)(k + 2)} = \frac{k(k + 3)}{4(k + 1)(k + 2)}$$
Adding $\dfrac{1}{(k + 1)(k + 2)(k + 3)}$ to both sides gives
$$\frac{1}{1 \cdot 2 \cdot 3} + \frac{1}{2 \cdot 3 \cdot 4} + \frac{1}{3 \cdot 4 \cdot 5} + \ldots + \frac{1}{k(k + 1)(k + 2)}$$
$$+ \frac{1}{(k + 1)(k + 2)(k + 3)} = \frac{k(k + 3)}{4(k + 1)(k + 2)} + \frac{1}{(k + 1)(k + 2)(k + 3)}$$
$$= \frac{k(k + 3)^2}{4(k + 1)(k + 2)(k + 3)} + \frac{4}{4(k + 1)(k + 2)(k + 3)} = \frac{(k^3 + 6k^2 + 9k + 4)}{4(k + 1)(k + 2)(k + 3)}$$
$$= \frac{(k + 1)^2(k + 4)}{4(k + 1)(k + 2)(k + 3)} = \frac{(k + 1)(k + 4)}{4(k + 2)(k + 3)}.$$
Therefore, if the equation holds for $n = k$, then it also holds for $n = k + 1$. This completes the proof.

Chapter 12

Section 12.1

1. Imagine wrapping your right hand around the z-axis with your thumb in the direction of positive z. The direction in which your fingers wrap is the same as the rotation of the x-axis to the y-axis. This is a right-handed system.

5. Imagine wrapping your right hand around the z-axis with your thumb in the direction of positive z. The direction in which your fingers wrap is the same as the rotation of the x-axis to the y-axis. This is a right-handed system.

9. Using the distance formula on page 688 yields

$$\begin{aligned}\sqrt{(\Delta x)^2+(\Delta y)^2+(\Delta z)^2} &= \sqrt{((-6)-(-2))^2+((-1)-3)^2+(0-0)^2}\\ &= \sqrt{(-4)^2+(-4)^2+0^2}\\ &= 4\sqrt{2}\end{aligned}$$

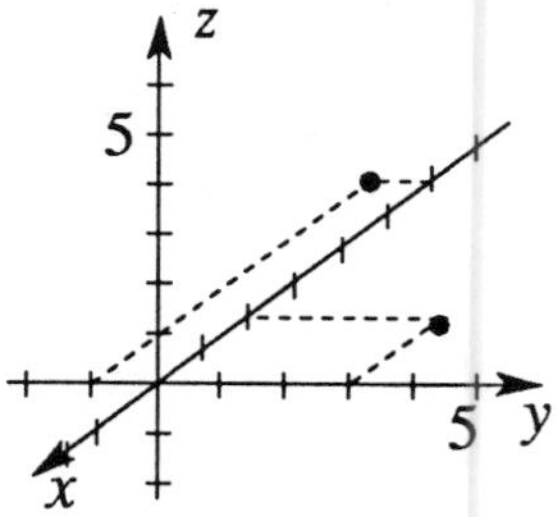

13. Examine the figure at right. The distance is equal to the value of the z-coordinate of the point, so the distance is 4.

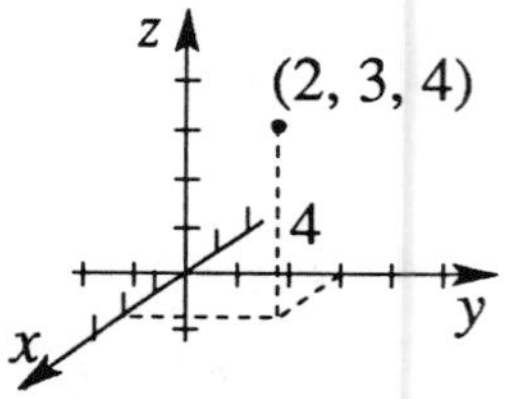

17. Examine the figure at right. The point that is on the same vertical line as (2, 3, 4) in the plane $z = -3$ is $(2, 3, -3)$. The distance is $4-(-3) = 7$

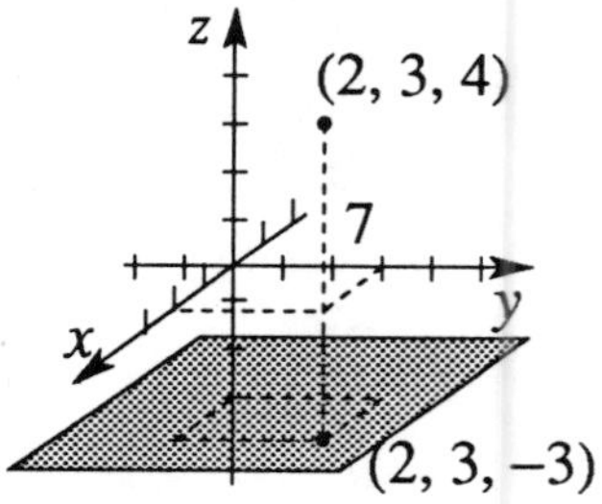

21. The set of points such that $y = -4$ is a plane perpendicular to the y-axis at $(0, -4, 0)$

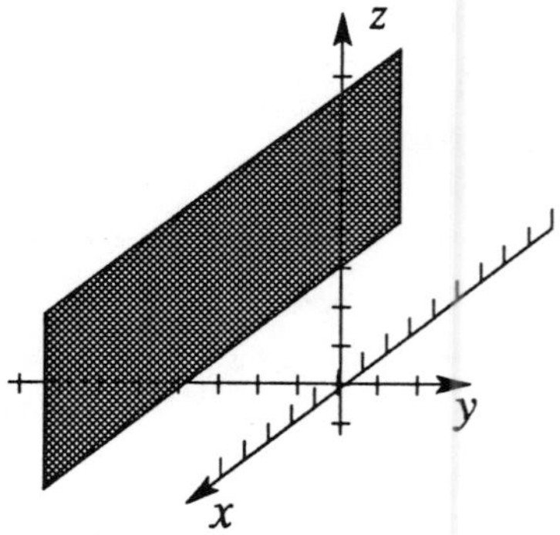

25. The set of points such that $x = 2$ and $y = 3$ is a line parallel to the z-axis passing through $(2, 3, 0)$.

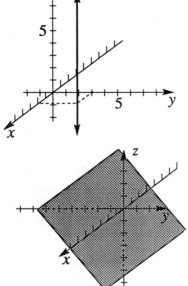

29. The plane passes through the given points and is parallel to the x-axis.

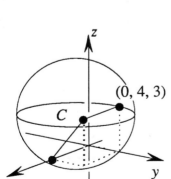

33. Compare $(x-2)^2 + (y-5)^2 + (z+1)^2 = 32$ with the general equation for a sphere on page 690. For this equation $(h, k, m) = (2, 5, -1)$ and $r = \sqrt{32} = 4\sqrt{2}$. Thus the center is at the point $(2, 5, -1)$ and the radius is $4\sqrt{2}$.

37. We need first to rewrite the equation in the form $(x-h)^2 + (y-k)^2 + (z-m)^2 = r^2$. We do this by completing the square on x, y, and z.

$$\left(x^2 + 2x \quad\right) + y^2 + \left(z^2 - 2z \quad\right) = -2$$
$$\left(x^2 + 2x + 1\right) + y^2 + \left(z^2 - 2z + 1\right) = -2 + 1 + 1$$
$$(x+1)^2 + y^2 + (z-1)^2 = 0$$

Compare this with the general equation for a sphere on page 690. For this equation, $(h, k, m) = (-1, 0, 1)$ and $r = 0$. Because the radius is zero, this "sphere" is merely the point $(-1, 0, 1)$.

41. Examine the figure at right. The radius from the center to $(0, 4, 3)$ is perpendicular to the yz-plane, and the radius from the center to $(5, 0, 0)$ is perpendicular to the x-axis. The center C therefore is $(5, 4, 3)$. The radius of the sphere is 5. Using the general equation for a sphere on page 690, we get

$$(x-5)^2 + (x-4)^2 + (x-3)^2 = 25.$$

45. The lengths of the sides are computed using the distance formula on page 688.

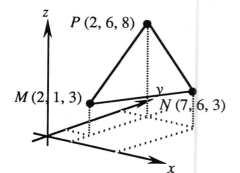

$$d(M,N) = \sqrt{(7-2)^2 + (6-1)^2 + (3-3)^2} = 5\sqrt{2}$$

$$d(N,P) = \sqrt{(2-7)^2 + (6-6)^2 + (8-3)^2} = 5\sqrt{2}$$

$$d(P,M) = \sqrt{(2-2)^2 + (1-6)^2 + (3-8)^2} = 5\sqrt{2}$$

Because $d(M, N) = d(N, P) = d(P, M)$, the triangle is an equilateral triangle.

49. We must first determine the lengths of the sides of the triangles to use the law of cosines. For the sake of discussion, let p, q, and r be the sides of the triangle as shown in the figure at right. The lengths of the sides are computed using the distance formula on page 688.

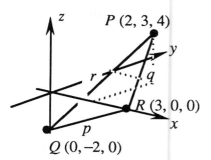

$$p = \sqrt{(3-0)^2 + (0-0)^2 + (0-(-2))^2} = \sqrt{13}$$

$$q = \sqrt{(2-3)^2 + (3-0)^2 + (4-0)^2} = \sqrt{26}$$

$$r = \sqrt{(0-2)^2 + (0-3)^2 + (-2-4)^2} = 7$$

Using the law of cosines, it follows that

$$\cos \angle RPQ = \frac{r^2 + q^2 - p^2}{2rq} = \frac{(7)^2 + \left(\sqrt{26}\right)^2 - \left(\sqrt{13}\right)^2}{2(7)\left(\sqrt{26}\right)} = \frac{31}{7\sqrt{26}}$$

$$\Rightarrow \angle RPQ = \cos\left(\frac{31}{7\sqrt{26}}\right) = 29.7°$$

$$\cos \angle PQR = \frac{p^2 + r^2 - q^2}{2pr} = \frac{\left(\sqrt{13}\right)^2 + (7)^2 - \left(\sqrt{26}\right)^2}{2\left(\sqrt{13}\right)(7)} = \frac{18}{7\sqrt{13}}$$

$$\Rightarrow \angle PQR = \cos\left(\frac{18}{7\sqrt{13}}\right) = 44.5°$$

$$\cos \angle QRP = \frac{q^2 + p^2 - r^2}{2qp} = \frac{\left(\sqrt{26}\right)^2 + \left(\sqrt{13}\right)^2 - (7)^2}{2\left(\sqrt{26}\right)\left(\sqrt{13}\right)} = -\frac{5}{13\sqrt{2}}$$

$$\Rightarrow \angle QRP = \cos\left(-\frac{5}{13\sqrt{2}}\right) = 105.8°$$

Section 12.2

1. The plane is the only one of the six that has positive x- y-, and z-intercepts. Thus III as the only possible candidate; its intercepts are $(4, 0, 0)$, $(0, 7, 0)$, and $(0, 0, 5)$.

5. The plane is the only one of the six that has positive x-intercept and y-intercept, but negative z-intercept. A survey of the list of equations yields IV as the only possible candidate; its intercepts are $(3, 0, 0)$, $(0, 3, 0)$, and $(0, 0, -2)$.

9. First we find the intercepts. To find the x-intercept, we let $y = 0$ and $z = 0$:
$$2x + 4(0) + 5(0) = 20 \Rightarrow 2x = 20 \Rightarrow x = 10.$$
The x-intercept is $(10, 0, 0)$.
To find the y-intercept, we let $y = 0$ and $z = 0$:
$$2(0) + 4y + 5(0) = 20 \Rightarrow 4y = 20 \Rightarrow y = 5.$$
The y-intercept is $(0, 5, 0)$.
To find the z-intercept, we let $x = 0$ and $y = 0$:
$$2(0) + 4(0) + 5z = 20 \Rightarrow 5z = 20 \Rightarrow z = 4$$
The z-intercept is $(0, 0, 4)$.
We plot these intercepts, connect them with line segments, and shade accordingly.

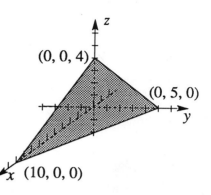

13. First we find the intercepts. To find the x-intercept, we let $y = 0$ and $z = 0$:
$$2x - (0) + 2(0) = 20 \Rightarrow 2x = 20 \Rightarrow x = 10.$$
The x-intercept is $(10, 0, 0)$.
To find the y-intercept, we let $y = 0$ and $z = 0$:
$$2(0) - y + 2(0) = 20 \Rightarrow -y = 20 \Rightarrow y = -20.$$
The y-intercept is $(0, -20, 0)$.
To find the z-intercept, we let $x = 0$ and $y = 0$:
$$2(0) - (0) + 2z = 20 \Rightarrow 2z = 20 \Rightarrow z = 10$$
The z-intercept is $(0, 0, 10)$.

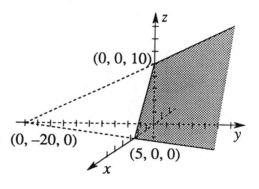

We plot these intercepts, connect them with line segments, and shade accordingly. Since the y-intercept is negative, the y-axis is extended in the negative direction to accommodate it. Dashed lines are used to indicate traces that are not in the first octant.

17. The graph of $y = 2$ is a plane that is parallel to the xz-plane and passes through $(0, 2, 0)$. The graph of $x = 5$ is a plane that is parallel to the yz-plane and passes through $(5, 0, 0)$. We draw in the traces and the line of intersection, and shade accordingly.

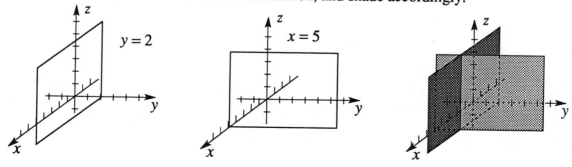

21. The plane $x + 3z = 6$ has x-intercept $(6, 0, 0)$ and z-intercept $(0, 0, 2)$. It is parallel to the y-axis. The plane $2y + z = 4$ has y-intercept $(0, 2, 0)$ and z-intercept $(0, 0, 4)$. It is parallel to the x-axis. We draw in the traces and the line of intersection, and shade accordingly.

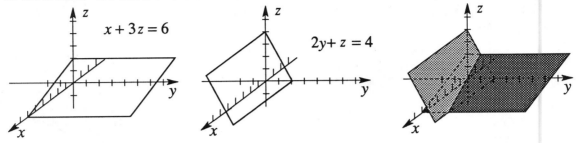

25. The plane $4x + 2y - z = 4$ has x-intercept $(1, 0, 0)$, y-intercept $(0, 2, 0)$, and z-intercept $(0, 0, -4)$. The plane of $x + y + z = 4$ has x-intercept $(4, 0, 0)$, y-intercept $(0, 4, 0)$, and z-intercept $(0, 0, 4)$. We draw in the traces, and shade accordingly.

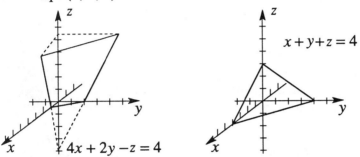

29. The graph of $y = 2$ is a plane that is parallel to the xz-plane and passes through $(0, 2, 0)$. The graph of $x + y = 4$ is a plane with x-intercept $(4, 0, 0)$ and y-intercept $(0, 4, 0)$; It is parallel to the z-axis. The graph of $2x + z = 6$ is a plane with x-intercept $(3, 0, 0)$ and z-intercept $(0, 0, 6)$; It is parallel to the y-axis.
We draw in the traces and the lines of intersection, and shade accordingly.

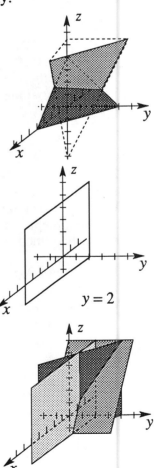

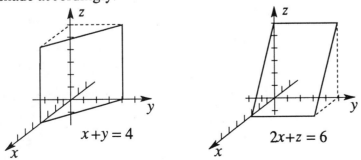

33. The plane $x+y+z = 4$ has x-intercept $(4, 0, 0)$, y-intercept $(0, 4, 0)$, and z-intercept $(0, 0, 4)$. The plane $2x - y + 2z = 6$ has x-intercept $(3, 0, 0)$, y-intercept $(0, -6, 0)$, and z-intercept $(0, 0, 3)$. The plane $x + y + 2z = 8$ has x-intercept $(8, 0, 0)$, y-intercept $(0, 8, 0)$, and z-intercept $(0, 0, 4)$. We draw in the traces and the line of intersection, and shade accordingly.

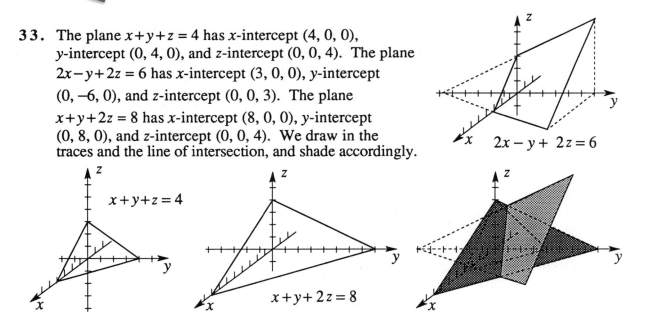

37. The point of intersection of these planes is the solution of the system of equations. We proceed by using the substitution method discussed in Section 10.1.

The first equation of the system $\begin{cases} x & = 3 \\ x + 3y & = 6 \\ x & + z = 2 \end{cases}$ gives us $x = 3$. We replace x by 3 in the

second equation to determine y:

$$x + 3y = 6 \Rightarrow (3) + 3y = 6 \Rightarrow 3y = 3 \Rightarrow y = 1$$

We replace x by 3 in the third equation to determine z:

$$x + z = 2 \Rightarrow (3) + z = 2 \Rightarrow z = -1$$

The point of intersection is $(3, 1, -1)$.

41. The point of intersection of these planes is the solution of the system of equations. We proceed by using techniques of Section 10.1 to find an equivalent system of equations that is upper triangular.

$$\begin{cases} 2x - y + 2z = 6 \\ x + y + z = 6 \\ 3x + 3y + z = 6 \end{cases} \quad \begin{matrix} E_2 \to E_1 \\ \Rightarrow E_1 \to E_2 \end{matrix} \begin{cases} x + y + z = 6 \\ 2x - y + 2z = 6 \\ 3x + 3y + z = 6 \end{cases}$$

$$\Rightarrow \begin{matrix} E_2 - 2E_1 \to E_2 \\ E_3 - 3E_1 \to E_3 \end{matrix} \begin{cases} x + y + z = 6 \\ -3y = -6 \\ -2z = -12 \end{cases} \quad \Rightarrow -\tfrac{1}{3}E_2 \to E_2 \begin{cases} x + y + z = 6 \\ y = 2 \\ -2z = -12 \end{cases}$$

From the third equation, we get $z = 6$, and from the second equation we get $y = 2$. Back–substituting these values into the first equation, we find x:

$$x + (2) + (6) = 6 \Rightarrow x + 8 = 6 \Rightarrow x = -2$$

The point of intersection is $(-2, 2, 6)$.

45. We seek a t such that $(-4t, 2t, t)$ is a solution to the equation $x+y+2z = 16$.

$$x+y+2z = 8 \implies (-4t)+(2t)+2(t) = 16 \implies (-4t)+(2t)+2(t) = 16 \implies 0 = 16.$$

This is a contradiction, so no such t exists. There is no point of intersection.

49. We proceed by using techniques of Section 10.1 to find an equivalent system of equations that is upper triangular and try to back-substitute.

$$\begin{cases} 2x +3y+4z = 6 \\ 2x +3y+4z =12 \\ x +y = 3 \end{cases} \quad \overset{E_3 \to E_1}{\implies} \quad \begin{cases} x +y = 3 \\ 2x +3y+4z =12 \\ 2x +3y+4z = 6 \end{cases}$$

$$\implies \begin{matrix} E_2-2E_1 \to E_2 \\ E_3-2E_1 \to E_3 \end{matrix} \begin{cases} x +y = 3 \\ y+4z = 6 \\ y+4z = 0 \end{cases} \implies \begin{matrix} E_1-E_2 \to E_1 \\ \\ E_3-E_2 \to E_3 \end{matrix} \begin{cases} x -4z =-6 \\ y+4z = 6 \\ 0 =-6 \end{cases}$$

The third equation of this last system, $0 = -6$, is a contradiction. This implies that the system is inconsistent. From the figure it appears that the planes representing the first and second equations, $2x+3y+4z = 6$ and $2x+3y+4z = 12$, are parallel. Because parallel planes do not intersect, there is no point of intersection.

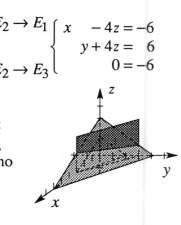

Section 12.3

1. a) $f(4, 1) = 2(4)-(1)^3 = 7$

   b) $f(1, 4) = 2(1)-(4)^3 = -62$

5. a) $g(0, 0) = (0)(\cos 0) = (0)(1) = 0$

   b) $h(0, 0) = \log_3((0)^2+(0)) = \log_3(0)$
   Because $\log_3(0)$ is undefined, $h(0, 0)$ is undefined

9. a) $h(3n, 18n^2) = \log_3((3n)^2+(18n^2))$
   $$= \log_3(27n^2)$$
   Using properties of logarithms, we get
   $$\log_3(27n^2) = \log_3(27n^2)$$
   $$= \log_3(27)+\log_3(n^2)$$
   $$= 3+2\log_3 n$$

   b) $3[h(n, 2n^2)] = 3[\log_3((n)^2+(2n^2))]$
   $$= 3[\log_3(3n^2)]$$
   Using properties of logarithms, we get
   $$3[\log_3(3n^2)] = \log_3(3n^2)$$
   $$= 3[\log_3(3)+\log_3(n^2)]$$
   $$= 3[3+2\log_3 n]$$
   $$= 9+6\log_3 n$$

13. Because the radicand of a square root must be nonnegative, the domain is the set of (x, y) that are solutions to $3x-y \geq 0$. First we sketch the associated equation $3x-y = 0$, using a solid line. Next, we select a test point, say $(4, 0)$ to determine the appropriate half plane:
$$3(4)-(0) \overset{?}{\geq} 0 \implies 12 \overset{?}{\geq} 0 \quad \text{True.}$$
The domain of the function is shown in the figure at right.

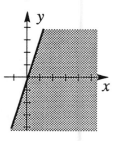

17. Because the radicand of a square root must be nonnegative, the domain is the set of ordered pairs (x, y) that are solutions to the inequality $\dfrac{x^2+y^2}{x^2-y^2} \geq 0$. Because $x^2+y^2 \geq 0$ for all (x, y),

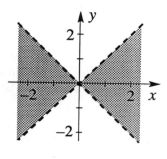

it follows that this inequality is true if and only if $x^2-y^2 > 0$. First we sketch the associated equation $x^2-y^2 = 0$ for this inequality. Solving this equation for y yields $y = \pm x$. We use dotted lines because this is a strict inequality. Next, we select test points in each of the regions defined by the lines:

$(3, 0) \Rightarrow (3)^2-(0)^2 \overset{?}{>} 0$ True. $(0, 3) \Rightarrow (0)^2-(3)^2 \overset{?}{>} 0$ False.

$(-3, 0) \Rightarrow (-3)^2-(0)^2 \overset{?}{>} 0$ True. $(0, -3) \Rightarrow (0)^2-(-3)^2 \overset{?}{>} 0$ False.

The domain of the function is shown in the figure at right.

21. From the figure, level curves appear to be circles. The traces in the xz-plane and yz-plane appear to oscillate up and down. Now, consider III. The level curves are $k = \sin(x^2+y^2)$. This implies that x^2+y^2 must be constant for each level curve, or, in other words, the level curves are of the form. $x^2+y^2 = r^2$. The trace in the xz-plane is $z = \sin(x^2+0^2) = \sin x^2$. The trace in the yz-plane is $z = \sin(0^2+y^2) = \sin y^2$. This makes III the likely choice for the figure in Problem 21.

25. The level curves are of the form $2x+3y = k$. Solving this for y yields $y = -\dfrac{2}{3}x+\dfrac{k}{3}$. The level curves are lines with slope $\dfrac{2}{3}$ and y-intercept $\dfrac{k}{2}$.

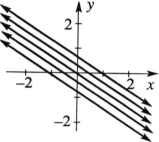

$k = -2 \Rightarrow y = -\dfrac{2}{3}x-\dfrac{2}{3}$ $k = -1 \Rightarrow y = -\dfrac{2}{3}x-\dfrac{1}{3}$

$k = 0 \Rightarrow y = -\dfrac{2}{3}x$ $k = 1 \Rightarrow y = -\dfrac{2}{3}x+\dfrac{1}{3}$

$k = 2 \Rightarrow y = -\dfrac{2}{3}x+\dfrac{2}{3}$

29. The level curves are $2\sin xy = k$, or $\sin xy = \dfrac{k}{2}$.

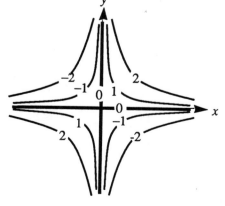

$k = -2 \Rightarrow \sin xy = -1 \Rightarrow xy = \dfrac{3\pi}{2} \pm 2n\pi$

$k = -1 \Rightarrow \sin xy = -\dfrac{1}{2} \Rightarrow xy = \dfrac{7\pi}{6} \pm 2n\pi$ or $\dfrac{11\pi}{6} \pm 2n\pi$

$k = 0 \Rightarrow \sin xy = 0 \Rightarrow xy = \pm 2n\pi$

$k = 1 \Rightarrow \sin xy = \dfrac{1}{2} \Rightarrow xy = \dfrac{\pi}{6} \pm 2n\pi$ or $\dfrac{5\pi}{6} \pm 2n\pi$

$k = 2 \Rightarrow \sin xy = 1 \Rightarrow xy = \dfrac{\pi}{2} \pm 2n\pi$

Each of these level curves is a family of hyperbolas. The figure shows one representative of each family.

33. The trace of this surface in the xy-plane is a semicircle of radius 4. Because there is no z variable in the equation, the surface is a cylinder parallel to the z-axis.

37. Consider the traces in the xy-plane, the xz-plane, and the yz-plane:

$z = 0 \Rightarrow$ trace is $x^2 + 9y^2 + 9(0)^2 = 36$ or $x^2 + 9y^2 = 36$; this is an ellipse

$y = 0 \Rightarrow$ trace is $x^2 + 9(0)^2 + 9z^2 = 36$ or $x^2 + 9z^2 = 36$; this is an ellipse

$x = 0 \Rightarrow$ trace is $(0)^2 + 9y^2 + 9z^2 = 36$ or $y^2 + z^2 = 4$; this is an circle

This implies that the surface is an ellipsoid (see page 713).

Its intercepts are $(\pm 6, 0, 0)$, $(0, \pm 2, 0)$, and $(0, 0, \pm 2)$

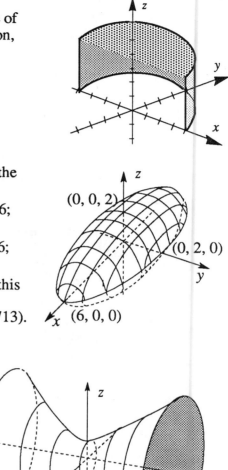

41. First, consider the traces in the xy-plane, the xz-plane, and the yz-plane:

$z = 0 \Rightarrow$ trace is $x^2 - y^2 + (0)^2 = 1$ or $x^2 - y^2 = 1$; this is a hyperbola

$y = 0 \Rightarrow$ trace is $x^2 - (0)^2 + z^2 = 1$ or $x^2 + z^2 = 1$; this is an circle

$x = 0 \Rightarrow$ trace is $(0)^2 - y^2 + z^2 = 1$ or $z^2 - y^2 = 1$; this is a hyperbola

This implies that the surface is a hyperboloid of one sheet (see page 713). Its intercepts are $(\pm 1, 0, 0)$ and $(0, 0, \pm 1)$

45. This equation can be rewritten as
$$(x-1) - z^2 - y^2 = 1.$$
This is a translation in the direction of the positive
x-axis of the paraboloid sketched in Problem 39.

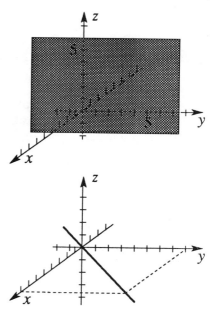

(1, 0, 0)

Miscellaneous Exercises for Chapter 12

---

1. Using the distance formula on page 688 yields
$$\sqrt{(\Delta x)^2 + (\Delta y)^2 + (\Delta z)^2} = \sqrt{(5-2)^2 + (4-2)^2 + (2-1)^2}$$
$$= \sqrt{3^2 + 2^2 + 1^2} = \sqrt{14}$$

5. The set of points such that $x = 4$ is a plane that is perpendicular to the *x*-axis at (4, 0 0)

9. The set of points such that $z = 0$ is the *xy*-plane. The set of points in the *xy*-plane such that $y = x$ is a line passing through the origin, the point (1, 1, 0), the point (2, 2, 0), and so on.

13. We need first to rewrite the equation in the form $(x-h)^2 + (y-k)^2 + (z-m)^2 = r^2$. We do this by completing the square on *x, y,* and *z*.
$$x^2 + y^2 - 8y + z^2 = 20 \Rightarrow x^2 + \left(y^2 - 8y \right) + z^2 = 20$$
$$\Rightarrow x^2 + \left(y^2 - 8y + \mathbf{16}\right) + z^2 = 20 + \mathbf{16} \Rightarrow x^2 + (y-4)^2 + z^2 = 36$$
Compare this with the general equation for a sphere on page 690. The center of the sphere is $(h, k, m) = (0, 4, 0)$ and the radius is $r = \sqrt{36} = 6$.

17. Examine the figure at right. The radius from the center to
(4, 0, 7) is perpendicular to the *xz*-plane, and the radius
from the center to (0, 4, 7) is perpendicular to the *yz*-
plane. The center therefore is (4, 4, 7). The radius of the
sphere is 4. Using the general equation for a sphere on
page 690, we get

$$(x-4)^2+(x-4)^2+(x-7)^2=16.$$

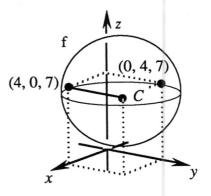

21. First we find the intercepts. To find the *x*-intercept,
we let *y* = 0 and *z* = 0:

$$2x+(0)+3(0) = 18 \Rightarrow 2x = 18 \Rightarrow x = 9.$$

The *x*-intercept is (9, 0, 0).
To find the *y*-intercept, we let *y* = 0 and *z* = 0:

$$2(0)+y+3(0) = 18 \Rightarrow y = 18.$$

The *y*-intercept is (0, 18, 0).
To find the *z*-intercept, we let *x* = 0 and *y* = 0:

$$2(0)+(0)+3z = 18 \Rightarrow 3z = 18 \Rightarrow z = 6.$$

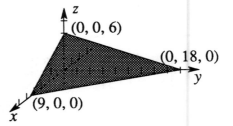

The *z*-intercept is (0, 0, 6).We plot these intercepts, connect them with line segments, and
shade accordingly.

25. The graph of *y* = 5 is a plane that is parallel to the *xz*-plane and passes through (0, 5, 0).
The graph of *z* = 3 is a plane that is parallel to the *xy*-plane and passes through (0, 0, 3).
We draw in the traces and the line of intersection, and shade accordingly.

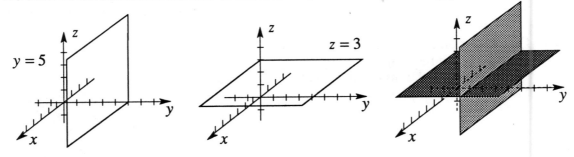

29. The graph of $x + 2y = 10$ is a plane with x-intercept $(10, 0, 0)$ and y-intercept $(0, 5, 0)$; it is parallel to the z-axis. The graph of $3x + 4z = 12$ is a plane with x-intercept $(4, 0, 0)$ and z-intercept $(0, 0, 3)$; it is parallel to the z-axis. We draw in the traces and the line of intersection, and shade accordingly.

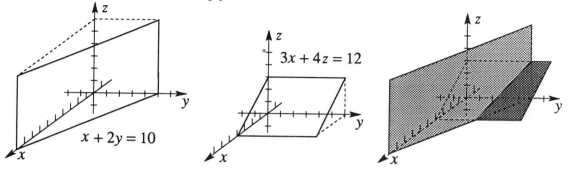

33. From the second equation, we get $y = 4$. Substituting this into the first equation allows us to determine x:
$$2x + (4) = 8 \Rightarrow x = 2$$
The value of z may vary, so the intersection is the line $(2, 4, t)$.

37. If the lines intersect, then for some t_1 and t_2,
$$(1 + 2t_1, \, 1 - t_1, \, 6 - t_1) = (10 + t_2, \, -3t_2, \, -1 - 3t_2).$$
This gives a system of three equations in two unknowns:
$$\begin{cases} 1 + 2t_1 = 10 + t_2 \\ 1 - t_1 = 3t_2 \\ 6 - t_1 = -1 - 3t_2 \end{cases}$$
Solving the top equation for t_2, we get $t_2 = 2t_1 - 9$. Substituting this into the second equation gives us $1 - t_1 = 3t_2 \Rightarrow 1 - t_1 = 3(2t_1 - 9) \Rightarrow t_1 = 4$. Now we use the value of t_1 to find the value of t_2: $t_2 = 2(4) - 9 = -1$. Using $t_1 = 4$ in the first line and $t_2 = 4$ in the second line each yields the point $(9, -3, 2)$. This is the point of intersection.

41. The domain of the inverse cosine function is $[-1, 1]$, so the domain of ϕ is the set of all (x, y) such that $-1 \le x + y \le 1$. This continued inequality is equivalent to the system $\begin{cases} x + y \ge -1 \\ x + y \le 1 \end{cases}$
The graph of the first inequality is the set of points that is above the line $x + y = -1$. The graph of the second inequality is the set of points that is below the line $x + y = 1$.

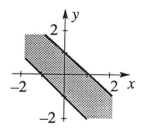

45. The level curves are of the form $x - \sin^{-1}y = k$. This equation can be rewritten $\sin^{-1}y = x - k$. From Section 6.4, we know that this is equivalent to $y = \sin(x-k)$, where $-1 \le y \le 1$ and $-\frac{\pi}{2} \le x-k \le \frac{\pi}{2}$, or $k - \frac{\pi}{2} \le x \le k + \frac{\pi}{2}$.

$k = 0 \Rightarrow y = \sin x, -1 \le y \le 1$ and $-\frac{\pi}{2} \le x \le \frac{\pi}{2}$

$k = \frac{\pi}{2} \Rightarrow y = \sin(x - \frac{\pi}{2}), -1 \le y \le 1$ and $0 \le x \le \pi$

$k = \pi \Rightarrow y = \sin(x - \pi), -1 \le y \le 1$ and $\frac{\pi}{2} \le x \le \frac{3\pi}{2}$

$k = \frac{3\pi}{2} \Rightarrow y = \sin(x - \frac{3\pi}{2}), -1 \le y \le 1$ and $\pi \le x \le 2\pi$

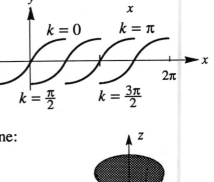

49. Consider the traces in the xy-plane, xz-plane, and yz-plane:

$z = 0 \Rightarrow$ trace in the xy-plane is $x^2 + y^2 - (0)^2 = 9$ or $x^2 + y^2 = 9$; this is a circle

$y = 0 \Rightarrow$ trace in the xz-plane is $x^2 + (0)^2 - z^2 = 9$ or $x^2 - z^2 = 9$; this is an hyperbola

$x = 0 \Rightarrow$ trace in the yz-plane is $(0)^2 + y^2 - z^2 = 9$ or $y^2 - z^2 = 9$; this is an hyperbola

This implies that the surface is an hyperboloid of one sheet (see page 713). Its intercepts are $(\pm 3, 0, 0)$ and $(0, \pm 3, 0)$

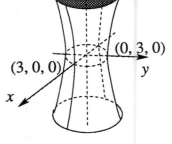